# VIRTUAL RECONSTRUCTION

# VIRTUAL RECONSTRUCTION

## A Primer in Computer-Assisted Paleontology and Biomedicine

**Christoph P. E. Zollikofer**
**Marcia S. Ponce de León**

Anthropologisches Institut
Universität Zürich-Irchel
Zürich, Switzerland

WILEY-INTERSCIENCE

A JOHN WILEY & SONS, INC., PUBLICATION

Published by John Wiley & Sons, Inc., Hoboken, New Jersey.
Published simultaneously in Canada.

No part of this publication may be reproduced, stored in a retrieval system, or transmitted in any form or by any means, electronic, mechanical, photocopying, recording, scanning, or otherwise, except as permitted under Section 107 or 108 of the 1976 United States Copyright Act, without either the prior written permission of the Publisher, or authorization through payment of the appropriate per-copy fee to the Copyright Clearance Center, Inc., 222 Rosewood Drive, Danvers, MA 01923, 978-750-8400, fax 978-646-8600, or on the web at www.copyright.com. Requests to the Publisher for permission should be addressed to the Permissions Department, John Wiley & Sons, Inc., 111 River Street, Hoboken, NJ 07030, (201) 748-6011, fax (201) 748-6008.

Limit of Liability/Disclaimer of Warranty: While the publisher and author have used their best efforts in preparing this book, they make no representations or warranties with respect to the accuracy or, completeness of the contents of this book and specifically disclaim any implied warranties of merchantability or fitness for a particular purpose. No warranty may be created or extended by sales representatives or written sales materials. The advice and strategies contained herein may not be suitable for your situation. You should consult with a professional where appropriate. Neither the publisher nor author shall be liable for any loss of profit or any other commercial damages, including but not limited to special, incidental, consequential, or other damages.

The publisher and the author make no representations or warranties with respect to the accuracy or completeness of the contents of this work and specifically disclaim all warranties, including without limitation any implied warranties of fitness for a particular purpose. This work is sold with the understanding that the publisher is not engaged in rendering professional services. The advice and strategies contained herein may not be suitable for every situation. In view of ongoing research, equipment modifications, changes in governmental regulations, and the constant flow of information relating to the use of experimental reagents, equipment, and devices, the reader is urged to review and evaluate the information provided in the package insert or instructions for each chemical, piece of equipment, reagent, or device for, among other things, any changes in the instructions or indication of usage and for added warnings and precautions. The fact that an organization or Website is referred to in this work as a citation and/or a potential source of further information does not mean that the author or the publisher endorses the information the organization or Website may provide or recommendations it may make. Further, readers should be aware that Internet Websites listed in this work may have changed or disappeared between when this work was written and when it is read. No warranty may be created or extended by any promotional statements for this work. Neither the publisher nor the author shall be liable for any damages arising herefrom.

For general information on our other products and services please contact our Customer Care Department within the U.S. at 877-762-2974, outside the U.S. at 317-572-3993 or fax 317-572-4002.

Wiley also publishes its books in a variety of electronic formats. Some content that appears in print, however, may not be available in electronic format.

*Library of Congress Cataloging-in-Publication Data:*

Zollikofer, Christoph P. E.
    Virtual reconstruction: a primer in computer-assisted paleontology and biomedicine / Christoph P.E. Zollikofer, Marcia S. Ponce de León.
        p.   cm.
        ISBN-13 978-0-471-20507-4 (cloth)
        ISBN-10 0-471-20507-9 (cloth)
    1. Paleontology–Data processing.   2. Paleontology–Graphic methods.   3. Biomedicine–Data processing.   4. Biomedicine–Graphic methods.   I. Ponce de León, Marcia S.   II. Title.
        QE721.2.D37Z65 2005
        560′.285′66—dc22

                                                                          2004017650

Printed in the United States of America

10 9 8 7 6 5 4 3 2 1

To my parents, Hans and Lotti
C. P. E. Z.

To the memory of my daughter, and to my parents,
Jaime and Rebeca, for being just who they are
M. S. P. L.

# CONTENTS

# PREFACE

In the biosciences—as in any other discipline of science—computers have become indispensable tools for data acquisition, processing, analysis, and visualization. It is fascinating that we can now navigate inside virtual organisms, for example, a patient's body, after this patient has already left the hospital. Or that we may take a rare human fossil specimen from the virtual bookshelf and inspect it on the computer screen, while the original remains locked in a safe repository. But how exactly do we transform organisms from physical reality into virtual reality? How can we accomplish interactive virtual tasks, such as reconstruction of fragmented and distorted fossil specimens or simulation of surgical interventions? How can we quantify the three-dimensional shape of organismic structures with their digital representations? And finally, how can virtual objects be printed out as three-dimensional hard copies?

This book bundles such questions within the concept of *virtual reconstruction*. It is written for bioscientists interested in computer science, but also for computer scientists interested in the biosciences. Its contents are borne out from a decade of research in computer-assisted paleoanthropology, and from lectures and courses in biomedical imaging, scientific visualization, and computational morphology. A great deal of the concepts and practical hints conveyed here emerged from collaborative work and innumerable discussions with colleagues in anthropology, zoology, paleontology, computer science, physics, radiology, pediatrics, and maxillofacial surgery.

Visit the book companion website at www.wiley.com/go/virtual reconstruction.

MARCIA PONCE de LEÓN
CHRISTOPH ZOLLIKOFER
Zürich
June 2004

# ACKNOWLEDGMENTS

Our first and special thanks go to Peter Stucki, director emeritus of the Multi-Media Lab at the University of Zürich. His constant enthusiasm on transdisciplinary research and his generous support were a major incentive to follow our own research approach.

A great deal of the concepts and practical hints conveyed in this book emerged from collaborative work and innumerable discussions with many colleagues in anthropology, zoology, computer science, physics, radiology, pediatrics, and maxillofacial surgery. These people were generous with their knowledge and time and provided access to key fossil specimens and clinical cases, the analysis of which forms an important part of this book. We would like to thank, in alphabetical order, Takeru Akazawa, Gea Bijl, Michel Brunet, Carlos Buitrago Tellez, Friedrich Carls, Cidalia Duarte, Francisco Estevez, Alexander Flisch, the late Walther A. Fuchs, Avishag Ginzburg, Lena Godina, Gerda Greuter, Almut Hoffmann, Isuf Hoxha, Beat Hümbelin, Hashime Ishida, Jean-Jacques Jaeger, Vitaliy Kharitonov, Osamu Kondo, Michael Krupa, Dan Lieberman, David Lordkipanidze, Thomas Lüthi, Rosine Orban, Ildiko Pap, Yoel Rak, Jürg Rickenmann, Patrick Semal, Chris Stringer, Ninoslav Teodorovic, Bernhard Vandermeersch, Peter Wyss, João Zilhão, and Christoph L. Zollikofer for fruitful and inspiring collaboration.

Special thanks go to Johannes Ledergerber for steady support and encouragement. Furthermore, we would like to thank all former students (many of whom are now esteemed colleagues) who helped crystallize critical points through constant feedback. These people are, in alphabetical order: Cyril Albrecht, Martin Bichsel, Martin Dürst, Thomas Fromherz, Adrian Hartmann, Michel Hafner, Mauro Iaccobacci, Abdelhakim Gezhal, Michael Path, Susanne Suter, and Jody Weissmann. We are very grateful to Beat Rageth and Enrico Solca, system administrators at the MultiMedia Laboratory, for always finding the time to give us a hand in the administration of huge amounts of biomedical data. We would like to extend our gratitude to Walther Fuchs for hints regarding sources in art history.

Earlier drafts of Chapters 3 and 4 were read by Frans Zonneveld, Jeffrey Zhao, Klaus-Ulrich Wentz, and Christoph L. Zollikofer. Their comments were invaluable in helping us to eliminate misunderstandings and in finding intuitive ways to elucidate the complexities of medical imaging technology. Kris Carlson provided valuable comments and proof-reading remarks from a physical anthropologist's and native English speaker's point of view. We are also most grateful to Erik Trinkaus, whose hospitable lab provided an ideal surrounding for writing up the initial chapters during a research stay at Washington University in St. Louis.

Writing books on empirical research always implies financial backing by institutions that support that research; we would like to thank the Swiss NSF, the L.S.B. Leakey Foundation, the W.A. Schultz Foundation, the Strategic Research Funds of the University of Zurich, and various industrial and private sponsors for continuous support.

We are greatly indebted to our editor, Luna Han, who encouraged us to put our experiences in computer-assisted paleoanthropology in a textbook, and who had the patience to see the project grow from virtuality into reality.

C.P.E.Z.
M.S.P.-L.

# INTRODUCTION

Reality leaves a lot to the imagination.

—John Lennon

Imagine you are sitting in front of a recently unearthed fossil specimen. Although glimpses of its morphology have been exposed during the recovery and they exhibit previously unknown anatomic features, most of it remains embedded in sediment matrix. Before it can reveal any further unique morphologies, extensive and time-consuming preparation with chisels, needles, drills, or acid baths is necessary. During this process, the specimen is exposed to a considerable risk of damage, such that one might prefer to avoid those regions that contain the most intricate structures. Likewise, the freed fossil might be too brittle to be replicated with traditional mold-and-cast procedures. Yet preparation and replication are only two of many steps in the long process of reconstruction. The specimen is probably fragmentary, such that isolated and dislocated parts must be repositioned and lost parts completed. Even worse yet, the specimen may have been distorted during the process of fossilization; distortions are usually not correctable, because the petrified morphology cannot be bent back into its original shape. Ultimately, anatomic peculiarities of the fossil must be described, measured, and compared with other fossil and extant specimens, so that these findings can be interpreted in terms of function, biomechanics, developmental change, and evolutionary modification.

How can one take advantage of the present-day three-dimensional imaging and computer graphics technology to conceptualize new computer-based tools, which render data acquisition, preparation, and casting less risky and anatomic reconstruction more reliable? Which specific tools do we need to explore fossil morphologies in a noninvasive manner, but as exhaustively as possible? And

*Virtual Reconstruction: A Primer in Computer-Assisted Paleontology and Biomedicine.*
By Christoph P. E. Zollikofer and Marcia S. Ponce de León.
Copyright © 2005 John Wiley & Sons, Inc.

finally, how can we extend our toolkit beyond the classic repertoire of rulers, calipers, and goniometers?

Imagine, on the other hand, you are sitting together with an accident victim and discussing an indispensable but complex maxillofacial surgical intervention. Today, clinicians routinely use medical imaging tools such as computer tomography (CT), magnetic resonance imaging (MRI), ultrasonography, and other noninvasive techniques in conjunction with computer graphics techniques to obtain a detailed representation of the patient's external and internal morphology. Medical imaging tools not only enhance diagnostic accuracy, they permit the detailed planning of interventions on the basis of patient-specific anatomic data as well.

What is the relationship between a real body and its reconstructed, virtual counterpart? What is exactly behind these technologies? And how do the answers to these questions constrain virtual planning and rehearsal of surgical interventions, or the reconstruction and analysis of fossil morphologies?

This book is geared toward those who ask these types of questions. Its principal aim is to provide a primer for concepts and techniques in three-dimensional data handling, which has tremendous potential in the biomedical sciences and organismic biology. We define "handling" here as the entire chain of tasks including data acquisition, processing, graphical representation, interactive manipulation, and morphometric analysis. Bioscientists are confronted with increasingly complex computing devices for each of these tasks, and with concomitant increases in software complexity. Most scientists, however, want to use computers to tackle specific problems. These scientists are concerned typically only *that* computer tools work, and less interested in exactly *how* they work. In E-mail communication, as one example, it suffices to know how to send, store, and receive messages; only the message content is of primary concern, not the Internet protocol or the chip architecture of your computer. However, basic knowledge of how Internet communication software works can be helpful with security issues.

One is led to believe typically that computers provide an objective, impartial, number-based approach to our world, and, in fact, it seems that the subjective content of an E-mail message is unrelated to the digital technology used to transmit it. In the biosciences, however, where we typically deal with three-dimensional organisms and their representations in the form of digital data in a computer, the choice of a specific data acquisition device or software tool, and the way in which these are applied, contributes to the results that we obtain. Computer "images" of real-world objects represent various and sometimes transformed aspects of reality. As we will show, the process that leads from *real* (i.e., physical) reality to *virtual* (i.e., computer) reality involves various stages of *reconstruction* much as the process of physical reconstruction of fossils, or of injured patients, involves. How reconstruction operates at each of these stages is the principal theme of this book. We all know that computer hardware changes rapidly, and that software is even more volatile. Nevertheless, concepts behind digital imaging and computer graphics are well-established. We concentrate on the "invariants" of computer tools. However, concepts can only be understood through practice. Accordingly, we provide frequent examples so that the reader has an opportunity to acquire hands-on experience

As a consequence of our own research, this book preferentially draws examples from physical anthropology and paleoanthropology (the study of human evolution based on the organismic remains of hominids) and from clinical applications, most notably maxillofacial surgery. We attempt to intersperse examples from other areas, notably paleontology.

Because computer hardware and software develop at a faster pace than printed books, it would be pointless to delve into product-specific features. These are covered amply in textbooks, software manuals, and online guides. Likewise, computer graphics and medical imaging are rapidly evolving research fields; research journals documenting new developments abound, and excellent textbooks cover basic concepts and methods, which are less prone to change. This book by no means replaces direct reference to such resources. Nevertheless, for the bioscientist who uses computer graphics and medical imaging primarily to solve problems in his or her own research field, it is increasingly difficult to comprehend what exactly happens behind a computer screen or inside a computer tomograph, and at which level of detail this working knowledge is essential, relevant, or at least beneficial during daily work.

This book is situated between the two poles of conceptual frameworks and practical application. To bridge the gap between theory and practice, we adopted a four-layer strategy:

- *Layer One: The main text*. It deposits foundations of the material treated in each chapter, introduces the basic concepts of each field, provides technical information, presents practical applications, and discusses potential issues for further research. The jargon in computer science is often metaphorical, which invites us to make extensive use of metaphors to promote a ready and intuitive understanding. Computers and computations are always considered from a bioscientific user's perspective. Because this perspective is quite different from that in computer science, the structure and sequence of material is different from dedicated textbooks treating similar topics. Most notably, we follow an inductive approach, which leads from practical bioscientific questions to practical computer solutions, rather than deducing the latter from computational theory.
- *Layer Two: Boxes*. Many readers may desire additional information, specific explanations, and hands-on instruction on how to proceed in a specific situation. Others may have tangential questions that arise while reading

the main text. And sometimes, it is simply easier to explain concepts in a schematic style rather than in written paragraphs. To address all of the above, we use boxes, autonomous single-subject units appositioned to the main text. Boxes contain occasional case studies that show how proposed concepts and methods are applied in "real-world" situations.

- *Layer Three: Appendixes.* For those readers who desire a more formal approach to the concepts and computational procedures referred to in text and boxes, the Appendixes provide such information in a "classic" linear algebra formulation. Furthermore, current standard data formats for text, images, biomedical images, and graphical objects are listed in this part of the book.

- *Layer Four: Companion Internet site.* A book on computer-assisted methods and applications would be incomplete if it did not use the Internet for dynamic and interactive visualization. Regarding our specific research subject, the companion site will encourage readers to explore and experiment with several provided sample data sets and applets. Many of the features whose exposition is static and black-and-white in the book can be better understood if interaction and colors come into play. Finally, because we do not wish to " reinvent the wheel," the companion site will contain a suite of useful links, which will direct the user to tutorials, applets, data bases, and technical definitions, that is, to the information that is vital in everyday problem-solving, but which changes rapidly over time. This site will be updated on a regular basis.

The eight chapters plus appendixes of the book are organized as follows:

- Chapter 1 gives a general introduction to the field of virtual reconstruction. As you will discover, the term "reconstruction" is not restricted to the recomposition of disintegrated fossil fragments or the surgical rebuilding of impaired tissue. Rather, it pervades computer-based work in biomedical sciences from initial (i.e., data acquisition) to final (i.e., data analysis) steps. The reader will become acquainted with different aspects of reconstruction and their implications for planning and executing each step. Furthermore, we examine workflow in a typical biomedical application and its differences and commonalities with related fields such as computer-aided design (CAD) and virtual reality (VR), and so on. Basically, this chapter erects the scaffold of "virtual reconstruction," and subsequent chapters fill in the construction material.

- Chapter 2 begins at the basis. It addresses a single issue—how biomedical data can be represented or structured, both in physical reality and as a logical concept. First, we examine the various types of data that can be acquired with current medical imaging technology and discuss relationships among volume data, surface data, vertex data, and data categories derived from these fundamental types. We explore the means by which a computer represents and handles data, both logically and physically, and present an overview of current data formats used to store text, image, volume, and texture information.

- Chapter 3 builds upon the conceptual framework and describes how data acquisition is carried out. After introducing basic notions used to characterize imaging devices (both 2D and 3D) and to assess quality of the images they produce, we turn our emphasis toward more recent generations of data acquisition modalities and devices, such as surface scanners and, of course, computer tomography (CT).

- Chapter 4 is dedicated to the typical next step, data processing in VR workflow. Here, we trace various approaches to the methods of processing two-dimensional (image) and three-dimensional (volume) data, to obtain data structures relevant for the desired analyses.

- Chapter 5 reveals to the reader how virtual reality and scientific visualization operate, and, furthermore, how these techniques can be utilized in virtual reconstruction. This chapter concludes the technical part of proposing available methods.

- Chapter 6 is dedicated to a subject that emerged from our own research in paleoanthropology and that, we hope, will radiate into neighboring areas. This is the reconstruction of incomplete, fragmented, or distorted fossil specimens. Most importantly, we will demonstrate how computer tools assist anatomy-based reconstructive work. From paleoanthropology, we will branch into general anthropology, surgical planning, and forensics, all areas where one also is confronted with problems related to incomplete anatomic information.

- Chapter 7 is dedicated to methods we use to return virtual objects to physical reality. Using CAD tools, engineers and architects can design machines and buildings on the computer screen. Rapid prototyping technology (RP) permits the automated conversion of these virtual objects into three-dimensional physical models, much as an image on a computer screen can be printed out as a two-dimensional paper copy. Hard-copying objects in the biosciences follows a *reverse engineering* approach. Fossil fragments or the individual anatomy of a patient already exist as real physical objects, and RP technology enables the user to return a virtual fossil reconstruction or a custom-designed implant to physical reality.

- Chapter 8 leads into a field —the acquisition of morphometric data—that is central to many biomedical studies. Morphometry is an active research area that has witnessed major advances stemming from new concepts in treating three-dimensional data statistically and rendering the results graphically. It is not possible to treat this topic with the same level of detail as the subjects of the previous chapters. However, we attempt to brief the reader on the logic behind morphometric analysis, and the principal characteristics of the new geometric-morphometric approach. This will enable you to make your own decisions regarding choices of the most appropriate method among the many competing morphometric approaches.

- The Appendixes provide a more formal, algebra-based approach to relevant issues in computer graphics and biomedical imaging. There is also an index for the main text and the boxes.

# VIRTUAL RECONSTRUCTION

**1**

Sometimes I dream about reality . . .

—Manu Chao in "Mr. Bobby"

## 1.1 A VIRTUAL REALITY CONTEST

A quartet of Greek painters—Timanthes, Androcydes, Eupompos, and Parrhasios—once invited their famous colleague Zeuxis to compete over their artistic skills. Zeuxis excelled with a still life of grapes, which he painted so lively and realistic that the birds who happened to fly by started to pick at the canvas. Zeuxis, swollen with pride, invited his colleagues to do better. When Parrhasios' turn came, and Zeuxis impatiently asked him to unveil his picture, he realized that he was deceived—the veil was the picture.

Narrated by Pliny the Elder[1] in less than a single sentence, this famous legend became a leitmotiv of Western art for at least 20 centuries. The less-known continuation is as follows: Zeuxis then painted a boy bearing a bowl of grapes. When the birds pecked at the grapes, he angrily admitted that he had failed, because the birds did not seem to be scared.

To appreciate the concept of *virtual reality* (VR), these stories are highly instructive, because they reflect various aspects of the creation of reality and its perception. First, Parrhasios did better because he was able to cheat humans, not only birds, whereas everybody recognized Zeuxis' picture as a picture. This tells us that virtual reality, that is, the creation of a world that is *perceptually equivalent* to physical reality, depends not only on technical devices and skills but also on the perceptive system itself. In fact, perception ultimately happens in the brain rather than in the sense organs.

---

[1]Roman historian, who died on the 24th of August, 79 A.D, during an eruption of Mount Vesuvius. Uncle of Pliny the Younger, author of the famous book *Historia Naturalis*.

---

On the other hand, the fact that Zeuxis did not succeed in his second attempt demonstrates the limits of perceptual equivalence. Birds "know" that immobile objects are normally harmless, whereas within this category small, colorful objects are likely to be edible. Hence, scarecrows can be visually extremely simple, but they must convey the perceptual illusion of someone shaking his arms, an effect that a picture couldn't produce at the time of Zeuxis.

Obviously, virtual reality is more than creating a good picture, and more than just perceptual equivalence. Technically, it can be defined as a computer-based environment in which the user interacts with geometric representations of real-world or model objects, utilizing tools and performing manipulations that emulate physical tools and actions while being immersed in this virtual world (Sherman and Craig, 2002).[2]

Hence, VR must create perceptual *and* interactive equivalence of at least some aspects of the physical world (Fig. 1-1). Perceptual equivalence requires that the computer generate and emit signals that emulate the physical world and match the characteristics of our sensory perception. For example, a 50-frames per second (fps) color film with stereo sound is perceptually more equivalent to the physical world than a 5-fps mono video. On the other hand, interactive equivalence requires that the computer provide tools to receive and handle the user's physical signals within a realistic response time and in a realistic way, that is, according to the laws of physics, notably mechanics. From this perspective, videoconferencing at 5 fps provides a perceptually more equivalent setting than watching a film.

When we inspect, manipulate, and modify objects in physical reality, perception and interaction build an immediate feedback loop. For example, if we pick up a pen from our desktop and start writing, we *see* how the text is being formed, we *feel* the pen's resistance between our fingers and against the paper, and we *hear* it scratching on the paper. Moreover, we literally keep an eye on the straightness and length of the lines we generate.

Transposing this everyday situation into a VR environment is not a simple task. Graphics tablets and pens provide a partial solution, but imagine the task of transposing the paper, pen, and hand into virtual reality, while the user is wearing stereo glasses and a data glove connected to a force-feedback device that simu-

---

[2] In the context of computer graphics, the terms *tool* and *action* are used in a relatively wide sense. Tool denotes any set of procedures (algorithms) that can be applied to a data set. Action denotes the act of applying a certain tool to a certain data set, typically by means of a graphical user interface. In a grammatical sense, user, tool/action, and data set are related to each other in the same way as subject, predicate, and object. It also bears mentioning here that the tool-making capacities of hominids at the beginning of this millennium largely focus on the development and implementation of *virtual* rather than physical tools.

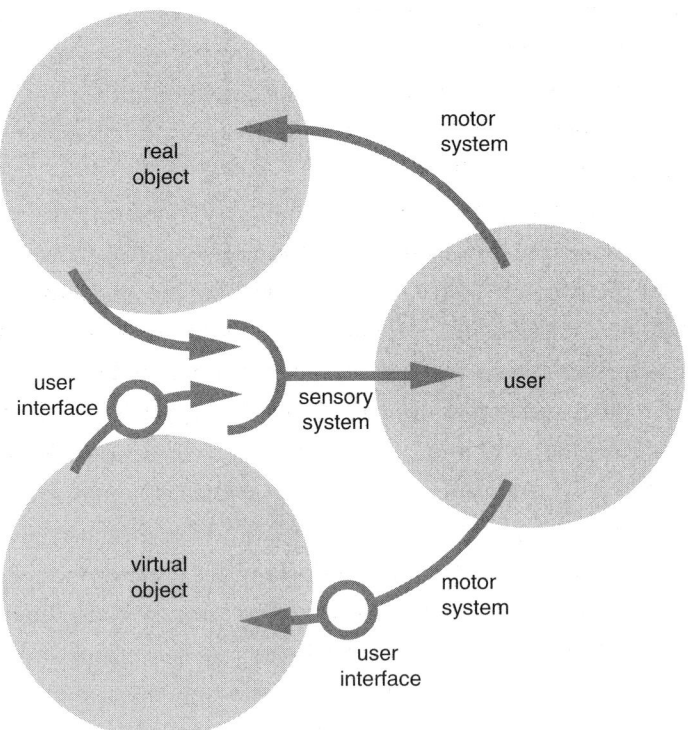

**FIGURE 1-1** Virtual reality (VR) and the human sensorimotor system. Humans perceive physical objects in physical reality with their sensory system (sense organs and sensory cortex of the brain) and modify them through actions or reactions of their motor system (motor cortex of the brain, muscles, and skeleton). A VR setting creates a parallel universe that emulates, transforms, and enhances reality. To be compatible with our sensorimotor system, VR must be perceptually and interactively equivalent to reality. This is achieved through user interfaces on the sensory and the motor sides of the system, for example, a computer display and a mouse or, in a more sophisticated setting, shutter glasses, a head movement monitoring device, and a data glove.

lates the physics of the pen's reaction force against the paper and against the user's hand. This is technically and computationally feasible, but the fundamental problem is the coupling between our sensorimotor system and the simulated reality: Perceptional equivalence in the real world relies on closed feedback loops. As soon as the pen touches the paper, reaction forces stop its downward trajectory, the ink starts flowing, and the written text becomes visible. As we are extremely sensitive to delays in the feedback loop, the illusion of reality—known as the *immersion effect* in a VR setting—is prone to crumble at the slightest incongruence.

Our sensorimotor system normally works sufficiently fast to keep track of real-time changes in the environment and to act and react adequately. What a delayed feedback signifies in real life can best be illustrated by a human trying to catch a fly. Because the visual system of a fly operates at much higher frame rates than that of a human, and a fly has less inertia but greater relative muscular power output than the human arm and hand, the fly almost always manages to break up the human feedback loop connecting visual tracking and catching.

Whereas the speed of reality sometimes derides the speed of our sensory system, the basic problem of VR is exactly the opposite: The speed of our sensory system is typically higher than the speed at which a VR system can emulate reality. For example, simulating the movements of the fingers of the virtual hand writing a text, detecting the collision between the paper and the pen, and rendering the entire visual scenery necessitate extremely demanding computations.

Is VR, as a form of re-creating reality, technically feasible at all? The answer is: Yes, at least in part, and especially in bioscientific applications. Fortunately, our brains are highly skilled in dealing with near-reality situations, and our sensorimotor system is used to handling metaphors of reality. For example, game stations in connection with joysticks do not represent the real, physical world. Yet computer game players get rapidly trained to handle these devices and soon will feel immersed into the virtual reality setting. Our proneness to illusion is probably an effect of our over two million-year history as tool users and tool makers. Tools, whether stone axes or computers, are mediators between our body and the outside world, and they provide a metaphoric access. In this sense, VR is an excellent tool for interacting with data, partly on a physical basis and partly on a metaphoric basis. Rather than perfecting the physical equivalence of VR tools, it is likely more sensible to take systematic advantage of our brain's ability to compensate with these shortcomings in computer interfaces, the mediators of a quasi-physical world of *enhanced*, or augmented, *reality* (Bowskill et al., 1997; Falk et al., 1999; Azuma et al., 2001).

## 1.2 VIRTUAL RECONSTRUCTION

How do these considerations bear on the topic of this book, virtual reconstruction? Let us discriminate between two paradigmatic settings of VR in scientific applications (Fig. 1-2). In the first, the user starts with a concept, a hypothesis, or a plan about how data or objects should appear. This can be an architectural plan of a building, a new design of a car, or a molecular structure, but it can also be an abstract model about how objects behave, for example, how planets circle around a sun, how plants ramify, how chemicals combust, or a set of instructions, like how an artificial creature behaves under specific circumstances. These concepts, plans, and models first must be realized, that is, converted into geometric models and graphical representations, then formulated as simulations of real-world processes, and finally visualized in a virtual scenery. This is the setting of *computer-aided design* (CAD), or *virtual construction*.

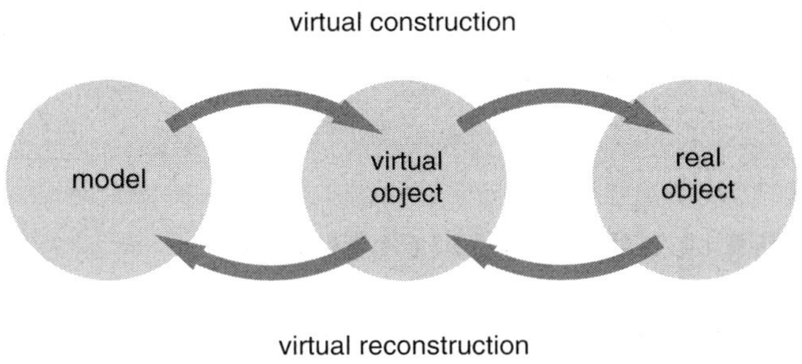

virtual construction

model    virtual object    real object

virtual reconstruction

**FIGURE 1-2**  Virtual construction and virtual reconstruction. During virtual construction (engineering and architecture), a mental model is transformed into a virtual object and then into a real object. During biomedical virtual reconstruction and reverse engineering, real objects are transformed into virtual objects, from which design principles (mental models) and replicas (physical models) can be drawn.

In the second setting, the user starts with real physical objects rather than mental concepts, for example, the human body, a fossil, a machine part, the surface of the earth. Because these objects already exist, the principal task is transforming them into geometric and graphical representations that can be handled subsequently in a VR environment. In industrial applications, this approach is known as *reverse engineering* (Ingle, 2001), because it aims to draw blueprints of preexisting objects. In biosciences, this is called *virtual reconstruction*.

Another major incentive behind virtual reconstruction is to enhance reality in a noninvasive way. Many essential tasks in the biosciences cannot be performed on real physical objects or they would at least expose these objects to unnecessary risks. Think of practicing a new and complex technique of surgical intervention, or the preparation of unique fossil fragments still embedded in sediment. These manipulations entail irreversible physical alterations, such that it is important to explore the objects and gain practical skills before the manipulations are performed. Virtual reconstruction meets this challenge by transforming physical objects into virtual objects and physical tools into virtual tools, so that the required tasks can be performed in a VR setting.

Yet, in many respects, virtual reconstruction goes beyond the mere replication of reality by opening new paths toward enhanced reality. Modern biomedical imaging methods, such as computed tomography (CT) and magnetic resonance imaging (MRI), permit noninvasive but pervasive acquisition of large data volumes, which can represent various physico-chemical properties of the organism. These data are not readily accessible to human perception, and thus they call for new means of visualization and exploration, new tools for manipulation and modification, and new methods for quantitative analysis.

How do we proceed during virtual reconstruction? We focus here on two specific subjects that form the empirical background of our own research. The first is computer-assisted paleontology—most notably paleoanthropology—and the second is computer-assisted surgery.

## 1.3 COMPUTER-ASSISTED PALEONTOLOGY

One of the major obstacles in many areas of paleontology is the lack of extensive fossil samples and the incompleteness of most specimens that have been found. In paleoanthropology—the area of anthropology pertaining to the fossil evidence of human evolution—specimens are even rarer, so that the difficulties encountered during the analysis of fossil morphology are multiplied. This is where *computer-assisted paleontology* (CAP) is beneficial (Zollikofer et al., 1998). Given the desire to gain maximum information from minimum material evidence, the primary challenge becomes enhancing investigative power while minimizing invasiveness of the methods used for data acquisition, fossil preparation and reconstruction, anatomic diagnosis, and morphometric analysis.

Following the paradigm of virtual reconstruction (Fig. 1-2), the basic premise behind CAP is to enable the user to perform the complete sequence of tasks involved in fossil analysis, but within a VR environment. The typical workflow of CAP is shown in Figure 1-3.

### 1.3.1 Data Acquisition

The first step in the reconstruction of a virtual fossil is the acquisition of three-dimensional data from the original specimen. Computed (or computer) tomography (CT) is the method of choice, as it provides X ray-based cross-sectional images revealing internal structures of solid objects in a noninvasive manner. Using a noninvasive technique to acquire such data is of utmost importance, given the paucity of fossil specimens. As a technique, CT was developed in the 1970s with the specific aim to improve volumetric radiographic techniques for clinical purposes. Today, CT has become a standard tool in medical diagnostics. Likewise, soon after its conception, CT was adopted for "fossil diagnostics," notably to reveal internal anatomic features and regions still covered by matrix (Jungers and Minns, 1979; Conroy and Vannier, 1984; Wind, 1984). Simultaneously, the technology was adapted and refined further for industrial applications, where it has been used extensively for reverse engineering and materials testing.

The process of CT imaging is radically different from classic radiography or photography. CT images must be *reconstructed* from radiographic projection data, which is a computationally intensive task. During the early development of CT,

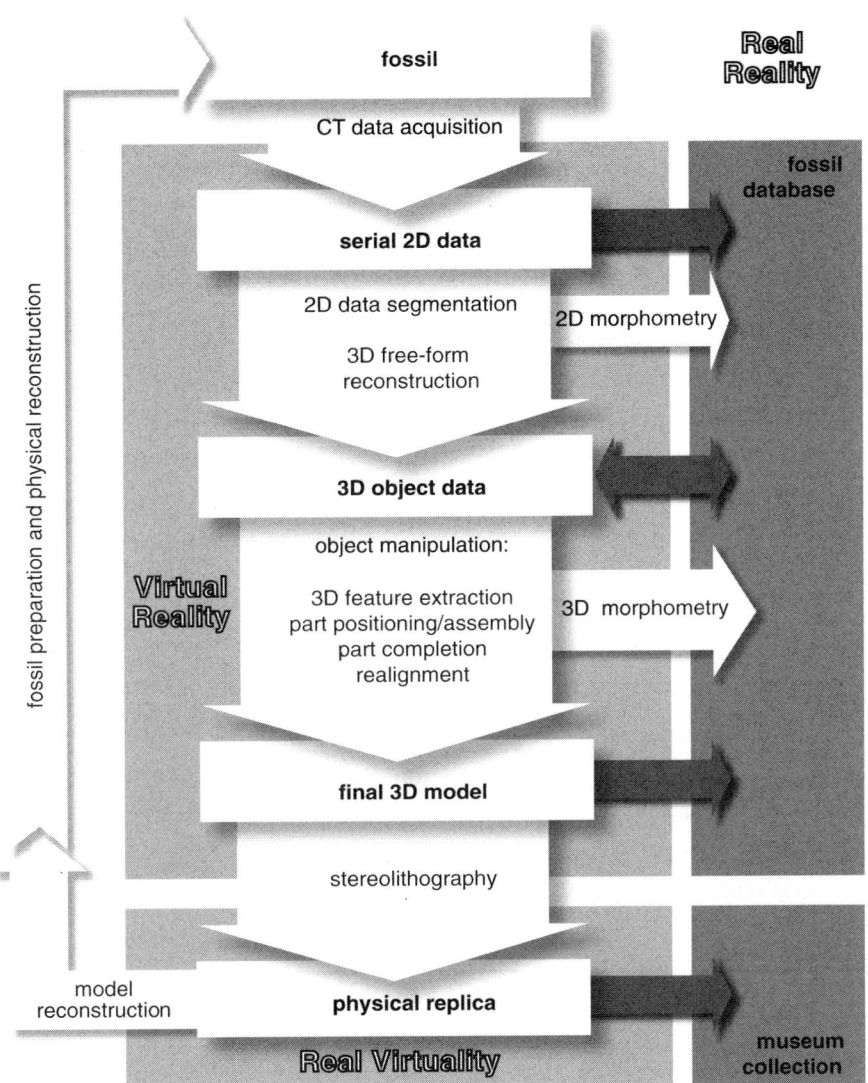

**FIGURE 1-3** Computer-assisted paleontology (CAP). Computed tomography (CT) is used to acquire serial cross-sectional image data from fossil specimens. All subsequent steps are performed noninvasively, in virtual reality (VR) through interactive computer graphics tools. Separating fossil from surrounding matrix is possible with image processing procedures. Image data are transformed subsequently into three-dimensional object representations, which can be manipulated interactively on the computer screen to plan and perform reconstructive tasks. After the virtual reconstruction of fossil morphology, extensive morphometric analyses can be performed in two or three dimensions, using comparative data from a database. Finalized virtual fossils can be transformed into physical models with rapid prototyping (RP) technology. Creating these "real-virtuality" (RV) objects presents a noninvasive alternative to conventional casting techniques. CAP procedures are iterative and interconnected; at any stage, data from "real" reality, VR, and RV can be compared to or combined with information from a database, and reconstructive or morphometric tasks can be refined (after Zollikofer et al., 1998).

therefore, acquisition and reconstruction of a single cross-sectional image took between 4 and 6 min.[3] With faster computer processors, and concurrent progress in X-ray detector technology, this time has been reduced to less than 1 s. Today, it is possible to acquire entire data volumes consisting of stacks of several hundreds of cross-sectional images in well under a minute.

### 1.3.2 Data Segmentation and Three-Dimensional Reconstruction

Once volumetric data of fossil specimens have been acquired, data segmentation procedures are applied and relevant object structures from the data set are extracted. For example, it is possible to free a fossil fragment embedded in sediment with a "computer chisel," and it is possible to extract a mineralized tooth crown from within its crypt because of the elevated X-ray density of enamel relative to the surrounding bone. After data segmentation, a geometric description and a graphical representation of the fossil fragments are created through three-dimensional reconstruction procedures. Up to this point, CAP followed the *reverse engineering* approach described earlier (i.e., a preexisting physical object is converted into a virtual object).

Manipulating fossil fragments on a computer screen as virtual objects instead of handling the real fossil specimens is of immediate practical and theoretical benefit. Whereas physical preparation and reconstruction are highly invasive and potentially destructive processes, the analogous computer-assisted manipulations are completely noninvasive. Additionally, although it is difficult or often impossible to reveal hidden structures in a real fossil without destroying it, CT volumetric data provide unlimited access to all external and internal anatomic structures from an infinite number of views.

### 1.3.3 Virtual Fossil Reconstruction

Using computer tools, it is possible to *reconstruct* fragmentary fossils in virtual reality (Zollikofer et al., 1995). This task can be compared to the assembly of a three-dimensional puzzle in which important pieces are unavailable. The underlying goals are to correct distortions that occurred during fossilization, to reestablish anatomic connections between preserved parts, and to replace missing parts by mirror-imaging preserved antimeres. To avoid the temptation of preconceived morphologies, each reconstructive step follows predefined anatomic criteria.

---

[3] The EMI *mark I* CT scanner, introduced in 1972, had a standard scan time of 285 s and a long scan time of 360 s, during which time the images were acquired and reconstructed (actually, two images were made simultaneously).

Strict criteria at each stage of reconstruction ensure reproducible results and a procedure accessible to other researchers. Furthermore, by performing alternative reconstructions and comparing them, the reliability of the process as a whole can be evaluated and a range of potential reconstructions is possible.

An additional benefit in virtual reconstruction compared to physical reconstruction is that each puzzle piece can be fixed in virtual anatomic space without the need of glue, plaster, or stabilizing elements. Virtual reconstruction is an iterative process in its essence, in which reconstructive attempts can be gradually refined without physically dismantling earlier versions and exposing the original specimens to further damage.

The finalized virtual reconstruction of a fossil marks an important step in the workflow of CAP; however, it is not the end point. Virtual fossils can be submitted to comparative morphometric analysis, used as a guide during physical preparation, and, last but not least, transformed into physical hard copies.

### 1.3.4 From Virtual Reality to Real Virtuality

*Real virtuality* (RV) denotes an environment in which a user interacts with physical models of three-dimensional objects generated or modified by computer-assisted procedures (Bresenham et al., 1993; Zollikofer and Ponce de León, 1995). Real virtuality replicas of virtual reality fossils can be created with rapid prototyping (RP) technology. RP was devised originally for the rapid production of physical models of CAD-generated industrial parts. Many RP technologies use the principle of stereolithography, where, analogous to the construction of topographic models from piles of cardboard layers, the object is constructed from automated superpositioning of cross-sectional layers. In this manner, it is possible to build models containing arbitrary geometric complexity, for example, including cavities, or parts within parts, in a single production step. Producing hard copies of virtual fossils via RP technology thus represents a valuable noninvasive alternative to classic mold-and-cast techniques. Being physical objects, RP models convey touch-and-feel sensory input and can therefore be handled, explored, and assembled in a more realistic fashion than virtual objects on a computer screen. For example, replicas drawn at different stages of fossil reconstruction can be used to monitor complex matching tasks, such as aligning masticatory occlusion between upper and lower dentitions.

### 1.3.5 Databases and Morphometry

Comparative morphometric analysis of reconstructed fossil specimens represents the final step of CAP. Morphometric investigations serve various aims, from functional and biomechanical analysis to the analysis of patterns of evolutionary and developmental differentiation. Nevertheless, a common denominator in morphometry is the search for patterns of covariation in the sample and the *reconstruction* of the underlying functional constraints and biological mechanisms.

The possibility of archiving and retrieving data at various stages of a virtual reconstruction opens new avenues of quantitative comparative analysis. Because virtual fossils are graphical objects, they provide an ideal basis for exploration of a wide variety of one-, two- and three-dimensional measurements. By supplementing the classic repertoire of rulers and calipers with virtual morphometric tools, spatial position of anatomic landmarks, as well as distances and angles between these landmarks, areas of muscular attachment, paranasal sinus volumes, surface curvatures, bone thickness distribution, moments of inertia, strain distribution, and so on can be measured. Making use of an extended database of digital fossils and the virtual desktop, it becomes possible to inspect fossil specimens that are literally separated by continents in side-by-side comparisons and to derive new measurement series for testing hypotheses without requiring reexamination of the original specimens.

## 1.3.6 Virtual Reconstruction in Space and Time

To summarize, CAP includes processes of virtual reconstruction at various stages. First, CT slices are reconstructed from X-ray projection data. Then, geometric object descriptions are reconstructed from the CT data volumes. Third, the computer-assisted reconstruction of fossil morphologies is made from the geometric object descriptions (e.g., fragmentary remains). Finally, morphologic variation in a sample is assessed by comparing fossil morphologies, with the aim of reconstructing function or patterns of developmental and evolutionary change.

Let us consider the last two stages—fossil reconstruction and morphometry—from a different perspective. Reconstruction must overcome the problems of spatial recovery of incomplete and disturbed morphologies. Fossils, however, are time windows into the course of evolution, and an important question is always, How did *time* affect fossil morphology? In living organisms, morphology represents a complex interplay between processes on evolutionary and developmental timescales. In fossils, an additional timescale should be considered, namely, diagenesis, which comprises all taphonomic and geological postmortem processes that alter expressed morphologies of an organism.

Fossil morphology, therefore, is the result of accumulation and integration of changes in form during three different scales of time (Fig. 1-4). First, developmental processes integrate form change over most of an individual's lifetime.

$$morphology = \iint (growth \otimes mutation)\, dt$$
$$\text{\small ontogeny, phylogeny}$$

$$fossil\ morphology = \iiint (growth \otimes mutation)\, dt$$
$$\text{\small diagenesis, ontogeny, phylogeny}$$

**FIGURE 1-4** Morphologies of living and fossil organisms. The morphology of a living organism can be understood as the double integral of morphogenesis over phylogeny and ontogeny. In the genesis of fossil morphology, diagenetic alteration acts as an additional integrator that obscures the original morphology. Fossil reconstruction signifies undoing the effects of this last integrator (after Zollikofer, 2002).

Second, in the evolutionary time scale, processes of mutation, selection, and adaptation integrate changes within the genome, which entail morphologic change at the species level. Third, geological processes create postmortem changes in the fossil morphology, such as degradation, alteration, or distortion. In this context, virtual reconstruction has the duty of "undoing" diagenesis by stepping backward through time until the time of death of the individual.

## 1.4 COMPUTER-ASSISTED SURGERY

Although computer-assisted paleontology represents an area of basic bioscientific research, concepts of virtual reconstruction can be applied in clinical environments as well. In fact, striking parallels exist between virtual reconstructions of fossils and virtual surgical interventions on patients, most notably in the areas of cranio-maxillofacial surgery and orthopedics (Fig. 1-5). For example, planning reconstructive surgical procedures after cranial trauma is similar to reconstructing a fossil skull from isolated fragments. Or the design of custom joint implants is based on computer graphics techniques similar to those used in the completion of missing parts in fossil specimens. Furthermore, surgical correction of cranial malformation has a counterpart in realignment of fossil specimens that have been distorted during the process of fossilization. RP models made from patient data permit surgical rehearsal and preoperative fitting of prosthetic parts, much as RP models made from fossil data help perform complex matching tasks.

Finally, with regard to morphometric analysis, a direct connection can be established between paleoanthropological and cranio-maxillofacial databases. An exact quantitative grasp both of the three-dimensional cranial morphology of fossil hominids and of cranial malformations in modern humans can yield deeper insights into the processes underlying craniogenesis and their evolutionary and functional significance.

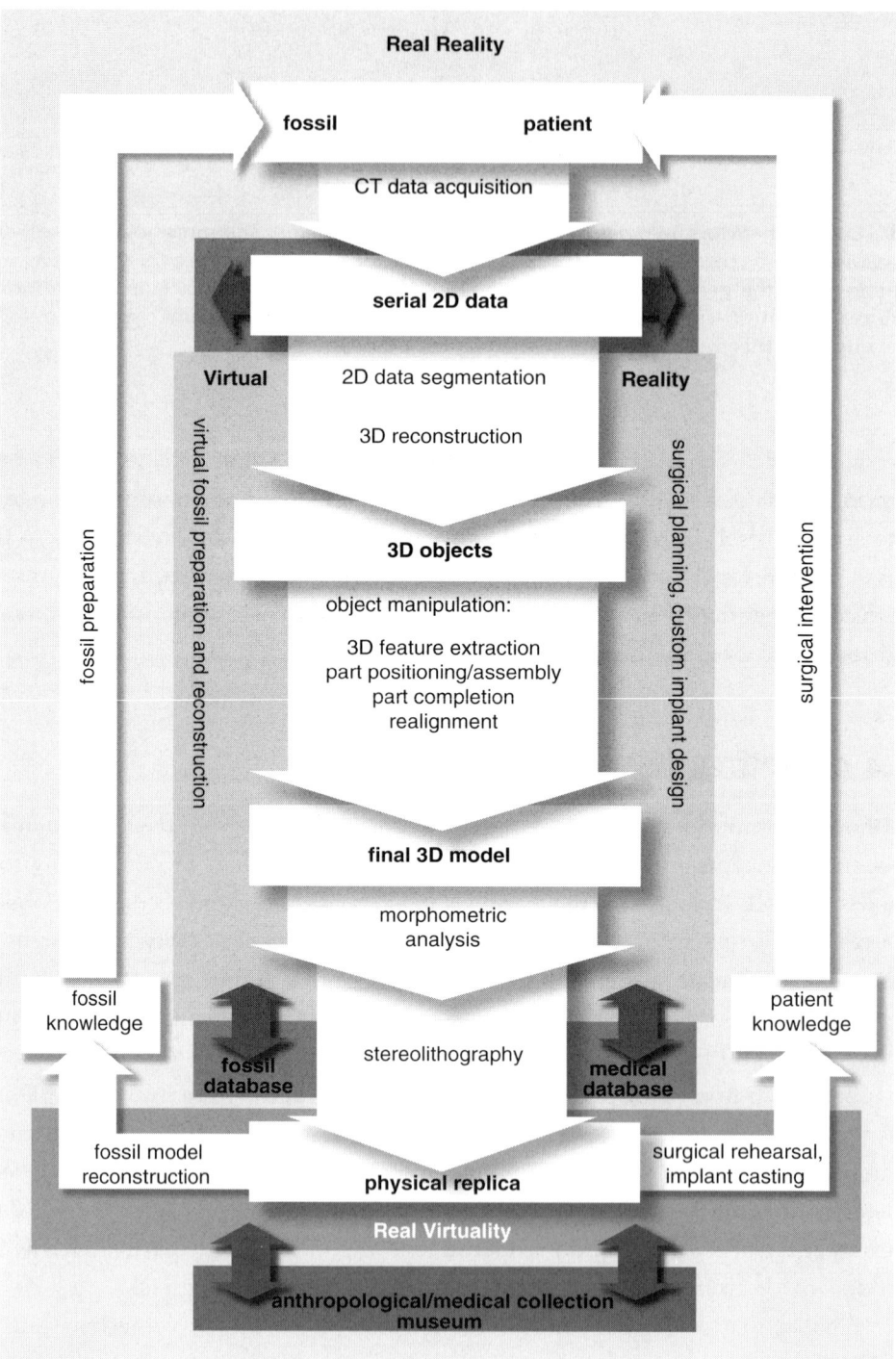

**FIGURE 1-5** Virtual surgery and CAP. Exploiting parallels between computer-assisted fossil reconstruction and surgical planning, patient data can be processed with computer graphics procedures originally developed for computer-assisted paleoanthropology (see Fig. 1-3). For example, CT-based hard/soft tissue separation is computationally analogous to distinguishing bone from matrix in virtual fossil preparation; each uses density differences to distinguish objects. Planning and simulation of surgical interventions involves the isolation and reposition of bony parts with computer graphics tools that are equivalent to those used during fossil reconstructions. The production of stereolithographic replicas is particularly important for surgical rehearsal and custom implant design/casting.

## 1.5 FURTHER READING

The fundamentals of virtual reality, real virtuality, and augmented/enhanced reality are treated in Drew (2003), Sherman and Craig (2002), and Ingle (2001). Feiner (2002) gives an inspiring overview of the potential of the field. Reverse engineering is treated from an industrial point of view by Ingle (2001), whereas the Visible Human data set represents the paramount example of biomedical reverse engineering (Spitzer, 1998).

# DATA REPRESENTATION

**2**

*Polonius:* What do you read, my lord?
*Hamlet:* Words, words, words.

—William Shakespeare, Hamlet, Prince of
Denmark, act II, scene II, line 199

## 2.1 WORLD FOOD ON A CHESSBOARD

In 1256, Ibn Kallikan proposed to put grains of wheat on a chessboard, one on the first square, two on the second, four on the third, eight on the fourth, and so on. The resulting number of grains is $2^0 + 2^1 + 2^2 + 2^3 + \ldots + 2^{63}$ and corresponds to the following 64-digit number in the binary system:

1111111111111111111111111111111111111111111111111111111111111111

This equals 18,446,744,073,709,551,615 and, in grains, amounts to 500-fold the world annual production of wheat.

This astounding number corresponds to the possible states of information that can be stored in a 8-byte computer word, as well as to the number of all conceivable $8 \times 8$-pixel black-and-white images. Most of these images are rather uninteresting, and only two of them represent the chessboard pattern.

In this chapter, the reader will be introduced to the binary system, bytes, computer words, pixels, images, and other terms used to describe the structure of computer data. We will look at these structures from two contrasting sides. From a bottom-up perspective, we examine how data are represented and manipulated in a computer, both physically and logically. This will familiarize readers with the basic concepts used to characterize computer hardware, data storage devices, and data formats. From a top-down perspective, we consider the various types of data that are encountered typically during computer-based biomedical applications.

*Virtual Reconstruction: A Primer in Computer-Assisted Paleontology and Biomedicine.*
By Christoph P. E. Zollikofer and Marcia S. Ponce de León.
Copyright © 2005 John Wiley & Sons, Inc.

We provide an overview of the structure of one-, two- and three-dimensional data and examine their relationships from the perspective of data acquisition, processing, and analysis (exactly how such data can be acquired, processed, and analyzed will be the subject of other chapters).

## 2.2 FACTS ABOUT DATA TO GET DATA ABOUT FACTS

### 2.2.1 Analog and Digital Data

The principal difference of a computer compared to other data processing devices is that all information is handled in digital form. The premise behind digital representation of information as opposed to analog representation of information is to separate the intensity of the information-bearing signal amplitude from the actual information the signal contains. Our nervous system utilizes this principle in long-distance information exchange. Transmitted information is contained in the temporal frequency of action potentials, known as spikes, rather than in the amplitude of spikes. In practice, the separation between signal contents and signal intensity is not straightforward. For example, knocking on a door tells the person on the other side that you want to enter, and the acoustic intensity of the signal reveals the urgency of your request. Knocking in Morse code, on the other hand, means transmitting a digital signal. Likewise, spoken language has mixed characteristics; shouting versus whispering can be as meaningful as the difference between yes and no. Written language, if we disregard calligraphy, is purely digital. No matter how much an old text has faded away over the years, as long as one can identify the characters, the textual information may be recovered.

But there is an additional potential problem in digital representation. As long as we do not know the meaning associated with a sequence of digits, it is impossible to make sense of the digital representation. As an example, it is possible to read Etruscan texts, but still impossible to understand them. For digital data, we need to know the *data format* to make sense of the digitally stored information.

This draws us into technical issues of digital data representation. Let us consider the storage of image data. Analog images can be "read" and visualized directly from the storage device without additional knowledge and special devices (think of 30,000 year-old cave paintings and of 150 year-old daguerreotypes). However, dust and scratches will degrade the analog signal, with time posing a substantial problem in the restoration of ancient pictures and films. Digital data storage does not encounter this problem. However, to read and visualize digital images, computer hardware and software are required for decoding the image data format, for reading it from the storage device, and for displaying it on a screen or a printout (think of your last attempt to open an image stored in an unknown format).

With these things in mind, one may jump to the conclusion that digital data storage is "forever," whereas analog storage is prone to deterioration and decay. Unfortunately, the opposite is true, not because the digital signal degrades more rapidly, but because encoding and decoding procedures and devices are short-lived. Data formats, both logical and physical, change with the design of computer hardware, computer software, and input/output devices, all of which experience rapid technological turnover.[1]

The primary aim of the following sections is to provide the reader with knowledge about the constants in the volatile world of digital data.

## 2.2.2 Bits, Bytes, and Words

The "atom of information" in the digital world is the *bit* (*b*inary dig*it*). It can assume only two discrete values, or states, 1 or 0. This polarity can also be thought of as yes/no, on/off, black/white, and so on. Physically, a bit is expressed as the presence or absence of some minimum amount of analog information, for example, in the form of electric charge, magnetization, or reflectivity at some specific location in a computer memory chip, hard disk, CD, and so on. Although these data storage media are two-dimensional objects, bits are always ordered logically along a single, long one-dimensional array. To store and retrieve information, it is necessary to specify each bit's position in the array. In computer terms, this is the physical *address*. Note that both the status of the bit (0 or 1) and its address (e.g., 3728703891) are integer numbers, that is, discrete information units.

Using a metaphor, we may consider the bits as yurts along one long road (Fig. 2-1). Each yurt has a number (the address) and a status telling whether its fireplace is lit (1) or not (0). Note that the size of the fire is irrelevant. The process of information storage can be envisioned as sending out a messenger to a specific yurt with instructions to light or extinguish the fire according to some superior instance (e.g., the computer user tapping on the keyboard). Likewise, the process of information retrieval can be compared to sending out a messenger to determine the status of a fireplace in a given yurt. Obviously, sending only one messenger to only one yurt on a long single-lane road is highly inefficient, with regard to speed and distance. Using the concepts of urban architecture, we may visualize how computer design can become more efficient:

- *Housing*: Minimize travel distances and general space requirements by reducing the size of the yurts, uniting them into larger building blocks, and

---

[1] Most of us are confronted with the trade-offs between digital innovation and decay on a daily basis. For example, try to read data on an old floppy disk or a large magnetic tape from the 1980s with modern computer technology.

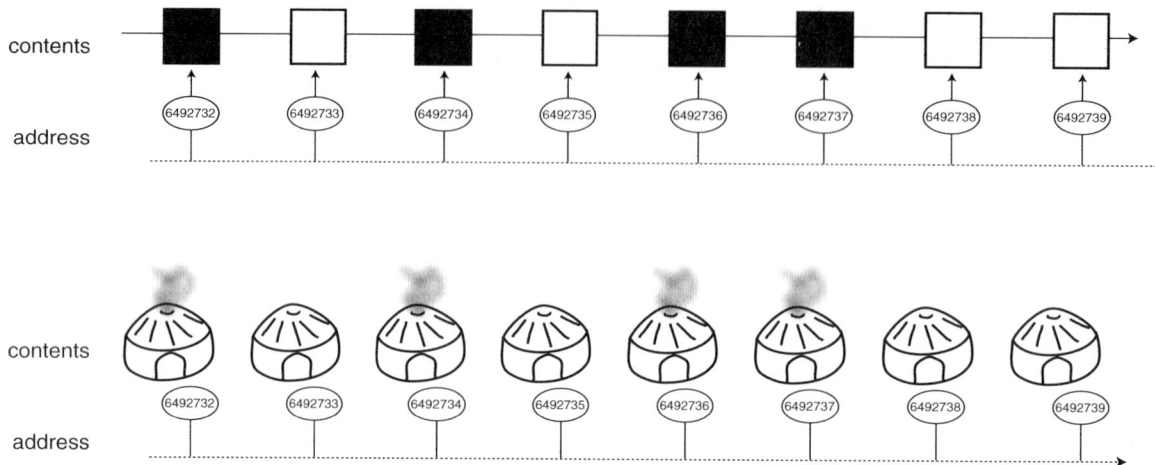

**FIGURE 2-1**    A metaphor of computer-style bit addressing.

constructing into the third dimension (while still conserving the linear addressing system, as in phone numbers).

• *Transportation*: Maximize the capacity of the road by sending out more than one messenger at a time along a multiple-lane highway, and by increasing the traveling speed of each individual messenger.

This thought experiment directly relates to computer chip design, which is the architecture of integrated circuits for digital data processing and storage. The first point concerns computer memory architecture. Over the past few decades, the storage capacity of digital devices has doubled, and space requirements have been halved, almost annually. The second point concerns processor architecture. In modern computers, bits are stored and retrieved at unfathomably high rates. For example, a 32-bit 1-gigahertz processor uses a 32-lane data highway, on which 32 messengers travel simultaneously one billion (1,000,000,000) times per second. Simultaneously, the amount of energy required to represent one bit has decreased, and the speed at which bit states are probed and/or changed has increased nearly to attain the limits of the physically feasible (but see Box 2-1).

Considering bits from a logical perspective leads to a subject of great practical relevance, namely, how information is *coded* in a computer. Bits make sense only if they come in sequences, together with a specific rule set that tells how the sequence is to be interpreted. For example, the integer number 139 is interpreted as

$$139_{10} = 1 \times 10^2 + 3 \times 10^1 + 9 \times 10^0 \tag{2-1}$$

The suffix 10 denotes that we use the decimal system. Computers use the binary system (base 2) to denote the same quantity as

**BOX 2-1**

**WHY ARE COMPUTERS SO SLOW?**

Compared with humans, computers are slower in carrying out everyday processing tasks such as recognizing objects in an image, despite "clock rates" of neurons (50 hertz) that are substantially slower in comparison to those of modern computers (on the order of gigahertz). The critical difference is the number of computing steps that can be carried out simultaneously. The neuronal system is massively parallelized. For example, the human retina consists of approximately 125,000,000 light receptor cells, which are "wired" to several layers of second-order neurons. These cells are highly interconnected and *simultaneously* perform data input/output, conversion, and processing tasks, before the signal converges to no more than 1,000,000 neurons, which send the information via the optic nerve to the brain. With these benchmark parameters, a very conservative estimate of the number of computations per second yields $1.25 \times 10^8 \times 50 = 6.25 \times 10^9$, compared to $1 \times 10^9$ in a 1-GHz processor. However, because neurons are organized in networks, the actual number of computations in the retina is much higher than this estimate.

So, the obvious answer to why computers are so slow lies in their parallel processing capabilities. Parallel processing is a highly effective strategy as long as computationally intensive tasks can be split up into several independent subtasks that can be handled by multiple processors. However, biocomputing tasks, such as object recognition, require highly interconnected solutions, such that a great proportion of the computing time gained through parallelization is spent in processor intercommunication.

$$10001011_2 = 1 \times 2^7 + 0 \times 2^6 + 0 \times 2^5 + 0 \times 2^4 + 1 \times 2^3 + 0 \times 2^2 + 1 \times 2^1 + 1 \times 2^0 \quad (2\text{-}2)$$

In both cases, the *position* of the digit within the string denotes the power to which the base is elevated, and the *value* of the digit represents a scaling factor of the respective power. In computer jargon, left- and rightmost bits of a number are therefore called its *most* and *least significant bits*, respectively. We will return to this notion below.

A sequence of 8 bits is known as an *octet*. However, this precise quantity is far less popular than *byte*, which today serves as a de facto synonym (byte originally corresponded to any short bit sequence that could be handled by a digital data processing device as an input/output unit). Bytes are widely used as basic information units of computer systems. One byte can represent $2^8 = 256$ different states of information, as given by the binary patterns $00000000_2$ to $11111111_2$. For example, this holds all integers between $0_{10}$ and $255_{10}$, which may code for the 256 gray values of a computer display, or for a set of 256 different text characters.

For the written representation of a specific bit pattern contained in a byte, the decimal system is not well adapted and the binary system is too cumbersome. Computer scientists use the hexadecimal notation. In this system, each digit can assume 16 different states, which are expressed with the numerical characters 0 through 9, followed by letters A to F. Accordingly, $139_{10} = 10001011_2$ is equivalent to

$$8B_{16} \qquad (2\text{-}3)$$

where "8" stands for $8 \times 16^1$ and "B" for $11 \times 16^0$. To indicate hexadecimal notation (hex-code, in computer jargon), the prefix "0x" is used rather than the subscript 16, such that we obtain

$$0x\ 8B \qquad (2\text{-}4)$$

With this notation, the binary pattern contained in one byte can conveniently be represented with just two digits. Accordingly, hex-code notation is a convenient way to represent binary data, irrespective of what the bit pattern actually signifies.

If we think of bytes are *characters* of a written language, one unit higher up in the hierarchy is the *word*. This is the byte string that the central processing unit (CPU) of a computer can handle in one processing cycle. For example, the 32-bit processor of a last-millennium PC understands and handles 4-byte words at rates of some 100 MHz (the 32-lane highway alluded to above), whereas the more recent generation of computers uses 64-bit processors to handle 8-byte words at rates of more than 1 GHz (i.e., 1 billion times a second). Evidently, the number of different states of information that can be represented with multibyte arrays increases exponentially with the number of bytes. A 4-byte (= 32-bit) word can represent $2^{32}$ = 4,294,967,296 different states of information, and a 8-byte (= 64-bit) word $2^{64}$ = 18,446,744,073,709,551,615 different states (the number of wheat grains in the chessboard legend).

Let us now consider the 4-byte word

$$1101\ 0001 \quad 0101\ 1110 \quad 1010\ 0010 \quad 1000\ 1011 \qquad (2\text{-}5)$$

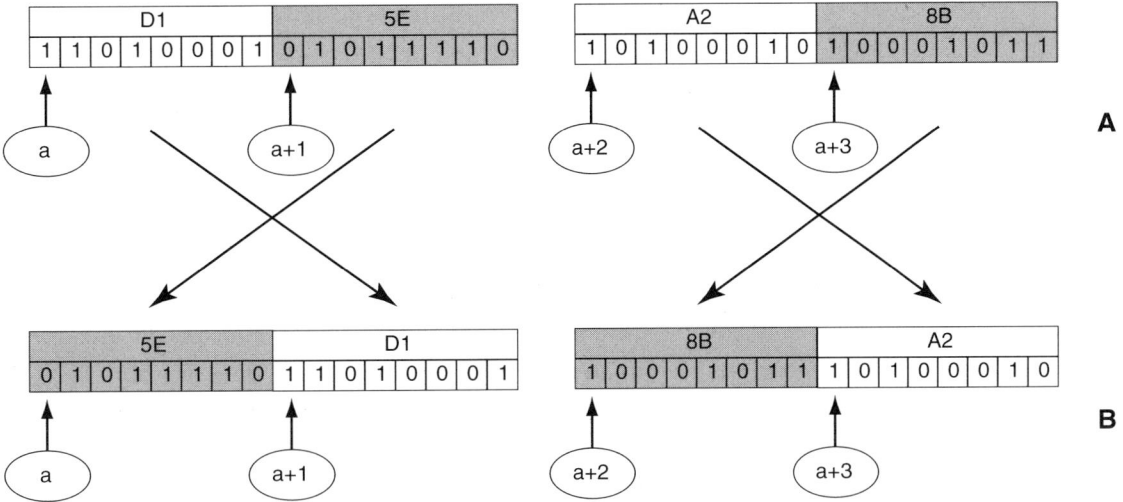

**FIGURE 2-2** Big-endian (**A**) and little-endian (**B**) representation of a 4-byte sequence.

which, in hexadecimal code, is

$$0x\ D1\ 5E\ A2\ 8B \qquad (2\text{-}6)$$

How is this chunk of information actually stored in computer memory, on a hard disk, a CD, or a magnetic tape? The straightforward answer is that it is stored in the same order as we wrote it down above (Fig. 2-2A).

For historical reasons, however, an alternative method is also in use (Fig. 2-2B). To understand the difference between these representations, some additional computer jargon is needed. In a sequence of 2-byte words, the left and right bytes are called the *most* and *least significant*, respectively, because they account for the numerically larger and smaller portions of the represented binary number. Accordingly, the big-endian representation maintains the ordering of the bytes from high to low significance, as when you say, as an English speaker, the number "25" (Sun SPARC, Motorola 68K, and PowerPC processors), whereas the little-endian representation inverts the byte order in the word, as when you say the number "15" (Intel processors).[2]

---

[2] The terms "big-endian" and "little-endian" were introduced into computer terminology by D. Cohen (1981), referring to an anecdote in Jonathan Swift's famous book *Gulliver's Travels* (1726). Here is Swift's original wording (Chapter 5): "Which two mighty powers have, as I was going to tell you, been engaged in a most obstinate war for six and thirty moons past. It began upon the following occasion. It is allowed on all hands, that the primitive way of breaking eggs, before we eat them, was upon the larger end: but his present

This seems an unnecessary complication; however, to use a biological term, it represents an "ancestral" feature of computers, because on early IBM PC processors, this representation was easier to implement in hardware (the most significant byte has the higher memory address). Reading data from both types of sources requires the ability to reverse the byte order, a process called *byte swapping*. Many application programs provide the option to store and retrieve data in either way.

The reader may now start experimenting with the DataFormat Applet provided in the Web Companion of this book. The Applet is conceived as a tool to explore various parameters that define how image data are organized, and how they can be represented on a computer screen. Throughout the following text, you may consult the Applet from time to time to investigate how each parameter influences the appearance of an image.

Let us now consider how digital data are stored physically by discriminating between two types of storage—permanent and temporary. Permanent means that data are not lost when you turn off your computer, whereas temporary data are lost. One problem of considerable practical concern is the rate at which stored data can be accessed and modified. At the low end is the digital magnetic tape, where the read/write (R/W) head must be positioned to a specified location by spooling through the entire length of the tape. In disk media such as computer hard disks and CDs, data are stored in sectors such that the R/W head can jump to specific locations (this produces the staccato sounds in your device when you read data from a CD). The fastest means of storing and retrieving data, although the storage is only temporary, is implemented in memory chips. In these chips, data are accessed through their logical address electronically rather than mechanically, such that any position in the memory can be reached at equal speeds (thus the name *RAM*: random access memory).

Another consideration in storage formats is the material form in which information is written down. This influences the speed of storage and retrieval. CD-ROMs possess physical pits that cannot be altered once they have been written (therefore the expression *ROM*, read-only memory). On tapes and hard disks, data reside as a relatively stable pattern of local differences in magnetization, whereas the fastest and most volatile form to store data is to represent bits as temporary charge differences in the RAM.

---

Majesty's grandfather, while he was a boy, going to eat an egg, and breaking it according to the ancient practice, happened to cut one of his fingers. Whereupon the Emperor his father published an edict, commanding all his subjects, upon great penalties, to break the smaller end of their eggs. The people so highly resented this law, that our histories tell us there have been six rebellions raised on that account; wherein one Emperor lost his life, and another his crown."

Each computer possesses a certain amount of built-in ROM, primarily for initiating the computer start-up. The ROM contains basic instructions that, during the booting phase, tell the computer how to load data from the hard disk into the temporary memory (the RAM). These data comprise large parts of the operating system (OS), which, once installed in the RAM, instruct the computer how to manage data, user instructions, and application programs. Each time you launch an application, it is loaded into the RAM as well. Computer programs essentially consist of instructions that tell the processor how to read, manipulate, and store data and how to deal with user interactions.

Computers tend to load as much of the application program instructions and data into the RAM as possible, because this temporary storage area provides the fastest access. If RAM space is filled, less-frequently used chunks of memory are "exported" temporarily to a reserved space on the hard disk—the so-called *virtual memory*—before the required information can be read from the hard disk into the RAM. Because access times to the hard disk are slower, "memory swapping" reduces the speed of the computer considerably. Especially in biomedical applications, where large data volumes must be accessed interactively, sufficient RAM space is crucial and even more critical than the latest graphics processor.

The size of computer storage areas and memories is measured in binary orders of magnitude but named with decimal terms. A kilobyte (KB or Kbyte), therefore, is not exactly 1000 bytes, but 1024, that is, $2^{10}$. Accordingly, a megabyte (MB) is $2^{20} = 1,048,576$ bytes, and a gigabyte (GB) $1,073,741,824 = 2^{30}$ bytes. Despite a lack of precise congruence, these quantities are reasonably close to decimal values $10^3$, $10^6$, and $10^9$.

### 2.2.3 Characters, Numbers, Pixels, and Voxels

Now that we have established an overview of the basic elements of data representation and storage, let us examine how digital data must be structured to represent one-, two-, and three-dimensional objects such as text, numbers, images, and volumes. The manner in which data are organized is the *data format*. In essence, this is a set of translation rules that instruct a computer program in interpreting a sequence of bytes. Data format works like the genetic code, which specifies how the molecular machinery of a cell interprets a sequence of nucleotides as amino acids (see Box 2-2).

We have already seen that a bit string of length $n$ can code for $2^n$ integer numbers. This is called a *binary data* format. But how can negative and/or real numbers (known as *floating-point numbers* in computer science) be coded? A floating-point number $x$ uses a factor $M$ (the so-called mantissa) and an exponent

## BOX 2-2

### DIGITAL DATA AND DNA

For the bioscientist, it is useful to compare the storage, representation, and retrieval of digital text data from a computer with the storage, representation, and retrieval of data from the nucleus of a cell. DNA stores genetic information as a linear sequence of nucleotides. In DNA, the latter are 4-state information units (A,T,G,C) rather than binary units (0,1), although it is sometimes convenient to consider them as binary units (AT vs. GC base pairs), or to conceive of each nucleotide as a 2-bit unit.

Reading data from a storage device, processing the information, and displaying the results on a computer screen is similar to reading DNA-based genetic information from a nucleus and displaying it as a protein in a cell. Both processes operate with a linear stream of bytes. Computer memory addresses are equivalent to specific recognition sites on the DNA, and the format specifications of digital data correspond to the genetic code. Whereas the genetic code serves as a dictionary to translate a sequence of nucleotide triplets into a sequence of amino acids (i.e., a polypeptide), the ASCII code specifies how sequences of bit octuplets are translated into text characters in a text string. Raster shifts during reading, corruption of the data, as well as use of the wrong code yield nonsensical information in both systems.

---

$p$ in the following way:

$$x = \text{sign} \cdot M \cdot 2^p \qquad\qquad (2\text{-}7)$$

$M$ has a value between 1.0 and 1.999 . . . , and $p$ is an integer number. Computers use 32 or 64 bits (4 or 8 bytes) to code single-precision and double-precision floating-point numbers, respectively. In both cases, the first bit codes for the sign. Following the sign bit, in a 32-bit number, the fractional part of $M$ (i.e., the digits after the decimal separator) is coded with 23 bits and the exponent $p$ with 8 bits. This allows the representation of numbers in the range of $\pm 10^{\pm 38}$ at about 7-digit precision. In a 64-bit floating-point number, $M$ and $p$ are coded with 52 and 11 bits, respectively, resulting in 15-digit precision numbers in the range of $\pm 10^{\pm 308}$.

Although text tends to be organized in two dimensions in a page and in three dimensions in a book, it is basically a continuous string of characters and thus closest in its actual structure to a sequence of bytes. A widely used format to trans-

| # | char | # | char | # | char | # | char | # | char | # | char | # | char | # | char |
|---|---|---|---|---|---|---|---|---|---|---|---|---|---|---|---|
| 0 | nul | 1 | soh | 2 | stx | 3 | etx | 4 | eot | 5 | enq | 6 | ack | 7 | bel |
| 8 | bs | 9 | ht | 10 | nl | 11 | vt | 12 | np | 13 | cr | 14 | so | 15 | si |
| 16 | dle | 17 | dc1 | 18 | dc2 | 19 | dc3 | 20 | dc4 | 21 | nak | 22 | syn | 23 | etb |
| 24 | can | 25 | em | 26 | sub | 27 | esc | 28 | fs | 29 | gs | 30 | rs | 31 | us |
| 32 | sp | 33 | ! | 34 | " | 35 | # | 36 | $ | 37 | % | 38 | & | 39 | ' |
| 40 | ( | 41 | ) | 42 | * | 43 | + | 44 | , | 45 | – | 46 | . | 47 | / |
| 48 | 0 | 49 | 1 | 50 | 2 | 51 | 3 | 52 | 4 | 53 | 5 | 54 | 6 | 55 | 7 |
| 56 | 8 | 57 | 9 | 58 | : | 59 | ; | 60 | < | 61 | = | 62 | > | 63 | ? |
| 64 | @ | 65 | A | 66 | B | 67 | C | 68 | D | 69 | E | 70 | F | 71 | G |
| 72 | H | 73 | I | 74 | J | 75 | K | 76 | L | 77 | M | 78 | N | 79 | O |
| 80 | P | 81 | Q | 82 | R | 83 | S | 84 | T | 85 | U | 86 | V | 87 | W |
| 88 | X | 89 | Y | 90 | Z | 91 | [ | 92 | \ | 93 | ] | 94 | ^ | 95 | _ |
| 96 | ` | 97 | a | 98 | b | 99 | c | 100 | d | 101 | e | 102 | f | 103 | g |
| 104 | h | 105 | i | 106 | j | 107 | k | 108 | l | 109 | m | 110 | n | 111 | o |
| 112 | p | 113 | q | 114 | r | 115 | s | 116 | t | 117 | u | 118 | v | 119 | w |
| 120 | x | 121 | y | 122 | z | 123 | { | 124 | | | 125 | } | 126 | ~ | 127 | del |

**FIGURE 2-3**  The ASCII text character format. The table indicates the (decimal) code of each character. The first 32 items (from 0 to 31) code for special characters (for example, #10 corresponds to "new line"); character #32 codes for a blank space.

late bytes into characters and vice versa is the *American Standard Code for Information Interchange* (ASCII). This *text format* comprises the usual alphabetic, numerical, and special characters that appear in a written text, as well as characters that control and communicate with the "printer," such as carriage returns and line feeds. As shown in Figure 2-3, the $128 = 2^7$ basic characters—alphabetic, numerical, and special characters—are coded in the first 7 bits of a byte. The eighth bit is used in various extensions of the ASCII code to include additional language-specific characters, such as umlauts, tildes, and accents. E-mail, as a highly text-oriented form of communication, makes extensive use of the ASCII code and its language-specific extensions. If you happen to encounter "spurious" characters, especially in non-English text, this normally indicates that the sender and receiver of the mail use different ASCII extension sets.

Whereas *character* is the term for the element of a text string, *pixel* (from picture element) denotes the element of an *image*. Note that *image* is a more general term than *picture*. An image is a two-dimensional data structure, and a picture is an image obtained by direct optical sampling from the physical space, for example, with a photographic camera.[3]

The concept of an image can easily be extended into the third dimension. Volume elements (*voxels*) are basic units of volume data sets. Volumetric data are often thought of as stacks of two-dimensional images and thus are referred to as three-dimensional images.

---

[3] The term *image* is also used to indicate that at least some data processing was involved in the creation of the data set. The acquisition of 2D or 3D medical data involves substantial data processing, thus the term "medical imaging."

### 2.2.4 Representing Gray Tones and Colors

Image data are stored as a linear sequence of bits. To interpret such an array in the proper dimensionality, instructions transforming the string of bytes into a two-dimensional pixel matrix or a three-dimensional voxel volume are necessary. Data formats that store images or volumes on a pixel-by-pixel (voxel-by-voxel) basis are called *bitmap formats*. The minimum information required to characterize the structure of an image is height and width, expressed through numbers of pixels, and *bit depth*, the number of bits used to represent one pixel (it is general convenience to use bit, not byte, numbers in this context). It is beneficial to view bit depth as the third dimension of an image, such as in a picture painted on canvas where various layers of pigment are stacked on top of each other (Fig. 2-4). For voxel data, three physical dimensions of the volume must be specified (length, width, depth), such that bit depth corresponds to the fourth dimension, which is easy to implement but harder to comprehend intuitively.

Let us now consider the representation of three types of image data: black-and-white, grayscale, and color. To represent a black-and-white image, a bit depth

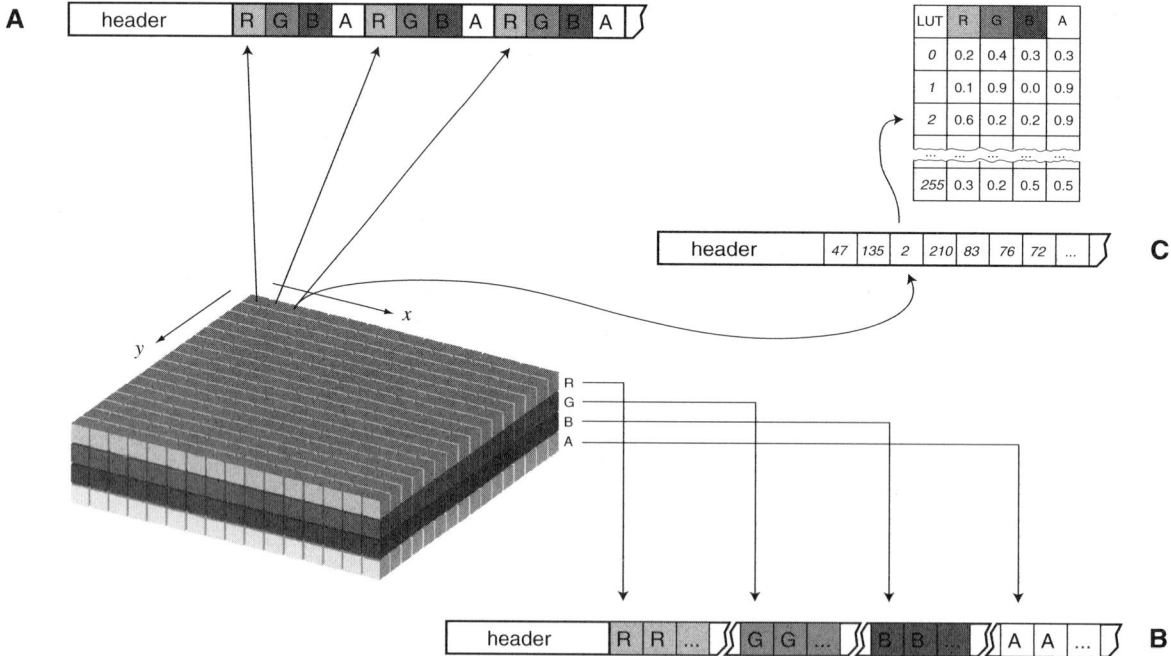

**FIGURE 2-4** Image data formats. The image is shown as a layered structure in which each pixel is represented by four components, R (red), G (green), B (blue), and A ("alpha", transparency). **A–C** depict three possibilities for organizing the data into storage formats. In **A**, the pixel data are stored sequentially, similar to characters in a text that start with the first pixel in the upper-left corner ($x = y = 1$) and follow $x$-rows. In this format, colors are *interleaved*, as R, G, B, and A components are stored pixel by pixel. In **B**, each channel R, G, B, A is stored separately. This is the *noninterleaved* format, in which all R entries come first, followed by all G entries, and so on. In **C**, all RGBA colors occurring in the image are given an individual identifier (an index) that is specified in a look-up table (LUT). The LUT is stored with the image, so that each pixel can be referred to by its color index.

of 1 suffices (i.e., one bit per pixel), as each pixel can assume only two possible states. Apart from the chessboard example in the introduction, black-and-white images are important in raster printing technology in which gray values are approximated more or less by dense patterns of black dots on a white background. Grayscale images contain additional complexity. Most computer screens display $2^8 = 256$ gray values, which range from black to white, and require 8 bits (1 byte) per pixel. By comparison, representing colors is more complex, whether as data, or on a computer screen. To appreciate the complexities in logical and technical requirements of color images, it is helpful to review briefly the organization of the human visual system and how computer display and color printing technologies are designed to match human color perception (Fig. 2-5).

The human retina contains several layers of neurons. The deepest layer contains the light-sensitive photoreceptor cells, which photochemically transform light into neural signals. There are two classes of photoreceptors—*rod* and *cone* cells—that are named after their morphology. At the molecular level, photoreceptor pigments act as probabilistic counting machines; they collect photons of different wavelengths with different probability. Rod cells are most prevalent, populating the entire periphery of the retina. They work at low light intensities, but they do not convey color information. The central region of the retina, known as the *macula*, allows object recognition by recording changes in the external environment during incessant "scans," or gaze. The macula is populated densely by three different types of cone cells—red (R), green (G), and blue (B) receptors— that are named according to their maximum sensitivity and that provide the input for color vision. Cones need relatively higher light intensities than rods to operate; this is why color vision functions poorly during the night or under dim light.

Because cones are restricted to the macula, color vision is confined to a relatively small high-resolution area within the retina. However, we do not perceive the periphery in our visual field as colorless and blurry, nor do we see the world as a pointillist picture of red, green, and blue dots that represent the distribution of the photoreceptors in the retina. Rather, the seamless perception of a colorful world is the result of high-level integration of color and spatiotemporal information. Strictly speaking, color vision is a *combination* of the inputs from adjacent red, green, and blue receptors. Respective "calculations" are carried out in the layered networks of retinal interneurons that receive information from the photoreceptor layer, as well as in specialized areas of the brain (from a developmental perspective, retinal interneurons migrated from the brain). During "calculations," signals from the three types of color receptors, $I_R$, $I_G$, $I_B$, are combined in two ways. The sum

$$I = I_R + I_G + I_B \tag{2-8}$$

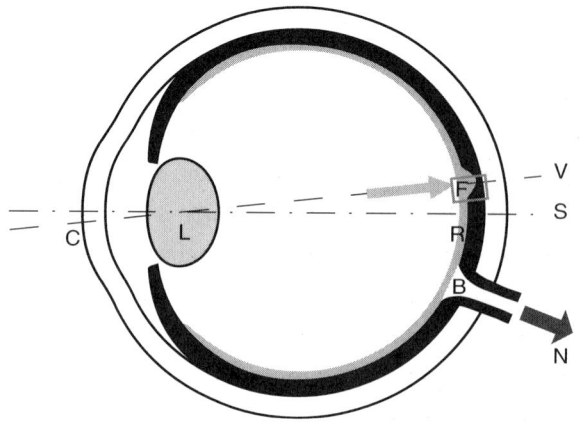

A

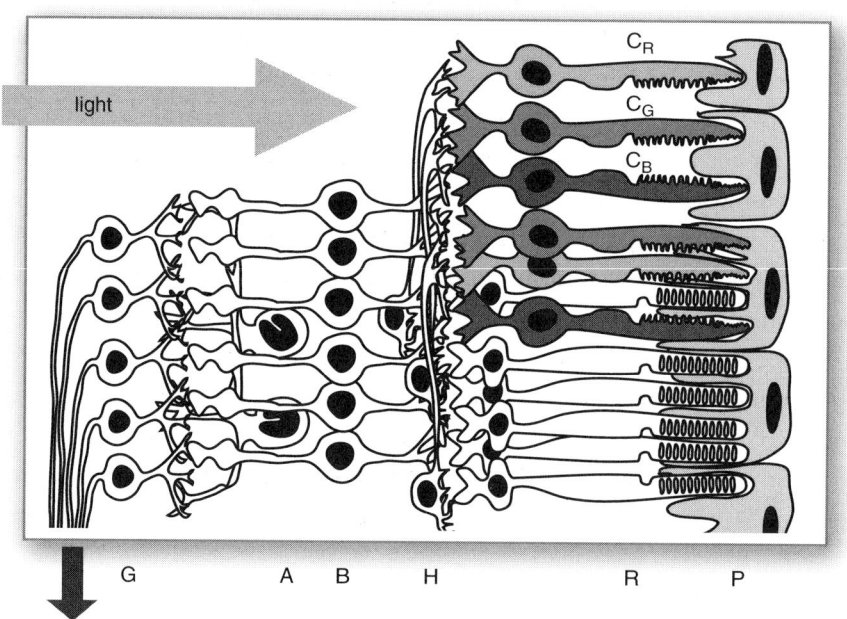

B

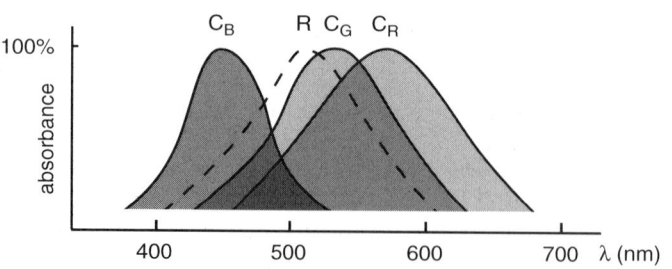

C

indicates the overall brightness. The *relative* contribution of each receptor type, $i_R$, $i_G$, $i_B$, is evaluated as follows:

$$i_R = \frac{I_R}{I_R + I_G + I_B}\,; \quad i_G = \frac{I_G}{I_R + I_G + I_B}\,; \quad i_B = \frac{I_B}{I_R + I_G + I_B}\,, \quad \text{with } i_R + i_G + i_B = 1, \quad (2\text{-}9)$$

Accordingly, whereas brightness is a sum, color perception is a proportion of the signal from each receptor type. This explains why object colors do not change when the overall ambient light intensity changes. Now, imagine that $I_R$, $I_G$, $I_B$ are three independent coordinates defining an *RGB color space*. Because color intensities are finite quantities, the full range of hues and intensities can be produced by any combination of $I_R$, $I_G$, $I_B$ in a color cube (Fig. 2-6).

Using geometry, $i_R + i_G + i_B = 1$ defines a triangular slice with edges R, G, B. This is the so-called color triangle (Fig. 2-6) that is often used to represent the range of color hues that a visual system can discern, or a computer screen can display, independent of brightness.

For reasons of perceptual equivalence, any technical device that displays colors for a human observer must take into account the *trichromatic* color scheme presented previously (see Box 2-3). The display of a color image on a computer screen is quite different from its printout version. On the computer screen, light is *emitted* in three primary colors (i.e., red, green, and blue) and blended within our visual system to produce a wide range of *additive colors*. Implementation of the RGB color scheme is achieved by a pointillist design of the monitor.[4]

**FIGURE 2-5** The human eye, retina and color vision. **A**: A schematic cross section through the left human eye, seen from above. Light is collected by optic elements (C: cornea; L: lens) and focused onto the retina (R). The central visual axis (V) departs slightly from the axis of symmetry (S) by leading to the fovea, that is, "pit" (F), a 1°-wide area of greatest visual receptor density. Neural signals from the retina travel to the brain via the optic nerve (N), which leaves the orbit at the blind spot (B). **B**: Retinal neuronal organization (magnification of the small box area around the fovea in **A**). Light-sensitive cone (C) and rod (R) cells are positioned along the entire retina and embedded in pigment cells (P), which separate them from each other (note the "inverted" design of the retina where light passes through the neuronal postprocessing layers before it reaches the sensory cells). Rod and cone cells transform light into neural signals, which undergo further processing in three layers of retinal cells—horizontal (H), bipolar (B), and amacrine (A). Subsequently, the preprocessed visual information travels to the brain via ganglion cells (G). The periphery of the retina is dominated by rod cells, which are responsible for colorless light/dark vision. The ability to see color is concentrated to the macula (an area 3° wide around the foveal pit) and based on a combination of signals from red ($C_R$)-, green ($C_G$)-, and blue ($C_B$)-sensitive cone cells. **C**: Relative light sensitivity (absorption) of the three types of cone cells ($C_R$, $C_G$, $C_B$) and of the rods (R). Note the overlapping cone curves; this is a precondition of color vision.

[4] The basic idea of pointillism, as promoted by Georges Seurat (1859–1891) and Paul Signac (1863–1935), was to paint tiny blobs of pure color that, when seen from a distance, were blended and perceived as additive colors.

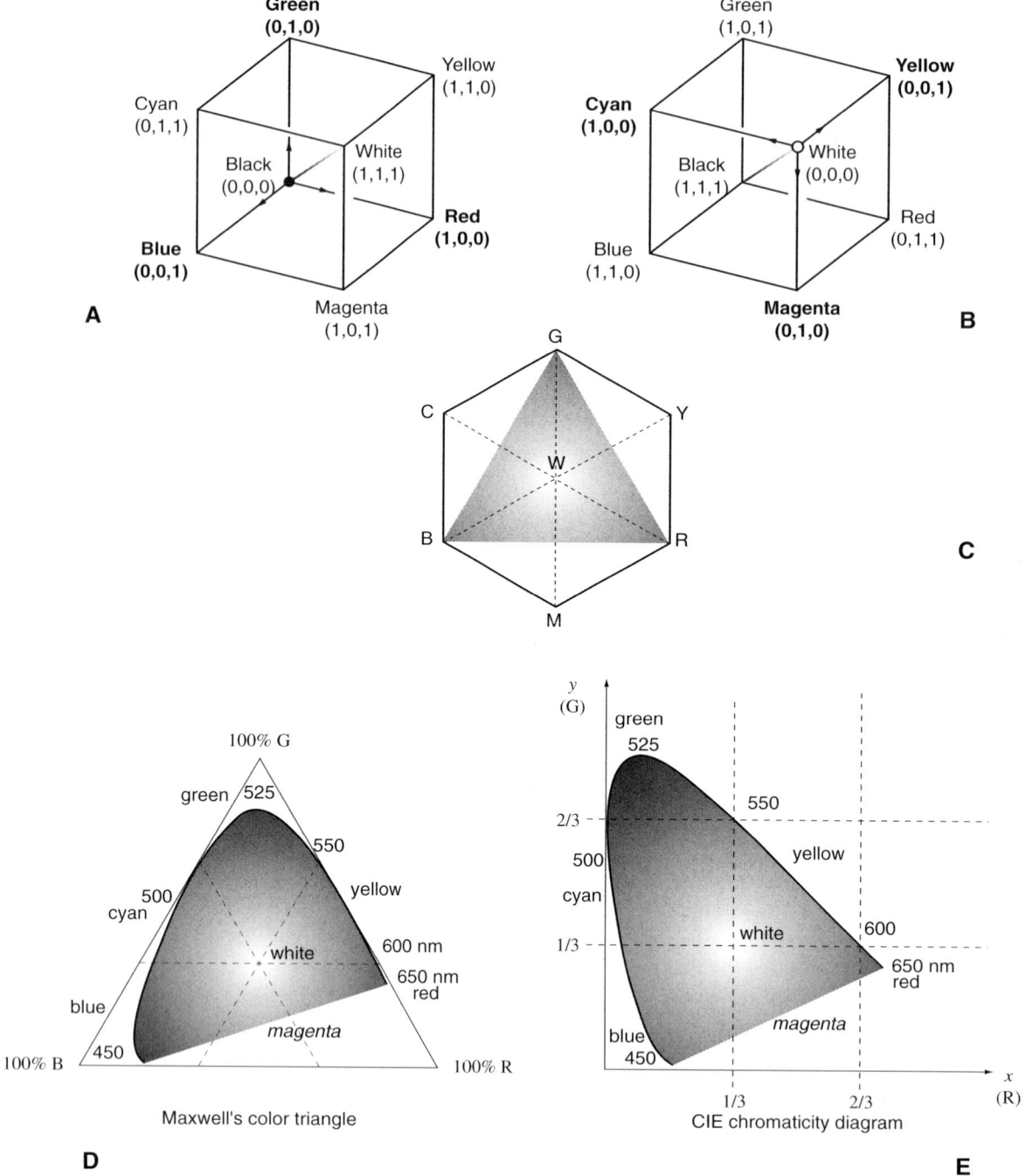

Maxwell's color triangle

CIE chromaticity diagram

**BOX 2-3**

## RGB VISION AND BEE VISION

Although we perceive a wide diversity of colors, this diversity is rather poor in comparison to the diversity of sounds. In fact, the full range of color represented in the RGB cube is equivalent to the full range of variation contained by a single three-tone accord (e.g., C-E-G), where the intensity of each component is varied from 0% to 100%. Whereas our visual system has only three color receptors to discriminate between different wavelengths, our auditory system possesses thousands of sound receptors, each tuned to an narrow frequency band (and thus wavelength range). It is much easier to fool our trichromatic color vision system in comparison to our "polychromatic" ears. In fact, "color deception for color perception" is the motto of every color rendering system, ranging from printing techniques to computer displays.

The spectral colors produced by the refraction of white light through a prism represent a physical continuum of wavelengths $\lambda$ between approximately 700 nm (red) and 400 nm (blue). Because each color corresponds to a specific wavelength, these are monochromatic colors. This spectrum can be *simulated* quite accurately on a computer screen by mixtures of R, G, and B, for example, by

**FIGURE 2-6**  Color cubes, hexagons, and triangles. Additive (**A**) and subtractive (**B**) color cubes are shown. The additive system (**A**) uses the principal colors red, green, and blue to define the color space; addition of these colors yields white, whereas their absence yields black (e.g., a black computer screen). In the subtractive system (**B**), the color space is constituted by cyan, magenta, and yellow. The origin of this system is "white," corresponding to the absence of printed colors on white paper. In both systems, gray values represent equal intensities of the three respective principal colors along the main diagonal. C–E: Colors can also be represented on a two-dimensional plane, if absolute intensities are disregarded. Looking along the main diagonal of the color cube yields a color hexagon (**C**). In Maxwell's[5] color triangle (**D** and grey area in **C**), the relative contribution of each color— R, G, and B—is plotted in a triangular coordinate system in which each point satisfies the equation R + G + B = 100%. **E:** The chromaticity diagram of the CIE (Commission Internationale de l'Éclairage) is a variant of Maxwell's triangle in which R and G axes are plotted at a right angle. Wavelengths depicted on the border of the diagram represent the perceptual location of pure (monochromatic) spectral light. Colors toward the edges of the triangle are not available because they represent physiologically impossible combinations of R, G, and B. On the other hand, colors appearing between blue and red (magenta, violet, purple) do not have monochromatic equivalents; they are mixtures of short-waved (blue) and long-waved (red) portions of the spectrum.

[5] James Clerk Maxwell, 1831–1879.

### BOX *2-3*

**RGB VISION AND BEE VISION (*Continued*)**

walking along the edges of the color cube from black over red to yellow, green, cyan, blue and then back to black. However, our visual system cannot differentiate monochromatic cyan ($\lambda$ = 510 nm) and the same color resulting from a blend of blue ($\lambda$ = 450 nm) and green ($\lambda$ = 540 nm), much less any other combination of colors. As long as the relative mixture of R, G, and B is the same, we perceive the same color, regardless of its physical constituents. On the other hand, hues in the vicinity of magenta (purple, mauve, violet) have no single-wavelength (monochromatic) counterpart—we may therefore conclude that colors exist only in our brains. Were we endowed through evolution with a tetra-, penta-, or polychromatic color vision system instead of a trichromatic system, computer display and printing technology would be even more technically demanding.

The human color vision system is only one of several evolutionary solutions to the problem of requiring color vision. Hymenopteran insects (e.g., ants, wasps, and bees) see the world quite differently. A bee also has three different types of color receptors; however, this trio is shifted toward shorter wavelengths so that red for us is invisible for the bee, and invisible ultraviolet for us constitutes the "blue end" of the bee's visual perception (Menzel, 1979). Additionally, bee photoreceptors work at much higher "frame rates" than human counterparts (Laughlin, 1981). Imagine how different flowers would appear to bees, but then realize that a bee flying by a high-tech computer monitor would see a rather odd slide show. To the bee, the screen would display a limited range of colors, and the screen images would change at a slow rate.

Stomatopods (mantis shrimps) have chosen another way. Their compound eyes exhibit 16 different photoreceptor types (Cronin and King, 1989; Marshall et al., 1991a, b). Evidently, evolution favored spectral visual acuity over spatial acuity in these animals.

Each pixel on the screen consists of three juxtaposed *subpixels* representing red, green, and blue, that is, the above variables $I_R$, $I_G$, $I_B$. This mimics the design in the human retina, and it can be verified by inspecting your computer screen with a strong magnifying glass.

Painting or printing with color pigments is quite different, as is best illustrated by the fact that superimposing pigment colors yields black rather than white. Pigments are *subtractive colors* that act as filters, when applied to a carrier medium such as paper, by *absorbing* (subtracting) certain wavelengths from incident light before it is reflected toward the observer. Figure 2-6 shows that subtractive colors are measured through a coordinate system with principal axes along the colors cyan (C), magenta (M), and yellow (Y). This system is diametrically opposed to the RGB system, such that each additive color has a subtractive counterpart. In practice, however, this correspondence is not exact; although every color in the color cube can be displayed effectively on a computer screen, the same is not possible for printed colors.[6]

Most notably, mixtures of cyan, magenta, and yellow pigments cannot produce the fully saturated hues that are represented by surfaces and edges of the color cube. Additionally, the overlay print of these three basic pigments normally does not yield fully saturated black. This problem typically is resolved by using four pigments rather than three pigments in printing technology: C, M, Y, and K (black).

The organization and structure of color image data files reflects a dichotomy between the RGB and CMYK systems in many ways. For an image stored in RGB mode, each pixel is represented by three color channels R, G, and B (the notion of a channel goes back to color video systems, where each color is transmitted on an individual channel). Normally, 1 byte (8 bits) is allocated per channel, resulting in a pixel depth of $3 \cdot 8 = 24$ bits. With 256 intensity values per channel, $256^3 = 16{,}777{,}216$ different colors are possible. Occasionally, a fourth 8-bit field is added to represent a transparency parameter (RGBA format), which controls how colors are treated during image stacking. In a CMYK image, the default pixel depth is 32 bits, which are subdivided into four 8-bit fields—C, M, Y and K. Among the storage arrangement possibilities for image data, two specific ones are worth mentioning. First, the R, G, B (and A) values of each pixel may be saved next to each other as a package. Second, each color channel may be saved as a separate layer (Fig. 2-4). The former and latter formats are also known as interleaved and non-interleaved modes of color image data storage, respectively.

At this point, the reader may experiment with the ColorApplet on the Web Companion. The applet permits mixing colors in RGB and CMY color spaces.

---

[6]Many applications issue an "out of gamut" warning when a color beyond the printable range is chosen.

### 2.2.5 Data Compression

Compared to text data, image data occupy considerable amounts of computer storage space because the number of pixels increases to the second power, and the number of voxels to the third power, for linear dimensions of an image. For example, the number of characters contained in this book ranges between 500,000 and 1,000,000. The entire text in ASCII format will occupy between 500 KB and 1 MB of storage space, which is about equivalent to one single RGB image measuring $512 \times 512$ pixels. With the steadily increasing capacity of storage devices, the problem of storage space can be overcome partially. However, during image data transmission, for example, via the Internet, transmission rates impose a real bottleneck, through which only relatively small data volumes can be funneled interactively. Data compression is therefore an important aspect of handling image data.

Compression algorithms reduce the actual amount of stored information by replacing redundant information with abbreviations and/or by removing insignificant information. To understand the potential benefits of compression, we continue with the example of the RGB image measuring $512 \times 512 = 2^{18}$ pixels. As seen above, its 24-bit pixels can represent $2^{24}$ different RGB colors. Considering the highly unlikely case in which every pixel in the image represents a different color, the number of actual colors ($2^{18}$) expressed as a fraction of the number of possible colors ($2^{24}$) is $2^{18-24} = 2^{-6} = 1/64$. In other words, only a maximum of approximately 1.5% of all possible colors can be represented in our image. In real images, this proportion is even smaller yet, because many pixels will have the same color. During data compression, these properties are systematically exploited.

One strategy in data compression involves eliminating bit patterns that occur more than once in the data set. With the technique called *run-length encoding*, series of repeated bit patterns are replaced by a single pattern and its number of repeats (Fig. 2-7).

Run-length encoding (RLE) of image data can be very efficient because a large percentage of adjacent pixels in an image have similar color values. A more sophis-

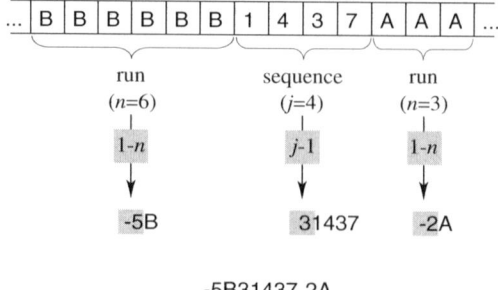

**FIGURE 2-7**  Principle of run-length encoding. Repetitive sequences of length $n$ are encoded by the number $1 - n$ and the repeated character; nonrepetitive sequences of length $j$ are encoded by the number $j - 1$ and the sequence of characters.

ticated technique requires searching for multiple, but not necessarily adjacent, occurrences of the same pattern in the data set. The principal task becomes defining a "dictionary" that associates each individual pattern encountered in the data set with an index entry. Compression is indexing the data set and storing it together with the dictionary. The most widely used algorithm that adopts this strategy is the LZW (Lempel–Ziv–Welch) compression scheme (Welch, 1984). RLE and LZW are *lossless compression* algorithms because every single bit of the original data set can be fully restored by reversing the compression procedure. The use of lossless compression schemes is mandatory whenever the state of every bit is significant (e.g., text data, numerical data, or image data used for quantitative analysis).

*Lossy compression* schemes are useful when precise quantities are less valuable than the overall visual context of an image; similarity to the original image is more important than identity. For example, digital photographs may contain large areas of relatively homogeneous color, in which local fluctuations are irrelevant. In these situations, it is sensible to replace a range of similar colors by an average color in order to reduce the total number of colors (the color palette) in the picture without losing spatial resolution. Because the amount of compression can be varied, it is possible to balance image quality against the volume of compressed data. Obviously, however, once the lossy compression scheme has been applied, the original image can no longer be fully recovered.

Let us return to the $512 \times 512$ pixel RGB image, which uses 24 bits of storage space per pixel. Pixel depth can be reduced to 8 bits in the following manner. We determine a palette of the $2^8 = 256$ most frequent colors within the image and replace less frequent colors by nearest neighbors in the palette. We then establish a *lookup table* (LUT) that numbers the palette colors from 0 to 255 and specifies the RGB value of each entry. This allows the coding of pixels with 8-bit color indices instead of 24-bit RGB values. The color-indexed image is saved with its LUT so that the actual color values at each pixel can be recalled from the table during display.

### 2.2.6 Some Common Image File Formats

Most image data come in standard formats that are characterized by suffixes attached to filenames. For example, image1.tif, image2.jpg, image3.gif are written in TIFF, JPEG, and GIF formats, respectively, which can be read by most standard Web browsers and image processing programs.[7]

An image data file consists of two parts, including a *header* field, containing information about the structure of the image, and the actual pixel field. The header

---

[7]Software components to read and write data in standard formats are often independent of the actual applications and come as so-called *plug-ins*.

typically contains a format identifier at the outset so "smart" applications can recognize the "endian-ness" of the bytes, as well as the format type, if no file extension is available. Following the format identifier, the header contains *image attributes*, that is, information on how pixel data were stored, how pixel data must be retrieved, and supplemental information like the physical size of the image (in real-world dimensions or as a printout), the author's name, the date and time of data acquisition, a description of the image contents, and so forth.

Why is there so much diversity in the way image data are formatted and stored? The principal answer is that image data are always context-specific. Parameters used during data acquisition, physics of the data acquisition device, media by which images are communicated and displayed, and applied compression schemes are highly diverse. For example, medical image data contain device- and patient-specific information that must be coded in special-purpose headers. And, as discussed above, images printed on paper are quite different from images displayed on a computer screen. A further reason for diversity in image data formats is that images represent considerable data volumes.

Arguably the most popular image data format, JPEG is named after the Joint Photographers Expert Group (http://www.jpeg.org) and is easy to recognize by its suffixes .jpg or .jpeg (capitalized letters are used alternatively). JPEG provides a so-called "baseline" lossy compression algorithm that allows the user to have flexibility in the degree of data compression. Furthermore, it is possible to select normal or progressive compression modes. In the first mode, an image is encoded at a fixed level of compression. In the second mode, multiple versions of an image are encoded at various levels of compression (and thus of spatial and color resolution). This is particularly useful in fast display of images over the Internet. A preliminary, low-resolution version of the image appears on the display in near-real time and is replaced gradually by more detailed versions. A JPEG header specifies parameters of compression, but it may also contain information on camera optics, exposure time, date, and so forth. This information is accessible with standards designed for archiving digital images.

That the JPEG format uses lossy compression has important consequences with regard to potential fields of application. In terms of image quality versus compression rates, JPEG yields the best results for color images that are designed for visual display. However, in quantitative analyses, JPEG compression is not desirable because unexpected alterations of the original data may occur. Furthermore, grayscale images with fine spatial detail, or black-and-white images that contain line drawings and/or text, do not benefit from JPEG compression because lossy compression introduces undesirable boundary effects, especially in image regions with high contrast.

The *graphical interchange format* (GIF) is used mainly to transport images via the Internet. GIF uses the 256-color look-up table approach described above to reduce storage requirements from 24 to 8 bits per pixel. Moreover, run-length encoding is used for further compression. In addition to compactness, the GIF format is useful for storage of image stacks, which can be rendered at a user-defined frame rate to create simple animations.

The *tagged image file format* (TIFF) is used primarily in storing image data with lossless compression schemes. The structure of a TIFF file is complex, but it provides a wide variety of options for saving user-specific information about an image. A tiff header contains three parts. The first part indicates the endian-ness of the byte ordering (an important aspect for compatibility between different types of hardware), the second part contains the tiff version number, and the third part indicates the location of the *image file directory*. This latter data structure contains all the required information for specifying image attributes such as length, width, and color representation scheme. In principle, however, any kind of information, even proprietary data, can be stored in this directory without affecting its readability by a different software on a different computer. This is known as *tagging*, in which each entry in the directory consists of an identifier (the tag), followed by a number that indicates the entry's length in bytes, and finally by the actual information (Fig. 2-8).

The tag-based header format has two major advantages over other formats. First, unknown entries can be skipped because length is known independently of content. Second, the header has no fixed length and can be expanded according to the actual requirements.

Medical images are acquired via a wide range of techniques, including computed tomography, magnetic resonance imaging, ultrasonography, and laser scan-

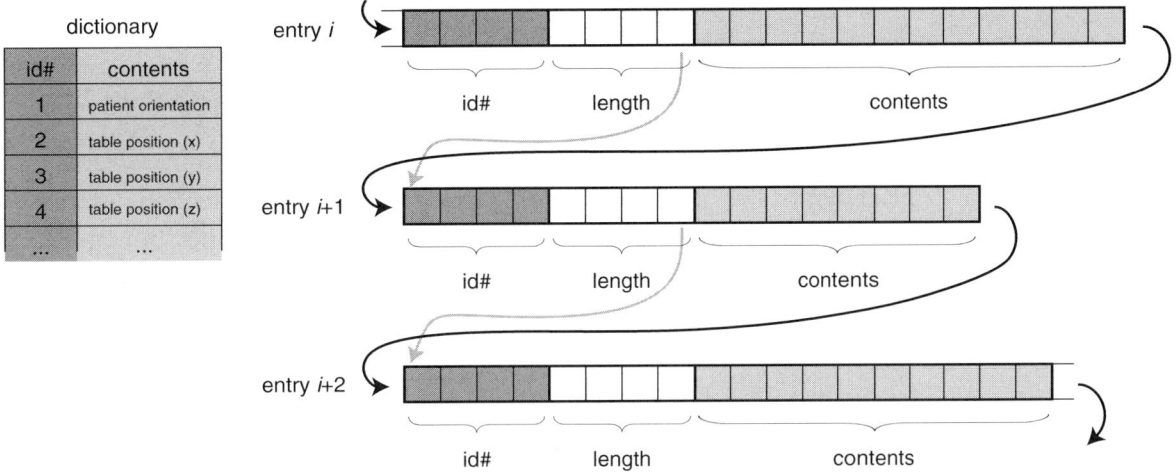

**FIGURE 2-8**  The structure of a tagged header in DICOM format. Each entry in the header consists of three elements—the identifier, the length field, and the field containing the actual information. With the dictionary, the identifier permits "translation" of stored information contents into an actual meaning. Optionally, information fields can be skipped (gray arrows), because their length is recognized independent of their contents.

ning. Accordingly, images document various physical properties, such as X-ray density, radio signal amplitudes, and sound pressure. Headers in medical imaging data files specify the exact circumstances and modalities under which data were acquired. For example, headers usually contain the date and duration of the medical examination, the parameter settings of the imaging device, patient-specific data, the physical dimensions of the image, and so on.

In the 1980s, a growing emphasis of medical imaging applications in clinical environments fostered device- and manufacturer-independent standardization in medical image data header information, which led to the development of hospital-wide *picture archiving and communication systems* (PACS). The American College of Radiologists (ACR) and the National Electrical Manufacturers Association (NEMA) devised the original standard in 1985; this was known as the *ACR/NEMA* image format. Further developments in imaging and networking technology extended the ACR/NEMA 2 format into the *DICOM 3* standard (Digital Imaging and Communications in Medicine; `http://medical.nema.org`), which today is nearly universal for all types of medical imaging data (Bidgood et al., 1997).

Like TIFF, DICOM uses a tag-based header format. However, each information item is identified by a name tag rather than by its position in the header. Taking into account technical progress, the entry list can be extended by including information from new imaging modalities. The disadvantage of the DICOM format, however, is that header information is not self-explanatory. Without special-purpose software tools, it is virtually impossible to extract useful header information from DICOM images. Box 2-4 indicates one means of skipping unknown DICOM headers to get direct access to actual pixel data of the image. This is helpful when displaying medical image data on a PC with standard image processing software.

### 2.2.7 Implicit Versus Explicit Representation of Object Data

Above in this chapter, we defined bitmaps as data formats that describe images pixel by pixel or volumes voxel by voxel. Digital data acquisition devices, such as cameras, laser scanners, and CT scanners, typically produce bitmap images. In these images, a color or intensity value is specified for each element of the data set. However, bitmaps do not specify which objects are represented, or which pixel belongs to which object. In other words, objects in the picture are represented implicitly rather than explicitly, and object recognition and quantization is the task of subsequent image processing steps (see Chapter 4). For example, the circle in Figure 2-10A is simply a set of black pixels on a white background from which our visual system extracts relevant shape information.

Let us now consider another way to represent a circle. Rather than *acquiring* data from a circle-shaped object, we *design* the circle with a CAD program

---

**BOX 2-4**

## THE DICOM MEDICAL IMAGE DATA FORMAT

---

The DICOM format uses tagged headers (see Fig. 2-8). Each entry contains three parts:

*The tag field.* A unique identifier indicating the type of information described. Information is classified hierarchically, using group tags and within-group element tags, the meaning of which is specified in the DICOM dictionary. Typically, group and element tags use 32 bits each.

*The length field.* An integer indicating the length in bytes of the actual data.

*The data field.* This field contains the data in ASCII or binary format, according to the tag-specific indications in the DICOM dictionary.

Figure 2-9 is an example of the specification of four DICOM entries.

---

and store the *explicit* object information. For example, we design and store the circle's radius, the coordinates of its center, and the width of the drawn line.[8] The explicit representation of the circle is not only more compact than its implicit form, but it has the additional advantage that it can be scaled without resampling the bitmaps. In fact, the relationship between the object and the pixels representing the object is fully flexible in the latter case, but fixed in the former case. In technical terms, the difference between explicit and implicit object representation is that between geometry-based *scalable vector graphics* and pixel-based *raster graphics* representation.

This brings us back to the dichotomy between biomedical and CAD applications. In biomedicine, we typically begin with the acquisition of implicit, raster graphics data to *reconstruct* the geometric object descriptions. In contrast, we *construct* three-dimensional object geometries in CAD explicitly by starting with a mental concept. In other words, biomedical reconstruction of objects follows a bottom-up approach, whereas CAD construction of geometric descriptions

---

[8]Similar arguments can be applied to 3D data structures. A data volume consisting of $N \times N \times N$ voxels may contain a sphere (represented, for example, by a set of white pixels against a black background), which, alternatively, can be described by its radius, $xyz$ coordinates of its center, and shell width.

| attribute name | tag (hex-code)<br>(group, element) | type,<br>multiplicity | attribute description |
|---|---|---|---|
| Pixel Spacing | (0028,0030) | ASCII, 1 | Physical distance in the patient between the center of each pixel, specified by a numeric pair - adjacent row spacing (delimiter) adjacent column spacing in mm. |
| Image Position (Patient) | (0020,0032) | ASCII, 3 | The $x$, $y$, and $z$ coordinates of the hand corner (center of the first voxel transmitted) of the image, in mm. |
| Slice Thickness | (0018,0050) | ASCII, 1 | Nominal slice thickness, in mm. |
| Slice Location | (0020,1041) | ASCII, 1 | Relative position of exposure expressed in mm. |

**Example header data for the entry** *"pixel spacing"*:

| 0028 | 0030 | 0000 | 0016 | 302e | 3133 | 3637 | 3138 | 3735 | 5c30 | 2e31 | 3336 | 3731 | 3837 | 3520 |
|---|---|---|---|---|---|---|---|---|---|---|---|---|---|---|
| group | elem. | hex. length | | contents in ASCII code | | | | | | | | | | |

This translates into:

| pixel<br>spacing | 22 bytes | 0 . | 1 3 | 6 7 | 1 8 | 7 5 | \ 0 | . 1 | 3 6 | 7 1 | 8 7 | 5 |
|---|---|---|---|---|---|---|---|---|---|---|---|---|

i.e., the pixel spacing is 0.13671875 mm in both $x$ and $y$ directions.

**Example header data for the entry** *"image position"*:

| 0020 | 0032 | 0000 | 0010 | 2d33 | 355c | 2d31 | 3432 | 5c2d | 3239 | 362e | 3520 |
|---|---|---|---|---|---|---|---|---|---|---|---|
| group | element | hex. length | | contents in ASCII code | | | | | | | |

This translates into:

| image<br>position | 16 bytes | - 3 | 5 \ | - 1 | 4 2 | \ - | 2 9 | 6 . | 5 |
|---|---|---|---|---|---|---|---|---|---|

i.e., the image is at position $x = -35$, $y = -142$, $z = -296.5$ mm.

**FIGURE 2-9**   Example of DICOM entry specification.

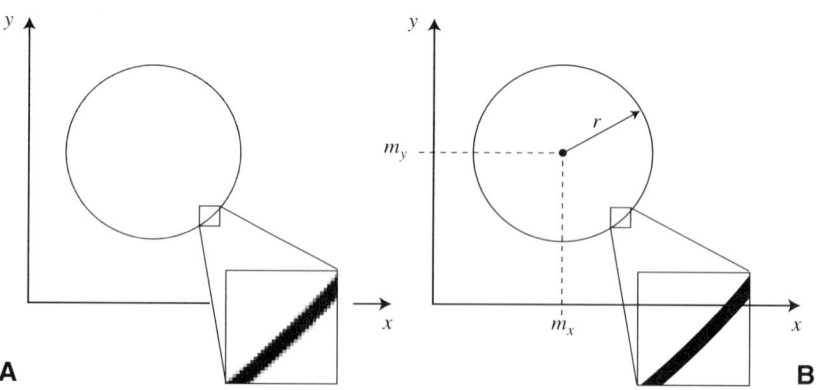

**FIGURE 2-10**   Implicit (**A**) and explicit (**B**) representation of a circle in an image. The circle in **A** is represented as a bitmap image, which becomes evident at higher magnifications. The circle in **B** is represented explicitly as a geometric primitive by specifying its radius $r$ and the position ($m_x, m_y$) of its centroid. This *scalable vector graphics* (SVG) representation permits arbitrary scaling of the object without losing resolution.

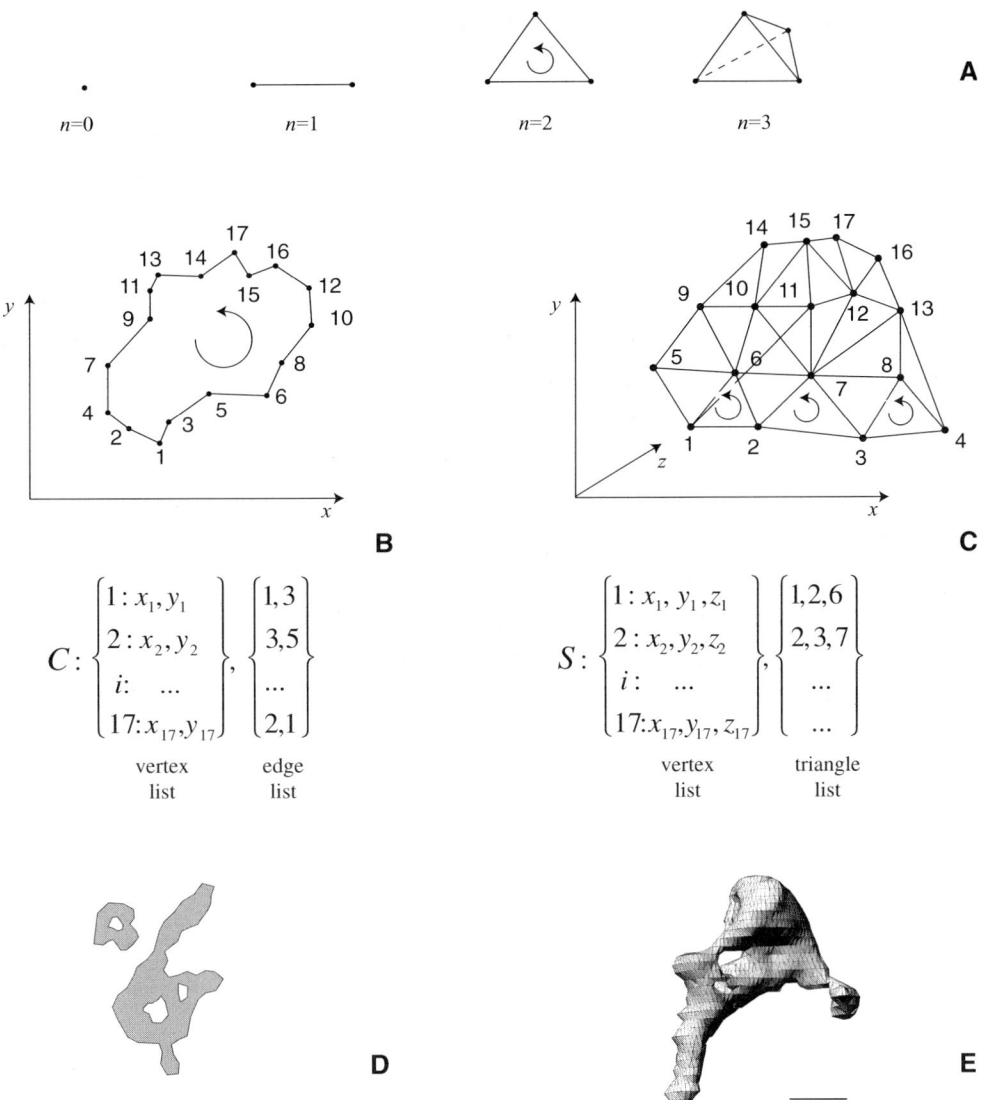

$$C: \begin{cases} 1: x_1, y_1 \\ 2: x_2, y_2 \\ i: \quad \dots \\ 17: x_{17}, y_{17} \end{cases}, \begin{cases} 1,3 \\ 3,5 \\ \dots \\ 2,1 \end{cases}$$

vertex    edge
list       list

$$S: \begin{cases} 1: x_1, y_1, z_1 \\ 2: x_2, y_2, z_2 \\ i: \quad \dots \\ 17: x_{17}, y_{17}, z_{17} \end{cases}, \begin{cases} 1,2,6 \\ 2,3,7 \\ \dots \\ \dots \end{cases}$$

vertex    triangle
list       list

**FIGURE 2-11**  Geometric simplexes, graphs, and free-form objects. **A**: Simplexes. The *n*-simplexes for $n = 0 \dots 3$ are the vertex, edge, face (triangle) and tetrahedron. **B**, **C**: Graphs. Graph *C* is an ordered set of edges (vertex pairs) specifying a contour, whereas graph *S* is an ordered set of triangles (vertex triples) specifying a surface. Note that ordering introduces directionality in edges and triangles and permits discrimination between "inside" and "outside" an object. **D**, **E**: Two- and three-dimensional free-form objects. Sets of several graphs are combined to define the geometry of complex objects. **E** shows the left anvil of a Neanderthal fossil (Le Moustier 1 specimen; see Ponce de León and Zollikofer, 1999), as reconstructed from 0.2-mm cross-sectional CT data (scale bar is 1 mm). **D** is derived from one such CT cross section.

follows a top-down approach. How does each operate? Let us first discuss the bottom-up approach. Objects of any geometric complexity are approximated with sets of so-called *n-simplexes*, or geometric primitives of dimension *n* (Fig. 2-11A). This sounds daunting, but it is just a formal description of familiar geometric entities, such as a point, or *vertex* (0-simplex), a line segment, or *edge* (1-simplex), a triangle, or *face* (2-simplex), and a tetrahedron (3-simplex). The principal task in geometric object description is establishing higher-order simplexes from sets of lower-order simplexes. Vertices can be obtained from bitmap data by specifying the position of pixels $(x,y)$ or voxels $(x,y,z)$, whereas edges are formed by vertex pairs, triangles by edge or vertex triplets, and tetrahedra by vertex or triangle quadruples. Subsequently, contours can be described as a set of edges and surfaces can be described as a set of triangles. The resulting structures are *free-form objects* because objects of arbitrary geometric complexity can be approximated by sufficiently large sets of simplexes. In computer graphics, the logical construct behind free-form objects is a *graph* (not to be confused with a graph that denotes two or more variables in a Cartesian coordinate system). In computer graphics, a graph consists of a set of simplexes with an associated list defining the geometric relations between them. For example, a graph specifies how vertices are interconnected to form line segments or triangles (Fig. 2-11B).

The most compact format for storing contour data is the *chain code* (Fig. 2-12A). It specifies an initial vertex position on the contour and then keeps track of each subsequent position as a relative direction (given that internode distances are constant). Formats of triangulated surface data are relatively more complex because triangles cannot be ordered linearly. A widely used standard is the *stl* format (from stereolithography) that was developed originally for exchange of free-form object data in rapid prototyping environments (Fig. 2-12B). An stl file provides a list of all the triangles composing a free-form object. Each triangle is specified with four *xyz*-coordinate triplets: The first triplet indicates the direction of the normal vector of the triangle surface, and the remaining three indicate the three-dimensional position of triangle vertices.

Let us summarize the bottom-up approach of object reconstruction. Free-form contours are built up from sets of line segments, or *polygons*, and free-form surfaces are built up from sets of triangles, or *polyhedra*. Polygons ("many edges") and polyhedra ("many faces") represent *discontinuous* surfaces, because they exhibit discontinuities between neighboring simplexes that do not necessarily occur in the original objects they represent.

How can we conceive of methods for the construction of continuous free-form objects?

## 2.2.8 Modeling Three-Dimensional Objects

To answer this question, we turn to the top-down approach of object construction. *Splines* are a special class of higher-order geometric primitives that describe smoothly curving *continuous* free-form objects. A spline is a function that defines

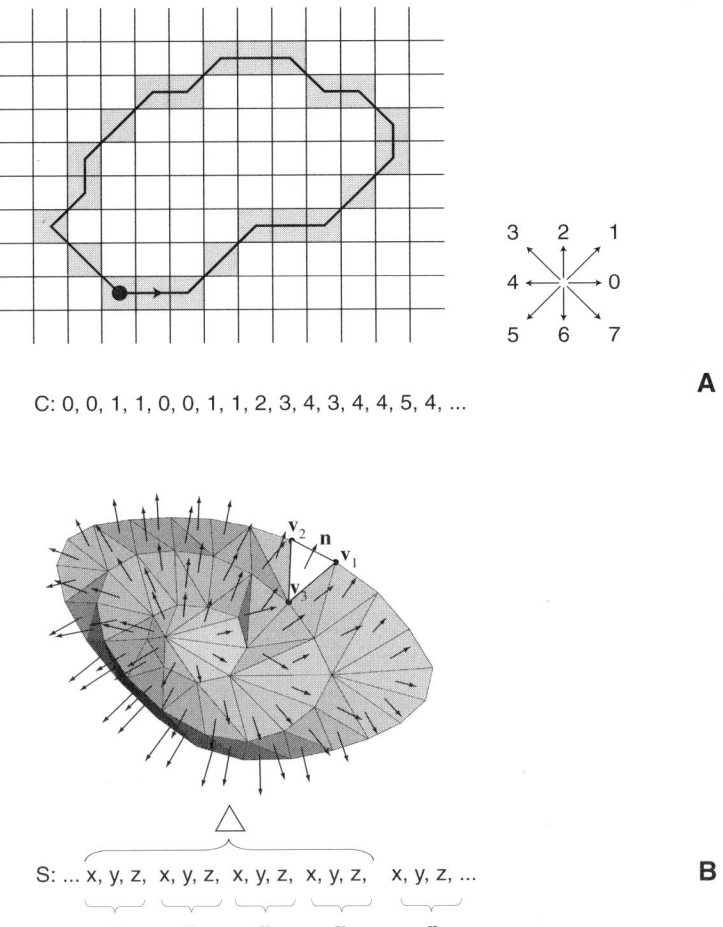

C: 0, 0, 1, 1, 0, 0, 1, 1, 2, 3, 4, 3, 4, 4, 5, 4, ...

**A**

S: ... x, y, z,  x, y, z,  x, y, z,  x, y, z,   x, y, z, ...

... **n**     $\mathbf{v}_1$     $\mathbf{v}_2$     $\mathbf{v}_3$        **n**   ...

**B**

**FIGURE 2-12**   Free-form object formats. **A**: The chain code of a 2D contour. **B**: The stl (stereolithography) format describes triangulated surfaces. Surface normal vectors have length 1; their direction results from the orientation of the triangles and determines the polarity between external and internal surfaces of the object.

the structure of an outline (in two dimensions) or a surface (in three dimensions) with the aid of *nodes* (or knots). If we think of the outline or surface as a rubber band or sheet, respectively, nodes correspond to the edges of a scaffolding structure attached to the outline or surface. Node positions define the final shape of the rubber structure (see Fig. 2-13). Because the rubber will accommodate its shape to the minimum energy state, the positions of the knots in a spline define the optimum shape of the outline or surface. In biomedical applications, spline functions are often used to parameterize free-form objects, that is, to approximate polygons and polyhedra with continuous lines and surfaces, respectively. The resulting combination of bottom-up object reconstruction and top-down object construction are discussed in detail in Chapter 4.

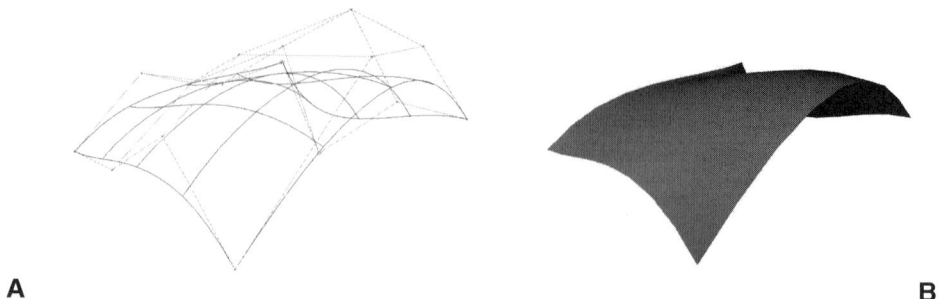

**A**                                                                                                                                **B**

**FIGURE 2-13**   Spline functions. A set of nodes representing edges of a grid structure (**A**) defines a smoothly curving surface (**B**).

## 2.3  A TAXONOMY OF BIOMEDICAL DATA

### 2.3.1  Perspectives on Data

A physician recording patient age, sex, body weight, and height, or a paleoanthropologist measuring the length, breadth, and volume of a fossil human braincase are both acquiring biomedical data. This is the typical approach to data sampling.

Here, we invert this typical perspective. We start with a three-dimensional, or volumetric, image data set that represents organismic structure, as it can be acquired with computer tomography (CT) or magnetic resonance imaging (MRI), and examine various types of data that can be derived from the original data volume (Fig. 2-14). In view of the reverse engineering approach often adopted in the biosciences, data acquisition begins with the dual aim of sampling quantitative information maximally and documenting and exploring the structure of an organism beyond given analytical goals. Thus we follow the traditional empiricist trail where observation precedes hypothesis-driven data acquisition. For example, we usually do not know how a fossil embedded in sediment actually appears; likewise, we do not know beforehand how all of the skeletal parts and soft tissues of a patient who suffered cranial trauma are affected by the injury.

### 2.3.2  Volume Data

Like any other physical object, an organism has as a particular spatial distribution of matter with associated physico-chemical properties, such as density, chemical composition, temperature, and color (Fig. 2-15). For the sake of simplicity, let us assume that data sampled from this organism occupy a cube with edge length $N$. Our sampling device has a spatial resolution of 1, so that data acquisition results in a set of $N^3$ voxels, each of which is characterized by position $(x,y,z)$ and an

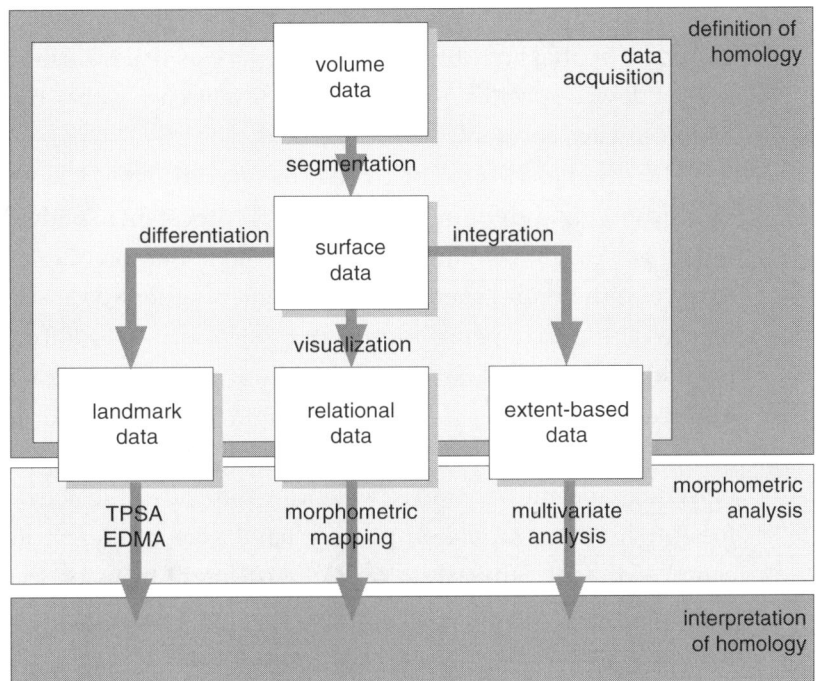

**FIGURE 2-14**   A taxonomy of biomedical data. After acquisition of volume data from biomedical objects, various types of 1D, 2D, and 3D data can be extracted. One important consideration is the recognition of homologous relationships between objects at various levels. Surface data denote boundaries between the object and its environment, or between subunits of the object. Further differentiation yields landmark data, which are the basis for subsequent geometric-morphometric analyses with methods such as thin plate spline analysis (TSPA) and Euclidean distance matrix analysis (EDMA). Integration yields more conventional data such as distances, surface areas, and volumes (extent-based data). Between surface, landmark, and extent-based data, there are data (e.g., curvature and thickness) that capture various relational properties between or within surfaces.

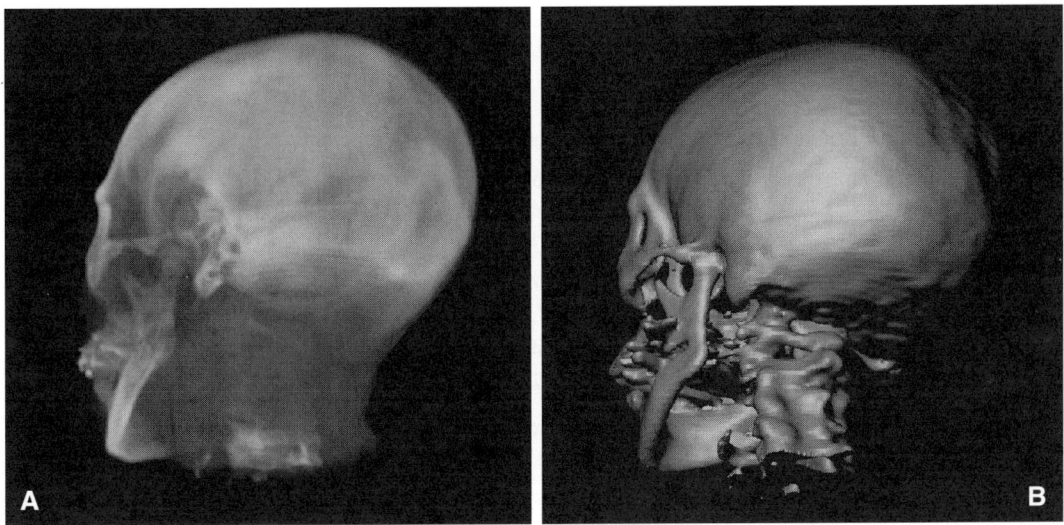

**FIGURE 2-15**   A volume CT data set of a patient. **A**: Skeletal structures within the head can be recognized effortlessly by our visual system. **B**: Quantitative characterization of these structures—for example, the skeletal surface—requires sophisticated image data processing steps.

associated physico-chemical property that is denoted by $I(x,y,z)$. Accordingly, the system of spatial coordinates $(X,Y,Z)$ that was used during data sampling represents a device-specific, external frame of reference rather than a "natural" system of coordinates.

The original volume of three-dimensional data lacks immediate biomedical significance and requires extensive preprocessing with biological reasoning, image processing, and image analysis. The definition of relevant data, such as the surface of the brain, skeletal parts within a patient data set, thickness of enamel in a fossil tooth, depends on the ability to identify and localize respective structures in the volume data set and to quantify them with adequate measuring tools. The criteria according to which the original data volume is segmented and classified must be formulated and provided by the user. For example, an anthropologist might be interested in representing the skeletal structures of a sediment-filled fossil, whereas a taphonomist may have a keen interest in the structure of the actual sediment. Similarly, a cardiologist might focus on the heart and vessels in patient CT data, whereas the same data set may be used by an orthopedic surgeon to assess osteoporosis in the vertebral column.

At the outset, identifying relevant data structures within the original data volume may appear straightforward because our visual system can perform this task effortlessly and efficiently. For example, we can recognize skeletal structures in the CT scan of a patient readily (Fig. 2-15). However, with computers, it is far from trivial to identify these same structures with the same reliability. This necessitates that the human visual system be replaced by the synthetic "visual" capacities of a computer, so that the task of data acquisition becomes a task of image processing and object recognition.

## 2.3.3 Surface Data

Because we dedicate Chapter 4 to this subject, we only outline here the mechanisms for finding solutions to object recognition. Objects are typically identified by their boundaries, which delimit them from the environment or from neighboring objects. We discussed above that boundary structures represent *contours* in two dimensions or *surfaces* in three dimensions. Surfaces arise through abrupt changes in local properties of the data volume. In computational terms, spatial discontinuities are detected by examining the first and second spatial derivatives of data volume. These measure local rates of change and "change in change," respectively.

Discontinuity-detecting procedures yield sets of potential surface vertices that must be ordered in a subsequent step to obtain an explicit geometric description of the surface to which they belong, such as a triangulated free-form object (cf. Fig. 2-12). Apart from computational issues, defining surface structures is a matter

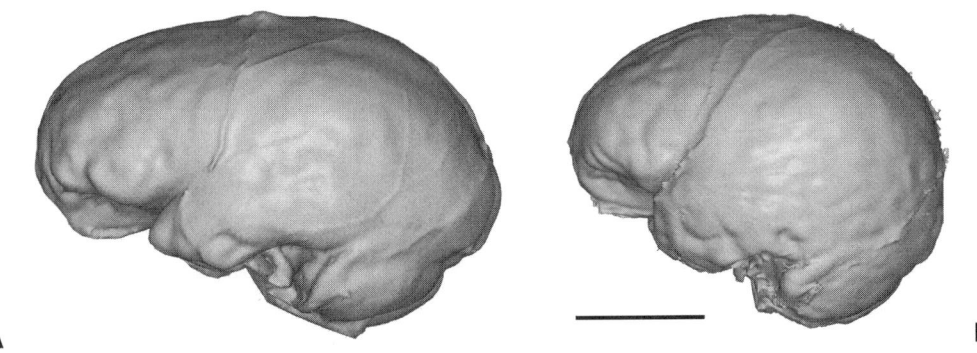

A

B

**FIGURE 2-16** Two virtual endocranial casts showing imprints of cerebral gyri and sulci, meningeal vessels, and venous sinuses. Correspondence relationships are defined between **A** and **B** at various levels of data types. The most comprehensive correspondence is established in endocranial volumes. But which point on surface **A** is developmentally or functionally homologous to a point on surface **B**?

of *biological correspondence*. In fact, biomedical object recognition always uses the concept of correspondence.[9]

We aim to characterize structures that are instances of biological object classes. For example, we presuppose that the surface of the brain of individual **A** is homologous to the surface of the brain of individual **B** (Fig. 2-16). Overall correspondence between boundary structures is recognized easily but remains rather unspecific because it does not indicate the point on surface **A** corresponding to a point on surface **B**. To solve this problem, we must establish more specific homologous relationships.

## 2.3.4 Landmark Data

Morphologic points of reference on which correspondence between specimens is established are called *anatomic landmarks*. These landmarks constitute a highly specific subset of object boundary structures because they identify convergence points for three or more distinct anatomic units, or extremal points on surfaces, such as spines, bulges, pits, and troughs (Fig. 2-17).

From the perspective of computational geometry, the landmarks can be identified as locations of maximal change in local surface properties. However, every extremal point is not necessarily a "good" landmark. As we have already noted in surface data, defining relevant landmarks ultimately depends on biological criteria and the constructed hypotheses. This relationship is of crucial importance because the connection between geometry and biology has direct implications for

---

[9] How biological correspondence is actually defined depends on the biological hypotheses stated at the outset of an analysis. We will return to this issue in Chapter 6.

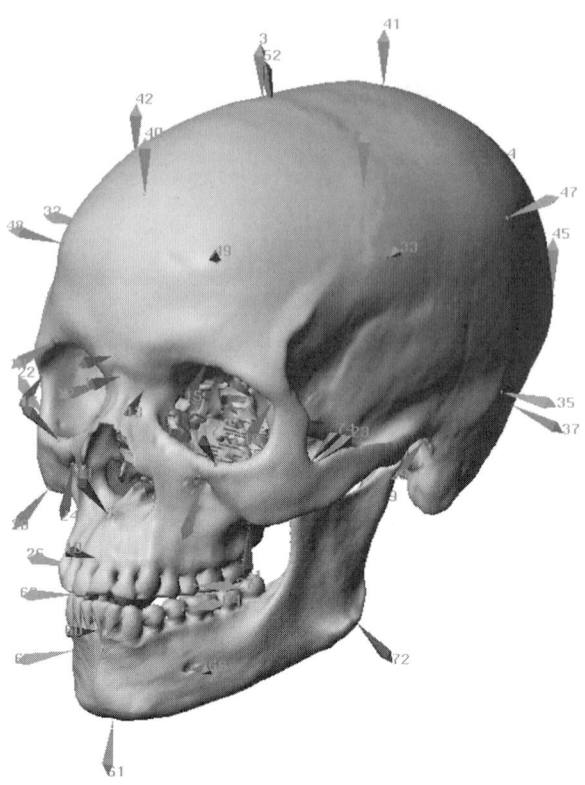

**FIGURE 2-17** Cranial landmarks identifying recognizable points of interspecimen correspondence.

the morphometric analysis of biomedical data (see Chapter 8). New imaging tools and new morphometric methods cannot replace this critical task—defining biological correspondence between different biomedical data sets.

To summarize, we derive surfaces from volumes and landmarks from surfaces through the computational procedure *differentiation*, that is, calculating rates of change at various levels of a spatial scale. From this perspective, landmark locations represent the most derived morphometric data and they stand for the most specific type of correspondence—point-to-point correspondence between organismic forms. However, these data are no different from morphometric data typically collected during most scientific work! Apparently, the complex task of differentiation is quite normal in practice. Nevertheless, whereas we usually measure interlandmark distances and angles, landmark constellations per se have become only recently amenable to morphometric analysis. We focus on this burgeoning subject in Chapter 8.

### 2.3.5 Extent-Based Data

In most classic morphometric analyses, weight (g), volume (mm$^3$), surface area (mm$^2$), and distance (mm) measurements are collected from various organisms and submitted subsequently to multivariate analysis. All of these measurements represent single values (or *scalars*) that characterize structure magnitude or size rather than its shape. From a computational point of view, these data are obtained by *integration* over boundary structures, or by appropriate subdivision of the original volume with data segmentation procedures. Extent-based data characterize quantity independent of object geometry. For example, measurements of body mass, brain volume, and body size reveal nothing about organismic shape. In summary, whereas landmark data are obtained through differentiation, extent-based data result from integration; the former characterize local correspondence, and the latter characterize global correspondence.

## 2.3.6 Relational Data

However, a large amount of potentially significant morphometric information exists between these two categories and remains unexploited by either landmark-based or extent-based methods. This information is underutilized for two reasons. First, landmarks are not dispersed evenly over biological forms; specific correspondence cannot be established in regions where identifiable landmarks are missing. Second, many relevant morphologic traits that characterize areas between landmarks are easier to recognize than to quantify; for example, tissue thickness or surface curvature cannot be quantified by landmark positions, nor can they be measured by surface areas or volumes (Fig. 2-18).

Therefore, a fourth type of data, one that captures localized morphometric properties without an explicitly defined homologous relationship, is useful. The salient property of this type of data—called *relational data*—is that local morphometric properties of an object are evaluated in relation to a given spatial neigh-

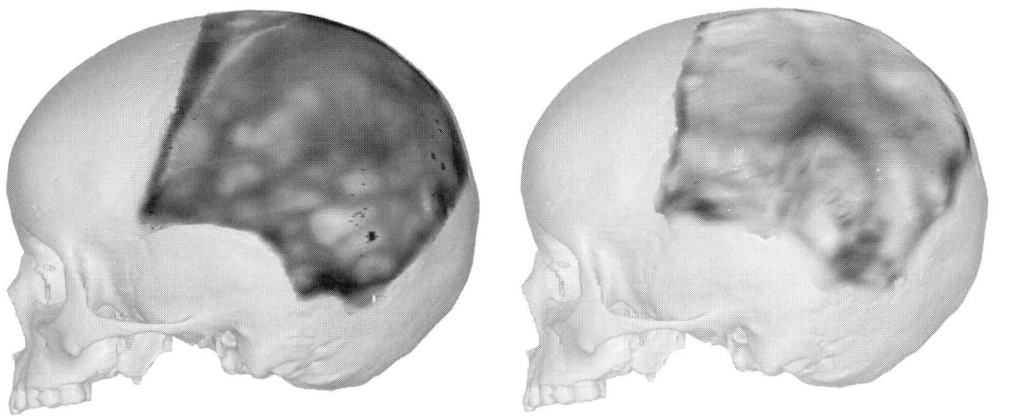

**A**  **B**

**FIGURE 2-18** Relational data. Bone thickness **(A)** and bone curvature **(B)** in the left parietal bone of a modern human child. Imaging localized morphometric properties is called *morphometric mapping*, or *functional imaging*.

borhood, or in relation to other data sets. For example, surface curvature expresses how much the orientation of the surface at a given point changes when we move in a given direction along a given distance. As another example, consider brain mapping, where brain activity is set in relation to cortical morphology. By virtue of their self-referential nature, relational data intentionally exclude the concept of correspondence. However, as mentioned above, detection of specific surface properties is a precondition for defining landmark locations. In this respect, the evaluation and visualization of relational data have the potential to characterize previously unrecognized biological correspondence, which can be resubmitted to comparative morphometric analysis.

## 2.4 FURTHER READING

The material of this chapter is presented in extended form in various introductory textbooks, for example, Long (1994), Simon (2001), Patt and Patel (2001), and Norton (2003). Topics concerning data format, color, and compression standards are best consulted on respective websites (see Web Companion). Issues of compression are amply covered by Hankerson (2003).

# DATA ACQUISITION

**3**

CAT-scan it before you publish in *Science*.
—Terry Harrison, November 19., 2001

## 3.1 DATA AND THE PHYSICAL WORLD

This chapter focuses on methods, technologies, and devices for acquiring three-dimensional digital data and representing physical objects with the data formats described in Chapter 2. A central question in the current chapter is, What is the relationship between a physical object under study and its representation with digital data?

Some 2500 years ago, Socrates pondered the relationship between objects in the physical world and their representation as a mental concept. He used the cave parable to illustrate underlying problems. This parable can be summarized as follows: A group of people living in a cave are tethered to chairs, forcing them to stare at the cave wall. Behind the seated people, a group of actors is portraying "reality". The cave is illuminated by torches that cast the actors' shadows against the walls. For lack of awareness, the spectators naturally perceive the shadows as the essence of reality. However, if the seated people were "freed" and led to the cave entrance, they would be blinded by the sunlight, but subsequently they would realize that the outside world is real. For Socrates, the sun represents the light of truth illuminating "ideas" (the real things), and the cave symbolizes the constraints of everyday perception (the shadow of reality).[1]

---

[1] As the reader may note, this story is highly polyvalent. It can be portrayed as an intellectual stimulant of deprivation experiments in twentieth-century experimental psychology. At the same time, it bears an astonishing resemblance to modern movie theater settings and, more generally speaking, it illustrates the potential impact of mass media on the public mind-set. And last, but not least, it reappears in the concept of a virtual reality "cave", where the user is immersed completely in a virtual environment.

*Virtual Reconstruction: A Primer in Computer-Assisted Paleontology and Biomedicine.*
By Christoph P. E. Zollikofer and Marcia S. Ponce de León.
Copyright © 2005 John Wiley & Sons, Inc.

From the perspective of data acquisition, we recall the situation within the cave. In modern terms, Socrates stated that the process of obtaining information from physical objects is not an immediate one but involves various stages of projection, mapping, or transformation. For example, our sense organs acquire data on specific properties of the environment and transform these into a sequence of neural signals. Likewise, a digital data acquisition device maps physical properties of an object as a sequence of bits in computer memory. As we will see, data acquisition devices, such as computer tomographs and laser scanners, literally follow the cave parable: The devices use a light source to generate projections and reflections of the objects.

As soon as a connection between a real physical object and its representation as digital data is established, several practical questions must be asked:

- *Which data acquisition device is most suitable to perform a specific task?* Data acquisition devices are tools to perform scientific tasks. As such, the choice of devices must incorporate specific aims of the analysis, not vice versa. However, this should not discourage the adoption of new, more sophisticated devices in new scientific endeavors. Medical diagnostics, for example, often requires acquisition of large volumes of tomographic data that reflect different structural and physiological properties of an organism. Only through subsequent exploration does the data volume yield a diagnostically relevant subset, such as a hidden fracture in the skull base or reduced lumen in a coronary vessel. On the other hand, for comparative analysis of stature in a group of patients, acquiring data for body weight and height may suffice.

- *Which properties of the object shall be recorded?* Devices are classified according to the type of recorded data. Adhering to the theme presented in Chapter 2 (Fig. 2-14), we categorize separately volume scanners (three dimensional, 3D), surface scanners [both 3D and two dimensional (2D)], and point data acquisition devices, such as 2D graphic tablets and 3D digitizers. A photographic camera is a 3D-to-2D scanner, and devices like balances and rulers are volumetric and linear integrators, respectively.

- *What is the metric relation between the physical object and its logical representation?* It is necessary to establish a link between the object and its representation by attaching a measurement scale to the data structure. For example, the size of each pixel in an image, measured in millimeters, and the speed of recorded video frames, measured in frames per second, are each a type of measurement scale. This information is contained typically in the header, preceding the actual data structures stored in a file.

- *What is the quality of the represented data with respect to the original object?* To answer this question two things must be considered. First, the limits of resolution at which the physical object structure can be represented must be determined. Second, measurement errors, that is, the accuracy and precision with which the data represent a given structure, must be ascertained. *Accuracy* is the degree to which a measured value equates the "true" value. Because true values are not known with absolute certainty, typically accuracy is expressed relative to values of standardized objects that are used to calibrate the data acquisition device following standard protocols. *Precision*, on the other hand, describes reproducibility of multiple measurements. *Resolution* represents the smallest interval that a data acquisition device can measure reliably.

A detailed explanation of these terms is given in Appendix A. In subsequent sections, we address these questions while emphasizing current three-dimensional data acquisition techniques, such as computed tomography, surface scanning and 3D digitizing. Beforehand, we return to the basics and discuss data acquisition devices that are more recognizable to all, such as the human eye and the photographic camera. This approach facilitates an insight into the potential and limitation of data acquisition techniques. Although principles of tomography are certainly more complex than principles of photography, the general terminology that assesses the performance of these data acquisition devices and the procedures that adjust the device parameters are largely similar.

## 3.2 VISION AND PHOTOGRAPHY AS DATA ACQUISITION: PERFORMANCE CONSIDERATIONS

The human eye and the photographic camera are data acquisition devices that transform three-dimensional scenes into two-dimensional images. Both are composed of two subsystems, the image-generating optical system, and the image-storing receptor system (Fig. 3-1). The overall performance of an eye or a camera depends on the performance of both subsystems. To obtain optimal results, the spatial and contrast resolution of the receptor system must be tuned to the spatial and contrast resolution of the optical system. These resolutions correspond almost perfectly in the fovea of the human retina, where the spatial density of the photoreceptors reaches theoretical limits of resolution for the cornea-lens system. In a digital camera, image pixels are typically larger, and contrast resolution lower, than allowed by the camera optics.

The process of data acquisition and the resulting data structures can be characterized similarly. In the eye, the optical system (cornea, crystalline lens, and iris) generates a retinal image that becomes a pattern of electric charge differences in photoreceptors (cones for color vision, rods for light/dark vision). Photoreceptors

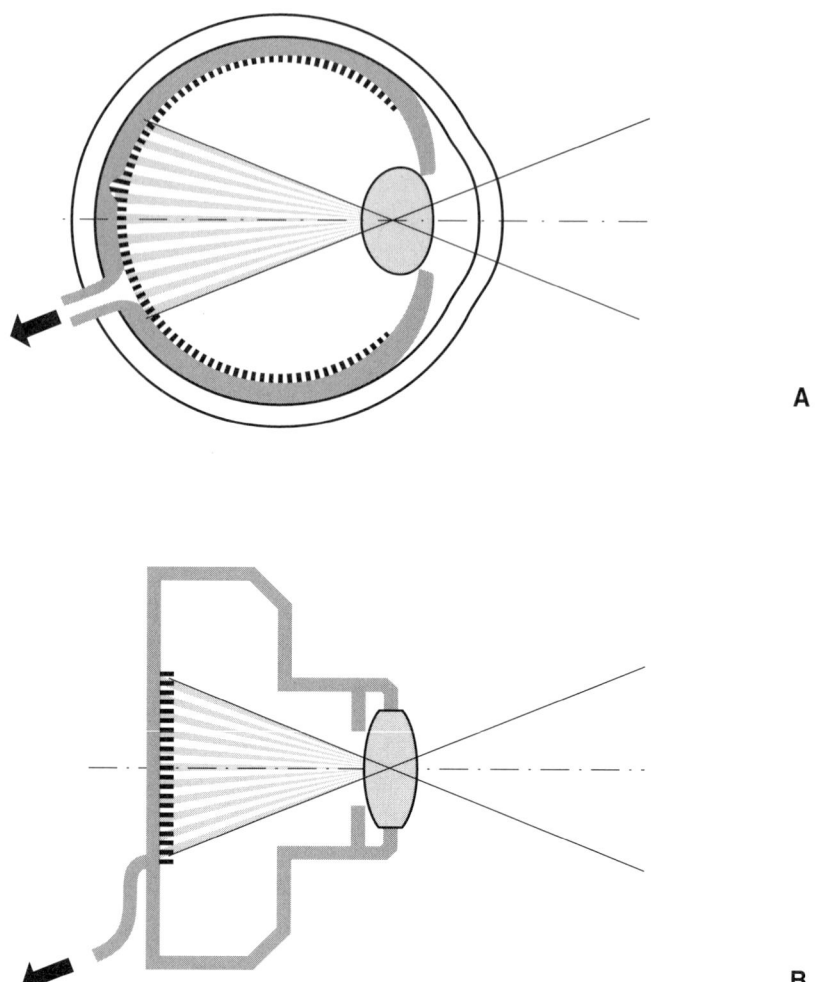

**FIGURE 3-1**  Human eye (**A**) and camera (**B**) as data acquisition devices. Both devices are composed of an image-generating optical system (gray) and an image-storing receptor system (black). To obtain optimum results, the spatial and contrast resolution of the receptor system (black-white pattern) must match the resolution of the optical system (gray-white pattern).

are the "pixels" in a retinal image. In the digital camera, the optical system (lenses and diaphragm) generates an image that becomes a pixel matrix encoding gray values or color values. Technically, this matrix is composed of light-sensitive charge-coupled devices (CCDs). CCDs are the "photoreceptors" in a digital image. Hence, in the eye and in the camera, the image is represented ultimately as a pattern of charge difference. These data structures are sent to the brain and computer, respectively, for further image processing (see Chapter 4).

To assess the performance of these data acquisition systems, and to evaluate the optimal combination of imaging parameters for a given imaging task, we may examine how "well" a system maps a given object onto its image, in other words, how "well" the image reproduces the original object. This is accomplished typically with standard objects of known dimensions and physical properties, such as black-and-white line patterns, because this permits direct quantitative comparisons between the object and its image. To assess the performance of measuring devices, we evaluate the following parameters:

- *The geometric properties of the system.* Because linear dimensions of line patterns are known, real-world dimensions of a pixel matrix can be expressed in millimeters per image pixel, or as angular degrees, independent of the distance between the object and the imaging plane.
- *The spatial resolution.* Spatial resolution is the minimum resolved distance under which two parallel lines (or two points) appear as separate objects. This "distance" is typically measured as a visual angle (for example, the visual diameter of the full moon is approximately 0.5 angular degrees). Spatial resolution can be evaluated by experimentally increasing the distance between the line pattern and the eye or camera until the pattern is no longer recognizable. Theory holds that spatial resolution depends on the diameter of the front lens in an optical system. For optimum performance, the resolution of the pixel matrix receiving the image should match the resolution of the optics. In the human eye correspondence evolved to near perfection, whereas digital cameras do not exploit typically optical resolution.
- *The contrast resolution.* This parameter refers to a minimum difference in image intensity that the system can resolve (see Fig. A-3 in Appendix A). Again, we must discern between contrast resolution in the optical system and contrast resolution of the pixel matrix that records the image. For the latter, contrast resolution corresponds to the possible number of coded gray levels (as we have seen in Chapter 4, a digital camera typically codes 256 gray values). Contrast resolution of the optical system is a more complex parameter because its definition changes at different levels of spatial detail. The so-called modulation transfer function (MTF) typically summarizes contrast and spatial resolution in the optical system, showing how well black-and-white (i.e., 100%-contrast) patterns are transferred by the system at any given spatial frequency (see Box 3-1).
- The *light sensitivity* should be considered on both sides of the imaging device. The amount of light captured by the optical system is a function of front lens area. The primary impetus behind building cameras with large front lenses is a gain in sensitivity rather than spatial resolution, because minimizing exposure time avoids blurring in moving objects and improves

## BOX 3-1

## DATA ACQUISITION AND TRANSFER FUNCTIONS

How well—in terms of spatial and contrast resolution—does an imaging system such as our visual system, or a camera, transfer real-world objects into data structures? The answer lies in testing the device with patterns exhibiting various combinations of spatial detail and contrast. This is best accomplished with sine wave patterns that have detail resolution represented by the spatial frequency $v$ (the inverse of wavelength $\lambda$) and contrast resolution represented by the modulation $M$, which measures the amplitude $A$ of the wave as a proportion of its mean intensity $\bar{I}$ (Fig. 3-2).

Subsequently, output modulation $M'$ of the system is measured as a proportion of the input modulation $M$ along a range of spatial frequencies $v$. Because of physical limitations, optical systems usually blur the input signal, such that the ratio $M'/M$ is always $\leq 1.0$. Plotting $M'/M$ as a function of spatial pattern frequency $v$ yields the *modulation transfer function* (MTF), which gives a detailed quantitative account of the imaging properties of a device. The MTF of the human visual system can directly be observed with the test image in Figure 3-3.

This image consists of a spatial sine wave whose frequency increases exponentially from left to right and whose amplitude decreases linearly from bottom to top. Although the image contrast decreases steadily at the upper end of the test image, we perceive an unequal course of attenuation corresponding to the bold line in Figure 3-4. Modulation transfer of the human visual system is most efficient at spatial frequencies around 10 cycles per degree (test this on yourself by grabbing a ruler, extending your arms, and looking at the millimeter ticks) and becomes less efficient as lower and higher ends of the frequency spectrum are approached. Contrastingly, efficiency in the MTF of technical systems (camera, radiographic projection, computed tomography) declines steadily only toward higher spatial frequencies. The system limit of *spatial resolution* is reached at the right end of each curve. Spatial frequencies become too high (accordingly, spatial detail too small) to be resolved even under optimum contrast conditions.

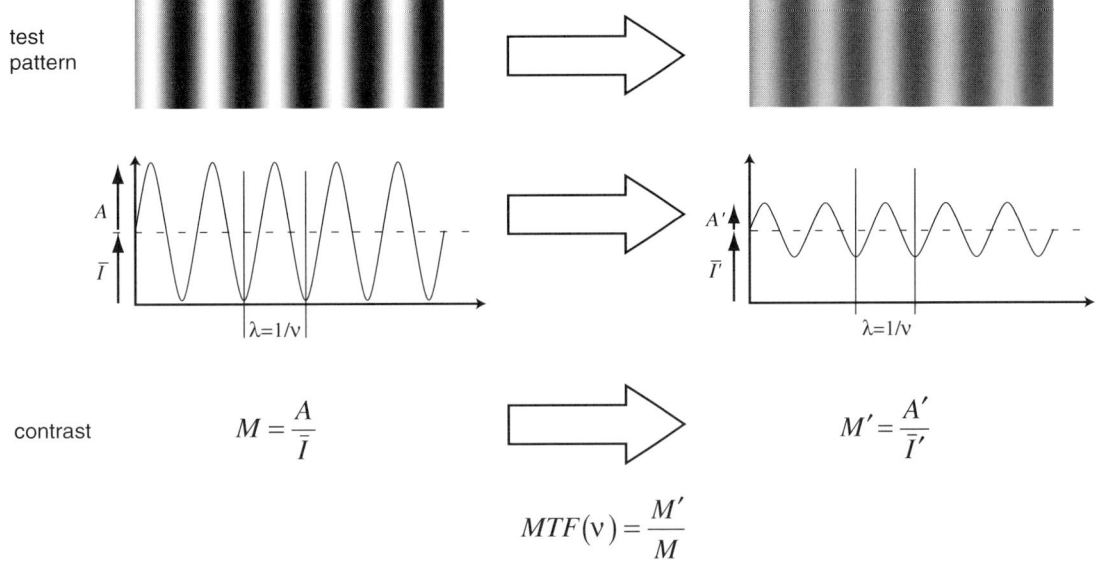

**FIGURE 3-2** A test pattern (left) used to measure modulation transfer (right).

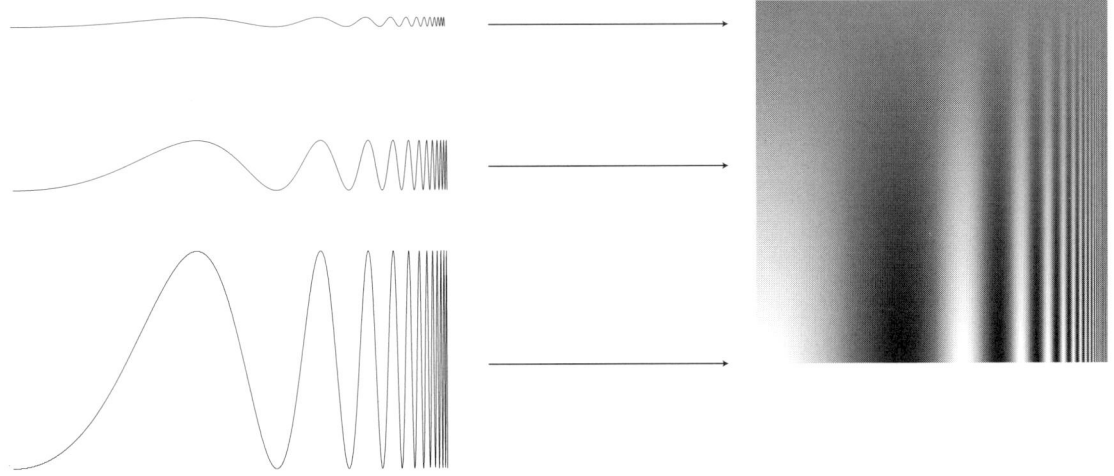

**FIGURE 3-3** Visualizing the modulation transfer properties of the human visual system. The curves (left) show the composition of the test image (right). Spatial frequency increases from left to right; contrast decreases from bottom to top.

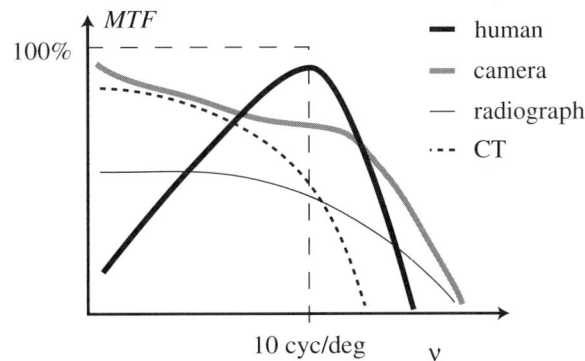

**FIGURE 3-4** Modulation transfer function of the human visual system and of various technical systems.

performance at low ambient light intensities. In comparison, we must consider sensitivity of light receptors in the retina or pixel matrix. Technically, this sensitivity can be enhanced in either of two ways. First, *response characteristics* (Appendix A) can be improved (more sensitive photopigments in neurons; more sensitive CCDs; more sensitive, coarser-grained, film in classic photography). Second, signals of neighboring receptors can be summed. The second method is used in the eye and the digital camera at the expense of spatial resolution (looking at graph paper, for example, observe how the spatial resolution decreases with decreasing ambient light intensity).

Now that we have assessed operational constraints of the eye and camera when functioning as imaging systems, we must discuss tuning the system for optimum results in specific imaging tasks. In human vision, tuning is performed dynamically at various levels of the system, such as through adjustment of the iris aperture, accommodation of the optics, adaptation of the photoreceptors to local lighting conditions, and neuronal postprocessing. Tuning imaging parameters of a camera is a similar process involving adjustment of the zoom factor (field of view), aperture, focus, exposure time, film sensitivity, and so on. Because trade-offs exist between all parameters, the definition of an optimum solution requires knowledge of the actual imaging task. This topic is amply covered in textbooks on scientific photography. After a short digression to the history of X rays, we will see that the same principles hold when a CT device is tuned to a specific imaging task.

## 3.3 COMPUTED TOMOGRAPHY

### 3.3.1 Frau Röntgen's Wedding Ring

In the nineteenth century electric phenomena fascinated professional scientists and amateurs alike. Specifically, the nature of the so-called cathode ray was vigorously researched. For example, it was well known that the flow of electric charge from the negative (cathode) electrode to the positive (anode) electrode elicited emission of visible light in a partially evacuated glass tube, or on a glass screen coated with fluorescent dyes. However, the material basis underlying this phenomenon remained unclear until late in the century. In 1887, the English physicist J. J. Thompson showed, via an ingenious experiment, that these rays represented the flux of a new elementary particle that was named the electron. The cathode beam could be deflected by a magnet, and its properties were independent of the materials used to construct the electron-emitting device.[2]

Wilhelm Conrad Röntgen, then of the University of Würzburg, also experimented with cathode rays and fluorescent screens. On November 8, 1895, Röntgen made one of those accidental, but nontrivial, observations that tend to occur in gently disorganized laboratories (i.e., where things that do not belong together are in close spatial vicinity). When Röntgen switched on his cardboard-wrapped gas-discharge tube, a nearby fluorescent screen started glowing in response to some invisible radiation emanating from the tube and penetrating the cardboard. This was the story behind the discovery of X rays, where "X" denoted the initially unknown nature of the radiation.[3] Röntgen placed various objects in front of the tube, casting their images on the screen and on photographic film.[4]

And this is how, shortly before X-mas 1895, the first medical-diagnostic X-ray image was produced: Frau Röntgen's left hand, complete with her wedding ring (Fig. 3-5). In the months following the discovery, X-ray technology was immediately adopted for medical and for paleontological diagnosis.

Soon it became clear that X rays represent short-wave electromagnetic radiation, that is, the high-energy complement of visible light. The mechanism by which X rays are generated occurs in three steps. First, a cathode emits high-speed electrons. Then, these electrons impact the anode. Finally, the resulting deceleration of the electrons resulting from the impact, and their interaction with nuclei of the atoms of the anode, elicit emission of X rays. Modern X-ray sources operate at high voltages (in the range of 90 to 140 kilovolts) and use rotating tungsten anodes to prevent melting under the impact of high-energy electrons. In principle, any device that is exposed to the impact of an electron beam, for example, the fluorescent screen of a TV monitor, produces X rays; however, most of these devices produce X rays at very low intensities, which are also filtered by lead atoms contained in the glass screen.

In medical diagnostics, X-ray imaging is an invaluable technique because it is relatively easy to handle and yields images with high spatial resolution but comparatively low contrast resolution. The major disadvantage, however, of classic radiography is that three-dimensional structures are imaged as projections; organs of the human body become transparent shadows superimposed on each other. For example, we do not know whether the wedding ring of Frau Röntgen was circular, quadratic, or triangular in shape because we know only one projection of it. Nevertheless, diagnostic background knowledge can overcome this limitation to

---

[2] You may repeat this experiment with your home cathode ray-fluorescent screen aggregate, namely, your TV monitor.

[3] In continental Europe, these rays are called "Röntgen rays," following a proposal of the anatomist von Kölliker on January 23, 1896, when Röntgen made a radiograph of von Kölliker's hand in public.

[4] A post-Romantic implementation of Socrates' cave parable.

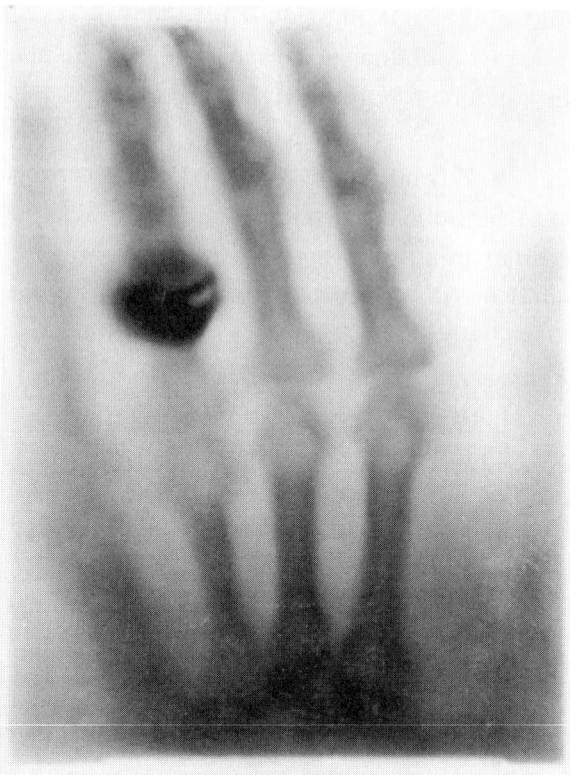

**FIGURE 3-5**  Frau Röntgen's left hand, as imaged by Wilhelm Conrad Röntgen on December 22, 1895 (reproduced by kind permission of the Deutsches Röntgen-Museum). The tube used to create this image was of an early type, in which the glass walls assumed the function of an anode. The lack of an actual focal spot of the X rays explains the blurred appearance of the image.

a certain extent. For example, most radiologists concerned with the topic suggest that Frau Röntgen's ring was probably an annular object.

Various methods have been devised to overcome the problem of projection and superposition. Typically, radiographs are taken from different directions and then examined simultaneously to evaluate the three-dimensional relationships between a suspected pathological structure and various organs. An alternative technique is stereoradiography. As in stereophotography, two projections made from slightly different viewing angles are combined into a single spatial image.[5] A third possibility, following logically from the first two, is allowing a computer to reconstruct the original three-dimensional structure from a series of X-ray pro-

---

[5] Fluoroscopy, the direct X-ray examination of patients, uses a similar principle. By interactively changing the orientation of the patient and/or of the plane of X-ray projection, physicians gain quasi-spatial insights into the patient's anatomy. The original method, which used fluorescent screens, has been largely abandoned because of the excess radiation exposure involved. Modern interventional radiography works with image amplifiers and low-dose radiation.

jections that are taken from various angles. This is the basic premise of computed tomography, or CT (Cormack, 1963, 1964; Hounsfield, 1973).

### 3.3.2 Radiographic Projections

What is the nature of an X-ray projection? X rays "collect" information about objects they traverse as they progressively attenuate along their paths. Let us denote with $I_0$ the signal intensity of an X ray before entering the object and with $I$ the intensity of the signal where it is received by an electronic detector (or radiographic film), after having passed through the object.

It can be recognized readily from the example shown in Figure 3-6 that the final intensity $I$ of the X ray is a multiplicative function of materials encountered along its path, as well as the length of the path. This is expressed in the Klein–Nishina formula of attenuation (Box 3-2), which states that $I$ is a power function

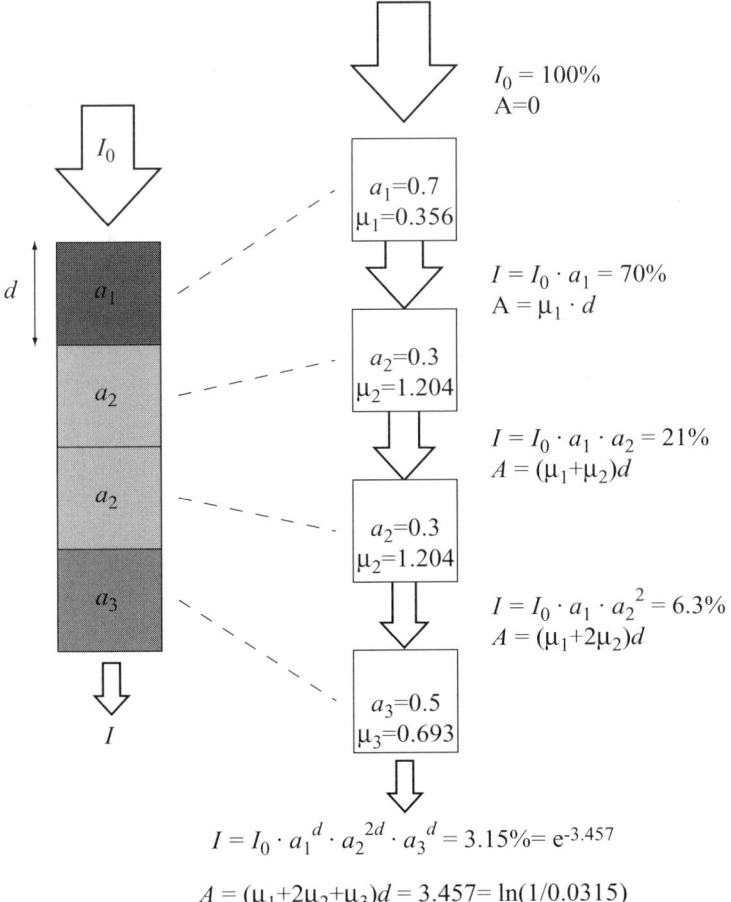

$$I = I_0 \cdot a_1{}^d \cdot a_2{}^{2d} \cdot a_3{}^d = 3.15\% = e^{-3.457}$$

$$A = (\mu_1 + 2\mu_2 + \mu_3)d = 3.457 = \ln(1/0.0315)$$

**FIGURE 3-6** An X ray passes through four voxels, each of which reduces its intensity $I_0$ by some proportion $a_i$. The final intensity $I$ is a multiplicative function of the transmission coefficient $a_i$, and an additive function of the logarithmic attenuation coefficient $\mu_i$.

## BOX *3-2*

### SUNGLASSES, X-RAY ATTENUATION, AND COMPUTER TOMOGRAPHY

Let us consider a light ray of intensity $I_0$ passing through an array of $d$ voxels (e.g., a stack of sunglasses), all having similar material properties and attenuating the light ray by the same factor $a$ (note that $a < 1.0$) (see Fig. 3-7).

The light ray leaves the first voxel with an intensity equal to $a \cdot I_0$, the second with an intensity equal to $[a(a \cdot I_0)]$, and so on, such that we obtain the final intensity $I$

$$I = I_0 \cdot a^d \tag{3-1}$$

In fact, $d$ can be treated as a continuous variable, if we think of infinitesimally small voxels. In "classic" notation with exponents, we express this equation with the "natural" base $e$ (Euler's number, 2.71828 . . .).

$$I = I_0 \cdot e^{\ln(a) \cdot d} = I_0 \cdot e^{-\mu \cdot d}, \tag{3-2}$$

where $\mu$ is known as the linear attenuation coefficient of a given material. In optics, Equation 3-2 is known as *Lambert and Beer's law of absorption* (light absorption in a sunglass means photon-catching by atomic electrons). The same formula can be used to describe the behavior of X rays on their path through a patient body. In radiology, Equation 3-2 is better known as the *Klein–Nishina formula of attenuation*. In an organism irradiated with X rays, attenuation results from X-ray scattering (Compton and Rayleigh scattering), whereas absorption (the sunglass effect) plays a minor role.

In a patient, $\mu$ typically varies along the X-ray path, that is, voxels have individual attenuation coefficients $\mu_i$. Furthermore, as depicted in Figure 3-8, path segments through each of the $n$ voxels have variable lengths $d_i$.

This yields

$$I = I_0 \cdot e^{-\mu_1 d_1} \cdot \ldots \cdot e^{-\mu_i d_i} \cdot \ldots \cdot e^{-\mu_n d_n} = I_0 \cdot e^{\sum_{i=1}^{n}(-\mu_i d_i)} \tag{3-3}$$

Note that, in logarithmic form, we obtain

$$\ln(I/I_0) = \sum_{i=1}^{n}(-\mu_i d_i) \tag{3-4}$$

## SUNGLASSES, X-RAY ATTENUATION, AND COMPUTER TOMOGRAPHY (*Continued*)

In computed tomography, the quantity $\ln(I_0/I)$, known as attenuation $A$, expresses the amount of X-ray intensity attenuated by the object. Because $\ln(I_0/I) = -\ln(I/I_0)$, we obtain

$$A = \ln(I_0/I) = \sum_{i=1}^{n}(\mu_i d_i) \tag{3-5}$$

This equation is important for the reconstruction of tomographic images because it shows that, on a logarithmic scale, we can sum, or integrate, linear attenuations along the ray path to obtain total attenuation.

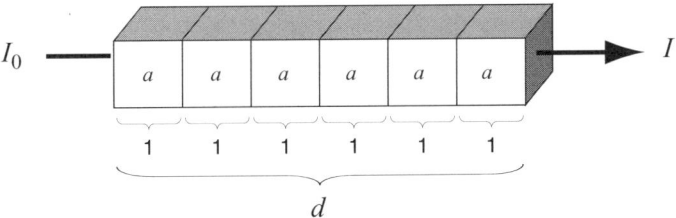

**FIGURE 3-7** A light ray passing through an array of voxels.

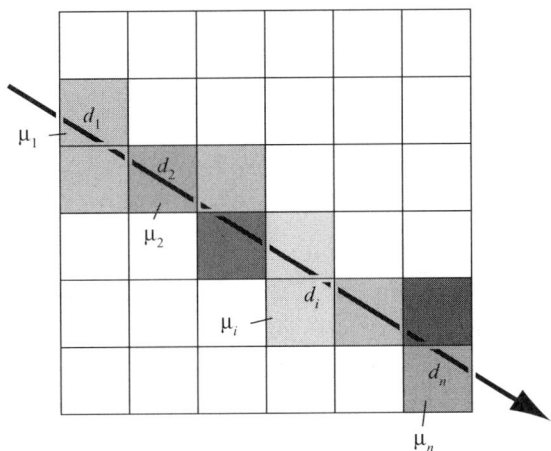

**FIGURE 3-8** An oblique ray passing a voxel field.

of local attenuation coefficients and distance traveled. Because we are interested in the relationship between incident and transmitted intensities rather than their absolute values, we use the logarithm of their proportion, $\ln(I_0/I)$, to define *attenuation A*. Using logarithms creates the additional advantage of transforming multiplications into additions and power functions into multiplications. Accordingly, we may state that the X ray *sums local attenuations* along its path through the voxel volume, where each local attenuation is the product of an attenuation coefficient $\mu_i$ (a material property of a voxel *i*) and the path length $d_i$ through that voxel. This sum is then projected on radiographic film or received by an electronic X-ray detector.

A fundamental question in computed tomography asks: How can local tissue densities $\mu(x,y,z)$ at voxel position $(x,y,z)$ within a three-dimensional organism be evaluated with attenuation values *A*? It is evident that more than one X-ray projection is necessary for solving this problem because, as in linear algebra, at least *n* equations are required to solve for *n* unknowns. The first step in a practical solution of this complex task is to subdivide it into several slices.

Whereas classic radiography works with two-dimensional projections of three-dimensional organisms, CT confines itself to one-dimensional projections of two-dimensional sections through an organism.

The principle of computed tomography is shown in Figure 3-9A (actual technical implementations are slightly different; see Fig. 3-14). A planar beam of parallel X rays is sent through the object under investigation and received by a one-dimensional detector array on the opposite side. Accordingly, the beam creates one virtual object slice of a given thickness. The source-detector ensemble rotates around the object to obtain projections from different points of view. As shown in Figure 3-9, each voxel $\mu(x,y)$ of the scanned object contributes to various projections. How can its actual value be evaluated from a combination of all available projections around the object?

The problem of reconstructing an *n*-dimensional object from the sum of its $(n-1)$-dimensional projections has a general, yet elegant, mathematical solution— the inverse Radon transformation—that predates computed tomography (Radon, 1917). Recall that an X-ray sums local attenuations along its path. If we think in infinitesimal terms (voxel size approaches zero, voxel number infinity), we pass from summation to integration. In other words, each projection along a ray path corresponds to a *line integral*. The Radon theorem states that the set of all line integrals taken around an object contains a complete description of that object. Mathematical methods by which the object can be recovered from its line integrals are complex (Natterer, 1986). In practical applications, an object must be reconstructed from a finite set of projections, with various approximation techniques. Thus CT images do not provide an immediate replication of reality; rather, they represent *reconstructions* of reality.

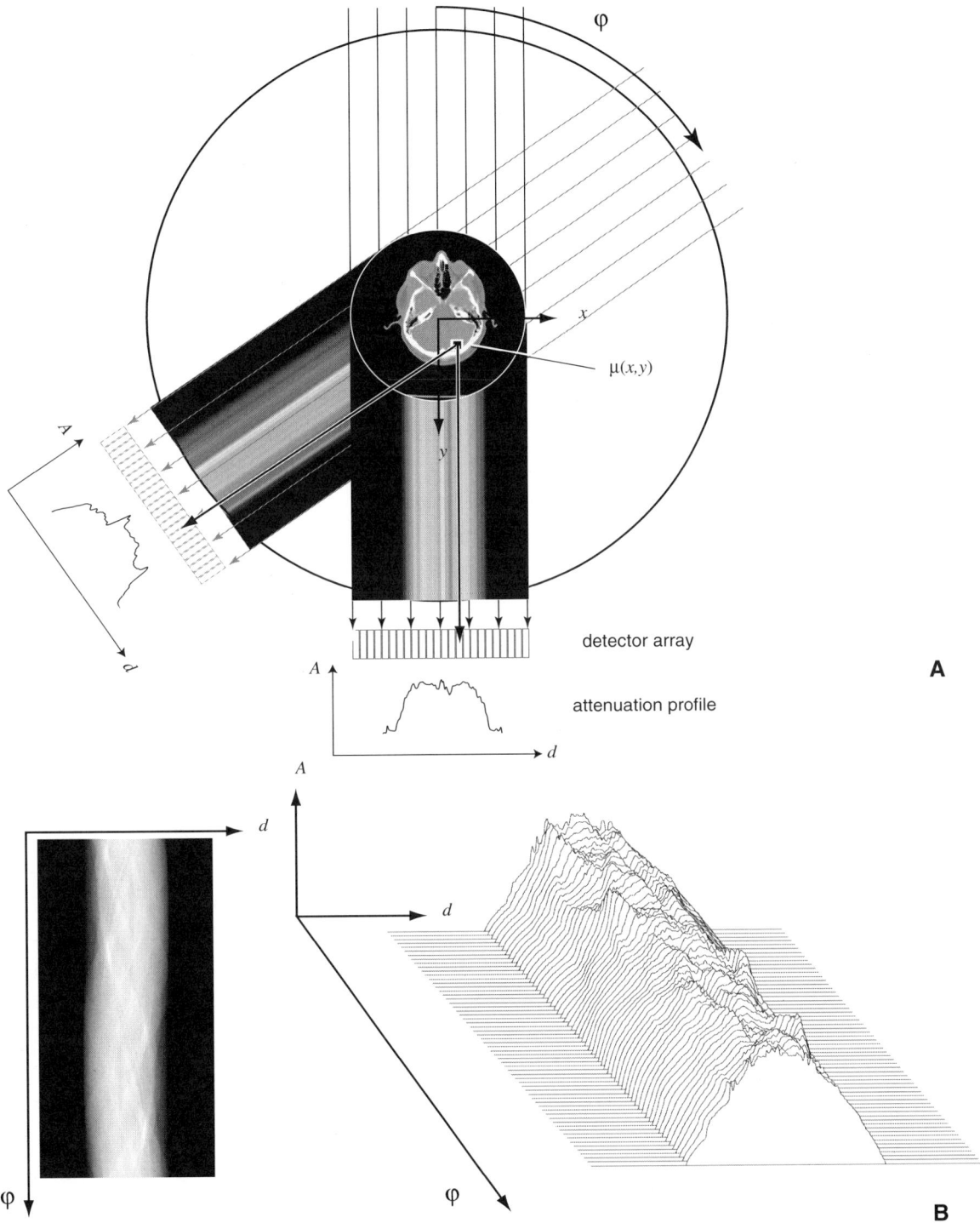

**FIGURE 3-9** Principle of computed tomography. **A:** Generating projections at various positions φ around the object (the picture shows a cross section through a human head. **B:** The sequence of projections, the so-called sinogram, represents the acquired attenuation data from one full turn of the CT gantry around the object.

Because CT imaging forms an essential part of computer-assisted biomedical applications, we will examine how cross-sectional images are reconstructed from projections in greater detail. This will enable us to assess the potential, as well as the limitations, of tomographic techniques and to determine the specific settings that are necessary to obtain optimum results.

### 3.3.3 Reconstructing CT Images

The fundamental idea of CT image reconstruction is to invert the process by which projections have been obtained. This method is called, appropriately, *backprojection*, because all projections are spread out in reverse direction along the ray paths that created them and then superimposed in the object's cross-sectional plane (Fig. 3-10). Recalling Equation 3-5, it is evident that we only know the summed attenuation $A$ along each ray path (i.e., the line integral), whereas individual contributions $\mu_i \cdot d_i$ are unknown. Accordingly, we assume an average value $\bar{\mu}$ for each individual attenuation coefficient

$$\bar{\mu} = \frac{A}{\sum_i d_i}$$

that does not vary along the ray path. The superposition of projections yields an adequate reconstruction of the original cross-sectional structures (the limited number of projections used in Fig. 3-10 results in a fuzzy image that exhibits streaklike artifacts).

The method of backprojection favors low-frequency portions of the image over high-frequency portions (recall from Box 3-1 that images can be considered mixtures of various spatial frequencies at different amplitudes). Accordingly, before backprojecting, projection data must be modified slightly by allotting more weight to the high-frequency portions than the low-frequency domains. As we will see in Chapter 4, functions that modify images (one-, two-, or three-dimensional) are called filters. The filter function for projection data is shown in Figure 3-11 as an instruction for amplifying spatial frequencies (similar to a transfer function; see Box 3-1). The filter function has the shape of a ramp (i.e. frequencies are amplified linearly) until a given cutoff frequency that marks the limit of spatial resolution of the CT scanner.

Compared to simple backprojection (Fig. 3-10), *filtered backprojection* yields reconstructions that are closer to the original cross-sectional image (Fig. 3-11).

In practical applications, various modifications of the ramp filter are used to enhance specific spatial frequencies in the reconstructed CT image. Medical CT

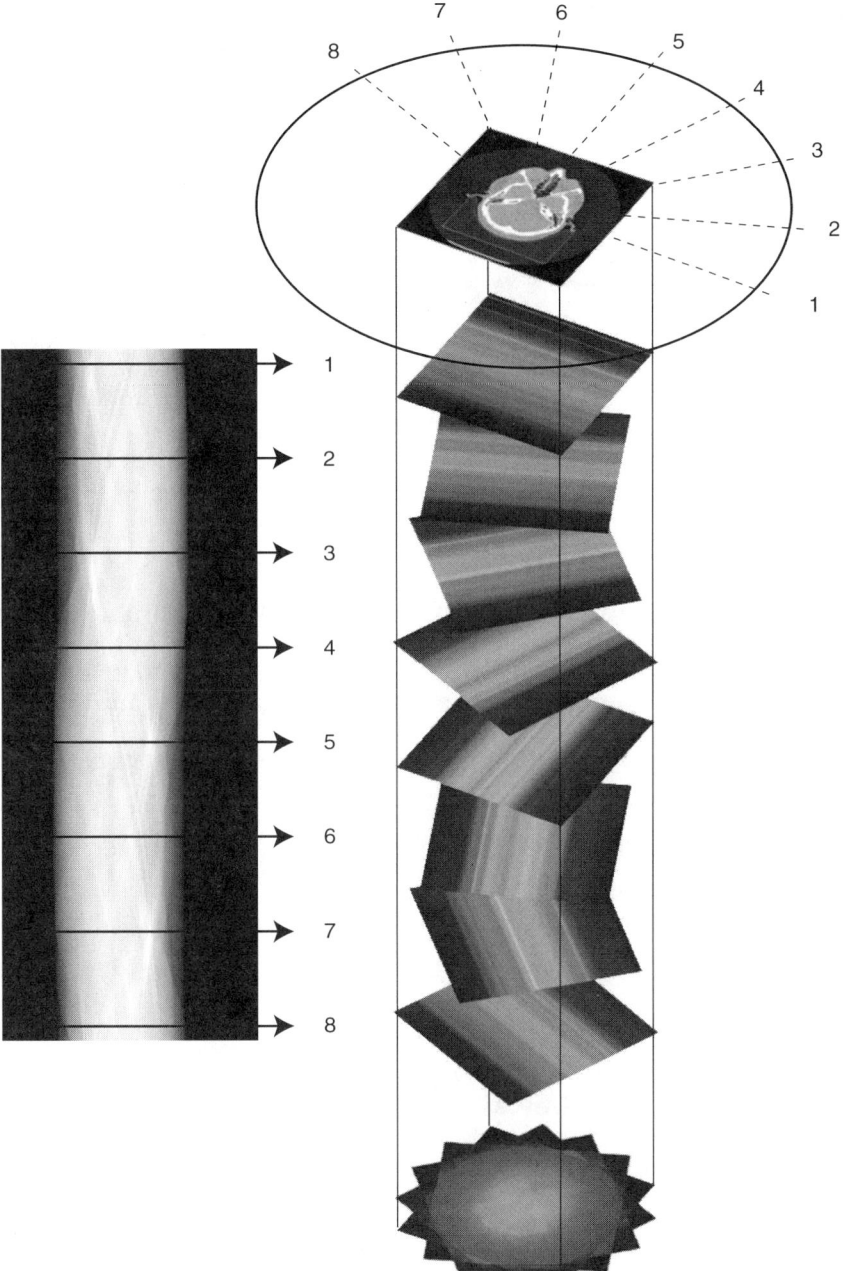

**FIGURE 3-10** Reconstruction of a cross-sectional image by backprojection of attenuation data. In the present example, only 8 projections (at angles 0°, 45°, 90°, etc.) are superimposed, yet a faint image of the cranial shape appears in the superposition at the bottom.

consoles typically provide a range of predefined so-called *reconstruction kernels* (or *filters*) from "hard" to "soft," or, alternatively, from "bone" to "soft tissue." In comparison to the ramp filter, hard and soft reconstruction kernels increase or reduce, respectively, the contribution of higher spatial frequencies during backprojection

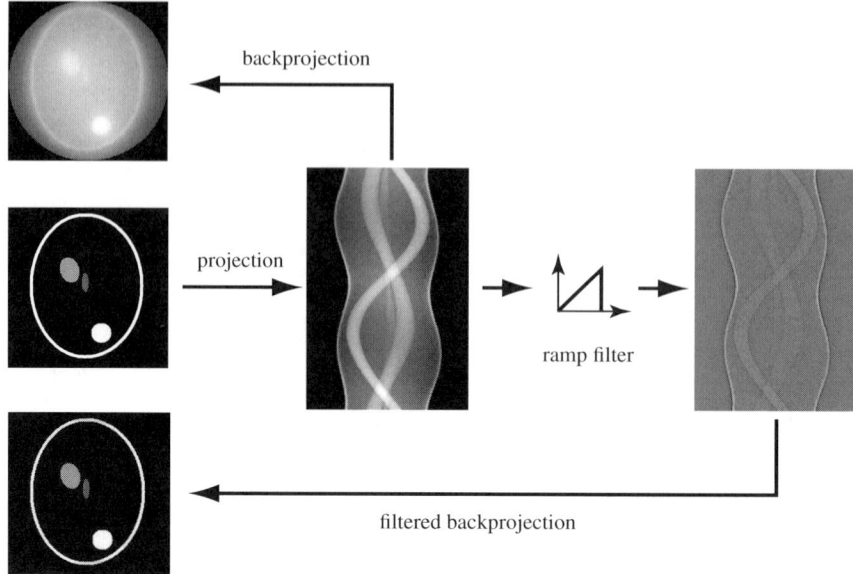

**FIGURE 3-11**   Filtered backprojection. A synthetic cross-sectional image (a so-called phantom) is projected in a CT simulator to obtain a sinogram. Direct backprojection yields a fuzzy image, in which the low spatial frequencies dominate. However, filtering the sinogram with the ramp filter, and then backprojecting it, yields a reconstructed image that is virtually identical to the original object.

(Fig. 3-12). A "hard" filter is used to enhance spatial detail (edges, i.e., density fluctuations over a short spatial range), as in trabecular architecture of cancellous bone, or bone fractures. A soft tissue filter does the opposite. It enhances long-range contrast and reduces image noise, which manifests as short-term fluctuation.

### 3.3.4 CT Scanning: Technical Considerations

Equipped with a theoretical backpack, it is now possible to explore the technicalities of CT scanning. The basic geometry of a modern medical CT scanner is shown in Figure 3-13. An X-ray source produces a fan-shaped planar beam, which is sent through the patient and received on the other side by an array of detectors. The plane of the CT section typically is called the $x$-$y$ plane, or image plane, and the reconstructed image is called a *CT slice*. The direction of table movement is parallel to the longitudinal axis of the patient and called the $z$-direction. The X-ray source and accompanying detector array are mounted in a wheel-like structure known as the *gantry*, which rotates around the patient to generate projections from various angles $\varphi$. The patient (or object) rests on a table that can be moved along the rotation axis of the gantry (the $z$-axis). Table movement permits the location

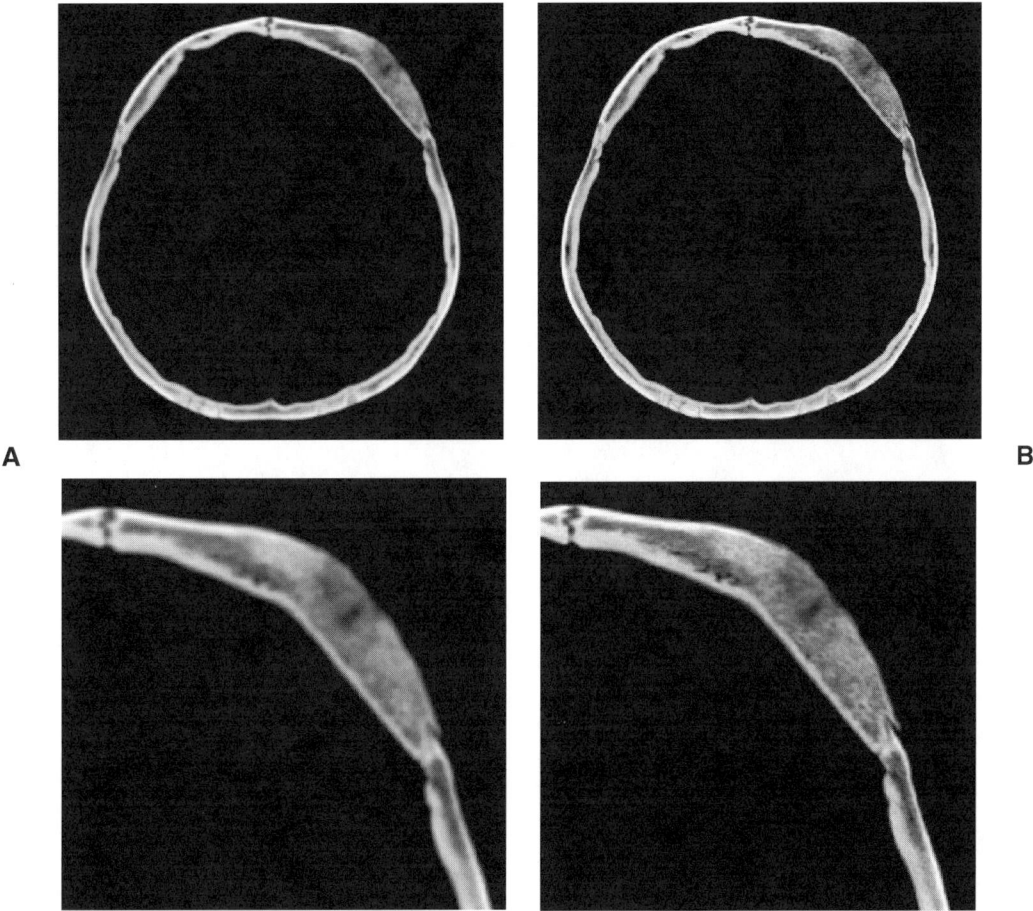

**FIGURE 3-12** CT image reconstruction with a soft tissue (**A**) and a bone filter kernel (**B**). Note the edge enhancement in **B**, revealing the fine structure of bone affected by a tumor (upper right portion of the image).

of specific cross-sectional positions and the option of performing serial cross sections. Projection data that are recorded by the detector array during a full revolution of the gantry are called a *sinogram*. Sinograms can be displayed conveniently as two-dimensional images, in which the horizontal axis represents detectors in the array and the vertical axis represents projections recorded at consecutive values of φ. The data represented by a sinogram are often called *raw data*, as opposed to reconstructed *image data*.

Image quality is a primary concern in CT scanning and is a notion related to various aspects of image accuracy, precision, and resolution discussed in Appendix B. We have seen that CT data acquisition consists of two consecutive steps, (a) acquisition of attenuation profiles (raw data) through exposure to X rays and (b) subsequent reconstruction of cross-sectional images through filtered

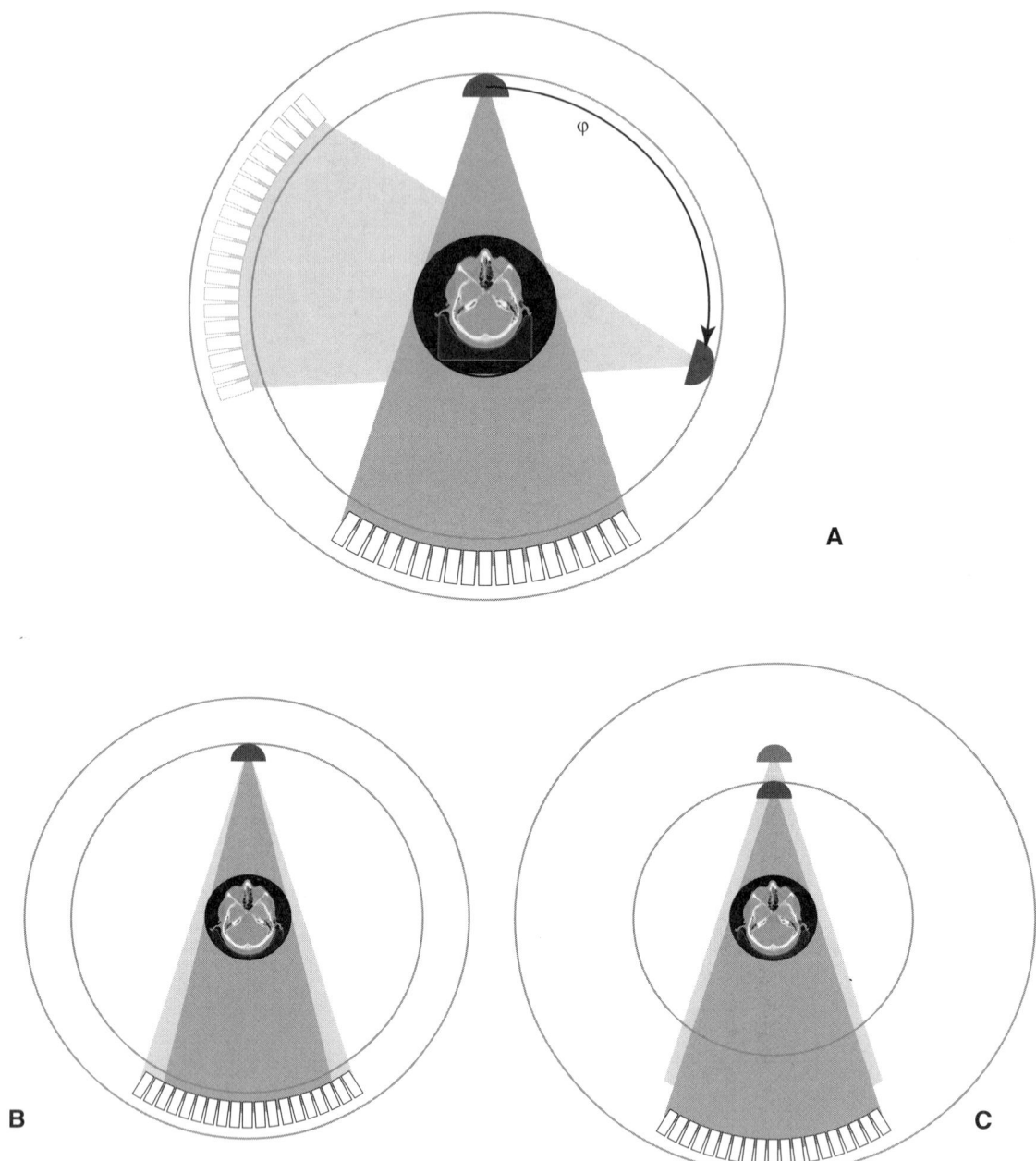

**FIGURE 3-13**  Medical CT device. **A**: Schematic drawing of a third-generation medical CT scanner. The X-ray tube generates a fan-shaped beam that passes through the patient and is recorded, on the opposite side, by an array of X-ray detectors. Tube and detector array are mounted in the gantry, which rotates around the patient to generate projections. **B**: In most devices, the scanned area can be adapted to the object diameter by reducing the fan angle. **C**: In some devices (Philips and Toshiba third-generation scanners), to achieve better spatial resolution, the X-ray tube can be moved toward the patient. This creates a projection onto the detector array that has a smaller field of view, resulting in geometric enlargement of the scanned patient region. During image reconstruction, special algorithms (called reordering and rebinning) convert projection data from the fan-beam geometry into a parallel-beam configuration.

backprojection. These two stages can be compared to two stages in photography where exposure of light-sensitive film or CCD and its development or digital post-processing represent two equally independent processes.

Quality in a photographic image is determined primarily by the parameters of the first step, namely, the choice of an adequate combination of lens aperture, focal length, exposure time, and film sensitivity, whereas the choice of the developing chemicals or the image-processing software modifies rather than enhances the image quality. Likewise, quality in a CT image is determined primarily by the parameters of the projection process, whereas variants of filtered backprojection only modify certain aspects of the reconstructed cross-sectional image. However, correspondence between the geometry of a CT device and resolution of the resulting CT image are removed further from each other than are optics or film of a camera and resolution of the resulting copy on photographic paper. This is because, in CT, there is a reconstruction process between acquisition and display.

### 3.3.5 Limitations of CT Data Acquisition

CT data acquisition has several limitations that arise from the finite size of detector elements, the finite number of projections available to reconstruct a cross-sectional image, and the limited amount of radiation. Furthermore, detector imperfections and deviations from the log-linear course of the Lambert–Beer law (see Box 3-2) also influence image quality in practical applications. Figure 3-14A depicts a cross section through four cranial fragments of the Le Moustier 1 Neanderthal specimen. Figure 3-14B is the same data set, but different window settings (i.e., display parameters) resolve the Styrofoam boxes holding the fragments in place. This image reveals concentric rings that indicate some degree of detector maladjustment (detectors must be calibrated occasionally to ensure equivalent responses to the same signal). Streaklike artifacts along relatively straight, contrast-rich object borders demonstrate the *partial volume effect*. As shown in Figure 3-14C, during projection 1, the object is projected only partially along its border onto the detector, which results in an apparently lower attenuation coefficient than during projection 2. Inconsistency between these values contributes to the observed artifacts. The dark bands between the bones result from beam hardening. The medical scanner used for this image was calibrated for beam hardening as it occurs in patients, but obviously not for unexpected beam-hardening effects of fossil bone.

Figure 3-15A demonstrates difficulties that arise during scanning of large fossil specimens with a medical scanner. A cross section through the skull of a Paleocene alligator (see Box 6-2) displays adverse partial volume effects and beam hardening, which create spurious density variations within the reconstructed object. While traveling through the object, low-energy portions of the X ray are

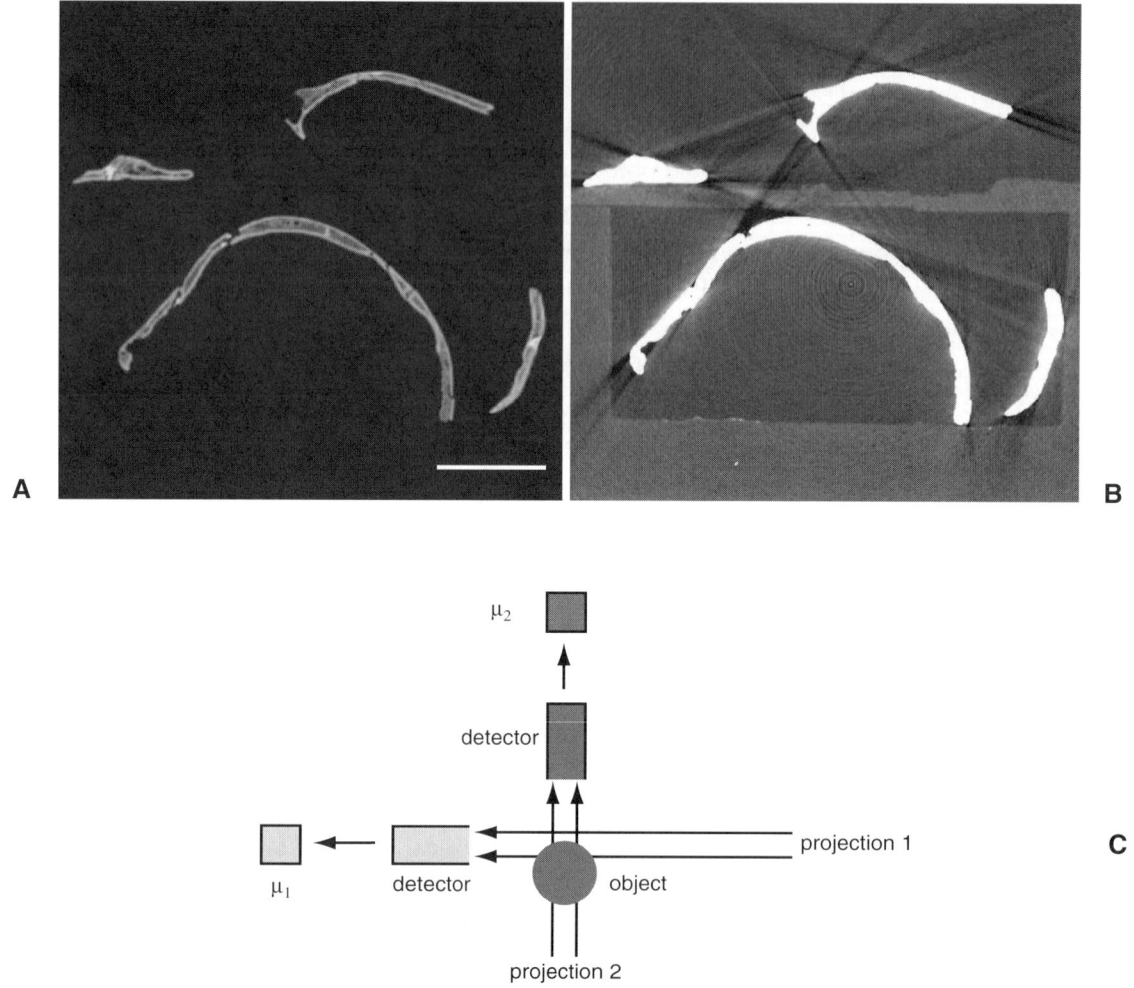

**FIGURE 3-14** CT scan of fossil cranial fragments (Le Moustier 1 Neanderthal specimen) demonstrating partial volume and beam hardening effects. **A** and **B** represent the same CT image data displayed with different contrast and brightness settings (scale bar is 5 cm). **C** illustrates partial volume effects. Projections 1 and 2 result in different inferred attenuation coefficients $\mu_1$ and $\mu_2$; $\mu_1 < \mu_2$ because only part of the object is projected onto the detector during projection 1.

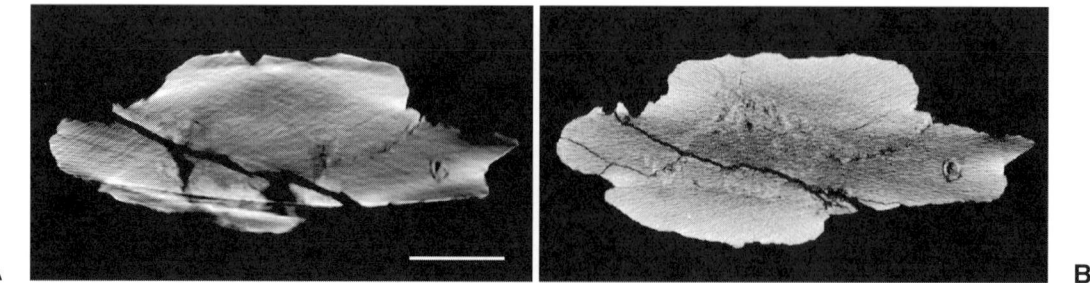

**FIGURE 3-15** Medical (**A**) and industrial (**B**) scans of a Paleocene alligator cranium. Note beam hardening artifacts in **A** and "frozen noise" streak artifacts in **B**. Scale bar is 10 cm.

more readily absorbed than high-energy portions. This yields a higher detector signal (less attenuation) than expected, such that the density of the reconstructed object is lower than the density of the actual object. Furthermore, contrast resolution is relatively low in this image because data were acquired at a slice thickness of 10 mm (this high aperture was necessary to obtain a reasonable detector response).

The same object was scanned again (Fig. 3-15B), but this time with an industrial scanner at higher X-ray energy levels. To improve resolution along the $z$-axis (i.e., in the scanning direction), slice thickness was reduced to 1 mm. Some streaklike artifacts and image noise indicate *frozen noise* effects, which originate from *photon starvation*. In photon starvation, some detectors operate at their sensitivity limits, thus producing attenuation signals with a high proportion of noise. The detector noise is backprojected across the image and thus "frozen" as an image artifact.

Figure 3-16A contains metal artifacts caused by dental fillings. Because of a high X-ray density, metallic implants completely block the X rays, resulting in zero detector signals. During image reconstruction, this type of signal deterioration induces starlike low- and high-density artifacts "radiating" from the artifact source. Various methods have been proposed to suppress such artifacts, primarily by applying special-purpose image processing algorithms to projection data (see Fig. 3-16B). These algorithms clip excessive attenuation in the sinogram and attempt signal restoration (Seitz and Rüegsegger, 1982; Kalender et al., 1987; Felsenberg et al., 1988; Path et al., 1998; Zhao et al., 2000).

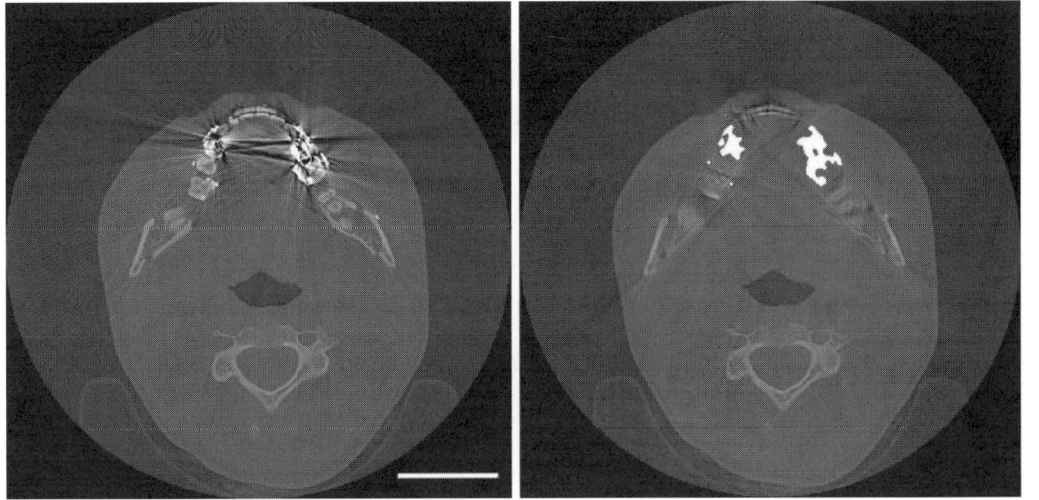

**A** **B**

**FIGURE 3-16** Metal artifacts caused by dental fillings (**A**) can be partially removed (**B**) with special-purpose software applied to CT projection data.

### 3.3.6 Slice-to-Slice, Helical, and Multislice CT

The CT techniques described to this point follow a two-phase step-and-shoot mechanism that alternates object positioning and data acquisition (Fig. 3-17A). In a typical medical scanner, each process lasts approximately one second. Thus the duty cycle of slice-per-slice data acquisition is roughly 50%, which is a relatively low value in a clinical environment that places a premium on time. This technique is known as *incremental scanning*; cross-sectional image data are acquired frame

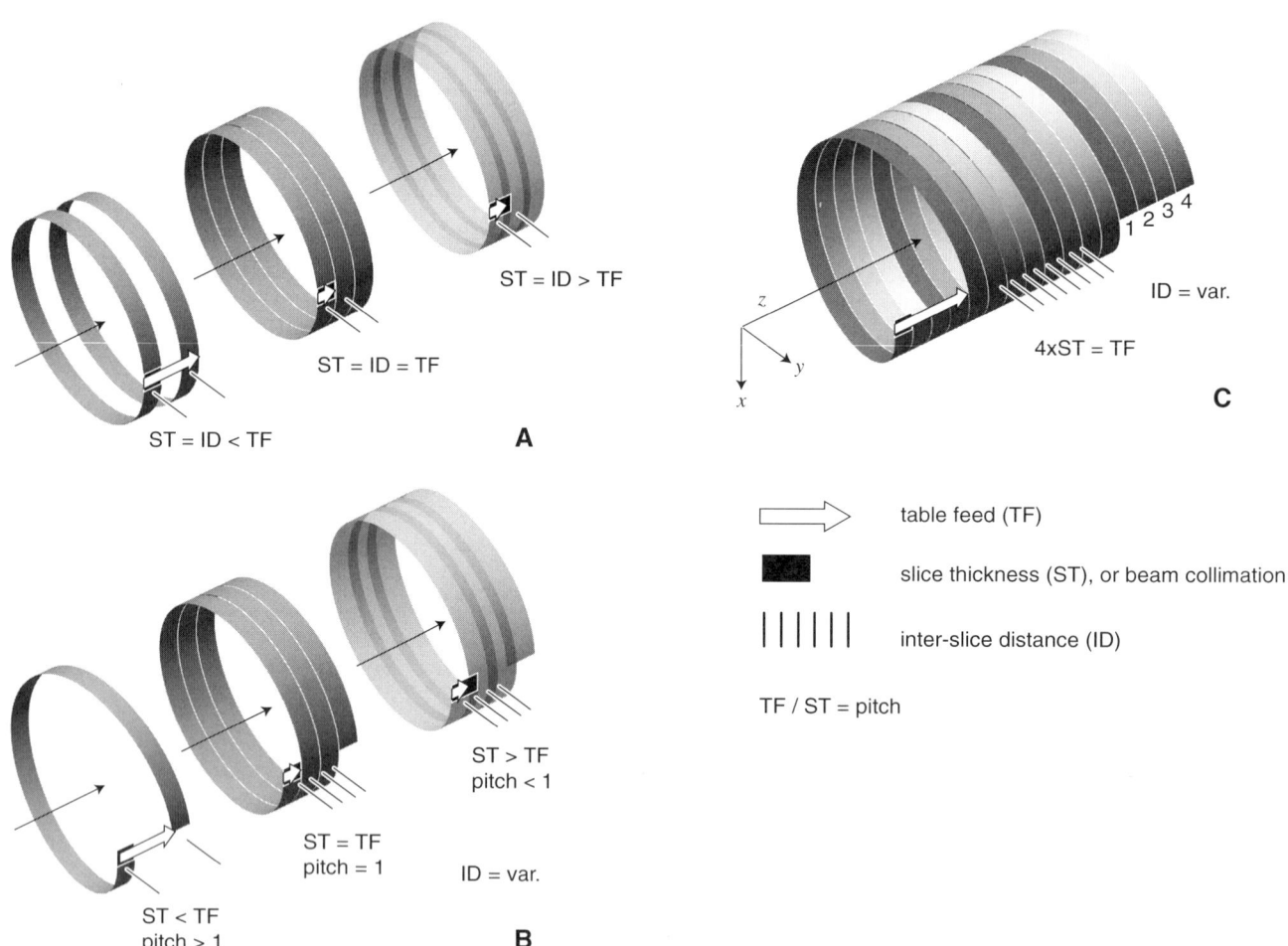

**FIGURE 3-17** Incremental, or slice-by-slice (**A**), helical (**B**), and multislice (**C**) CT scanning. **A**: During incremental scanning, the gantry rotates around the object at fixed table positions $z$, such that the table feed is identical to the interslice distance. **B**: A helical scanner acquires projection data while the object is moved continuously along the $z$-axis. Slices can be reconstructed retrospectively at arbitrary $z$ positions, but the projection data must be interpolated along the helix. Results are free of distortion when helix pitch (the length of one turn of a spiral divided by its width, or the table feed per full rotation per beam width) is 1:1, or smaller than 1:1. **C**: A multislice scanner uses a series of detector rows to perform helical scans. This technique permits dense volume data sampling (large $z$ intervals with low pitch values; for terminology see Silverman et al., 2001).

by frame, that is, slice by slice. In 1989, the technique of *helical* (or spiral) *computer tomography* (HCT) was introduced (Kalender et al., 1990; Soucek et al., 1990; Kalender, 1994). In HCT, instead of performing CT images at fixed table positions, the table is moved through the rotating gantry at a constant speed, resulting in a spiral-like path of the detector array around the patient. Because data acquisition and patient positioning are performed simultaneously, the duty cycle approaches 100% (Fig. 3-17B). This technique is extremely valuable in a medical setting because it permits volume CT data acquisition of the entire abdomen or thorax while the patient withholds breathing. An additional advantage of HCT is that overlapping reconstructions can be made without adding X-ray dose.

A salient feature of HCT is that data acquisition is no longer restricted to fixed cross-sectional positions. During data acquisition, not only is the projection angle $\varphi$ incremented, but so is the position along the $z$-axis. Accordingly, a full set of projection data per fixed image plane is no longer available but must be interpolated from the raw data volume. This is advantageous because it is possible to reconstruct cross-sectional images at any $z$ position retrospectively and without loss of quality. However, this also can be disadvantageous, because interpolation techniques necessarily introduce some error, which reduces image quality relative to incremental CT image quality. Interpolation artifacts appear as helical irregularities (spiral artifacts) on the object surface, primarily in areas where object morphology changes substantially relative to the $z$-axis (Fig. 3-18) (Wilting and Timmer, 1999). Nevertheless, conscientiously adjusting HCT device parameters during data acquisition and image reconstruction almost equates the accuracy of this technique with accuracy of sequential CT. Therefore, HCT is especially well-

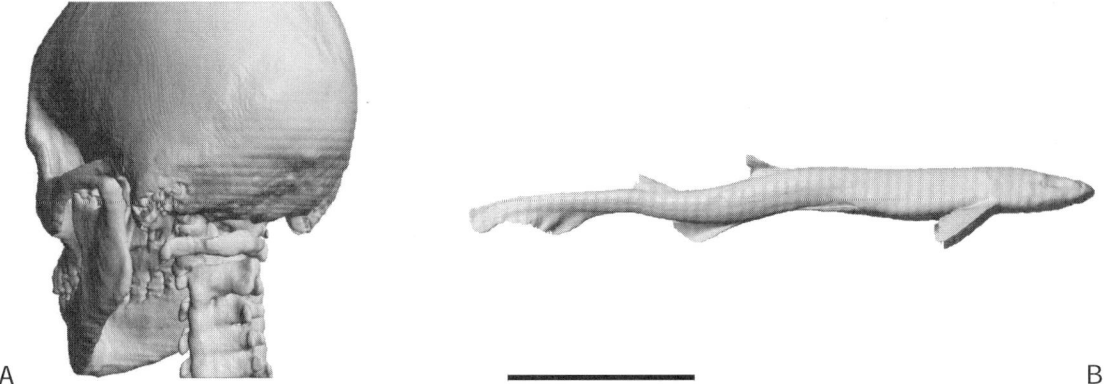

A                                                                                        B

**FIGURE 3-18** Helical scanning artifacts. **A**: Artifacts in this patient CT scan made with a one-detector-array helical scanner appear as stripes, notably in the occipital region, where the object geometry changes rapidly in directions parallel to the scanning plane. **B**: This multislice helical CT scan of a shark exhibits spiral-shaped artifacts on the specimen's surface resulting from inappropriate data acquisition parameters (pitch >> 1). The scale bar for both pictures is 100 mm.

suited for the rapid acquisition of large data volumes from extended skeletal samples.

As *single-slice scanners*, sequential and helical CT devices acquire data with a one-dimensional array of detectors. Recently, *multislice scanners*, in which multiple detector rows are arranged one after the other along the z-axis, thus resulting in a two-dimensional detector structure, have been developed (Fig. 3-17C) (Cody et al., 1998). The major advantage in multislice scanning is that data can be acquired simultaneously from 4 to 64 contiguous slice locations along the z-axis. This lowers scanning time, increases resolution along the z-axis, and requires less X-ray energy because emitted radiation can be collected more efficiently. In CT images acquired with sequential scanning, the z dimension of a voxel is typically larger than the x and y dimensions. Multislice scanners promote the acquisition of *isotropic voxel data* (i.e., volumetric data sets in which voxels have equal dimensions in x, y, and z directions). This facilitates reconstruction of arbitrary cross sections through the data volume that have the same image quality as slices in the original scanning plane. With multislice helical scanners, quantitatively reliable volume data can be acquired with the same guidelines as single-slice HCT.

### 3.3.7 Industrial and Micro Computed Tomography

Medical CT scanners are adjusted to the dimensions and material properties of the human body. As a general rule of thumb, an isolated tooth establishes the limits of spatial and density resolution of a medical CT scanner. Dense fossil specimens, as well as specimens considerably larger or smaller than a human body, often require analysis with an industrial tomograph (Fig. 3-19). Industrial tomography is used primarily for materials testing and reverse engineering purposes, for example, to detect fatigue fissures in gas turbine blades or to recover the three-dimensional structure of a machine part for which blueprints are no longer available. Because radiation intensity is not a typical concern in industrial tomography, higher X-ray intensities, smaller detector apertures, longer sampling intervals, and an adaptable source-detector geometry are all available options for improving spatial and contrast resolution. Overall, dense material and objects with nonstandard dimensions can be examined more accurately with industrial tomography than with conventional medical CT scanning.

An application of special interest because it permits the noninvasive analysis of fossil bone tissue at the histologic level is *micro computed tomography* ($\mu$CT, or microCT). The technique was developed for tomographic assessment of alterations in human cancellous bone, especially in connection with osteoporosis (Rüegsegger et al., 1996), as well as for nondestructive three-dimensional analy-

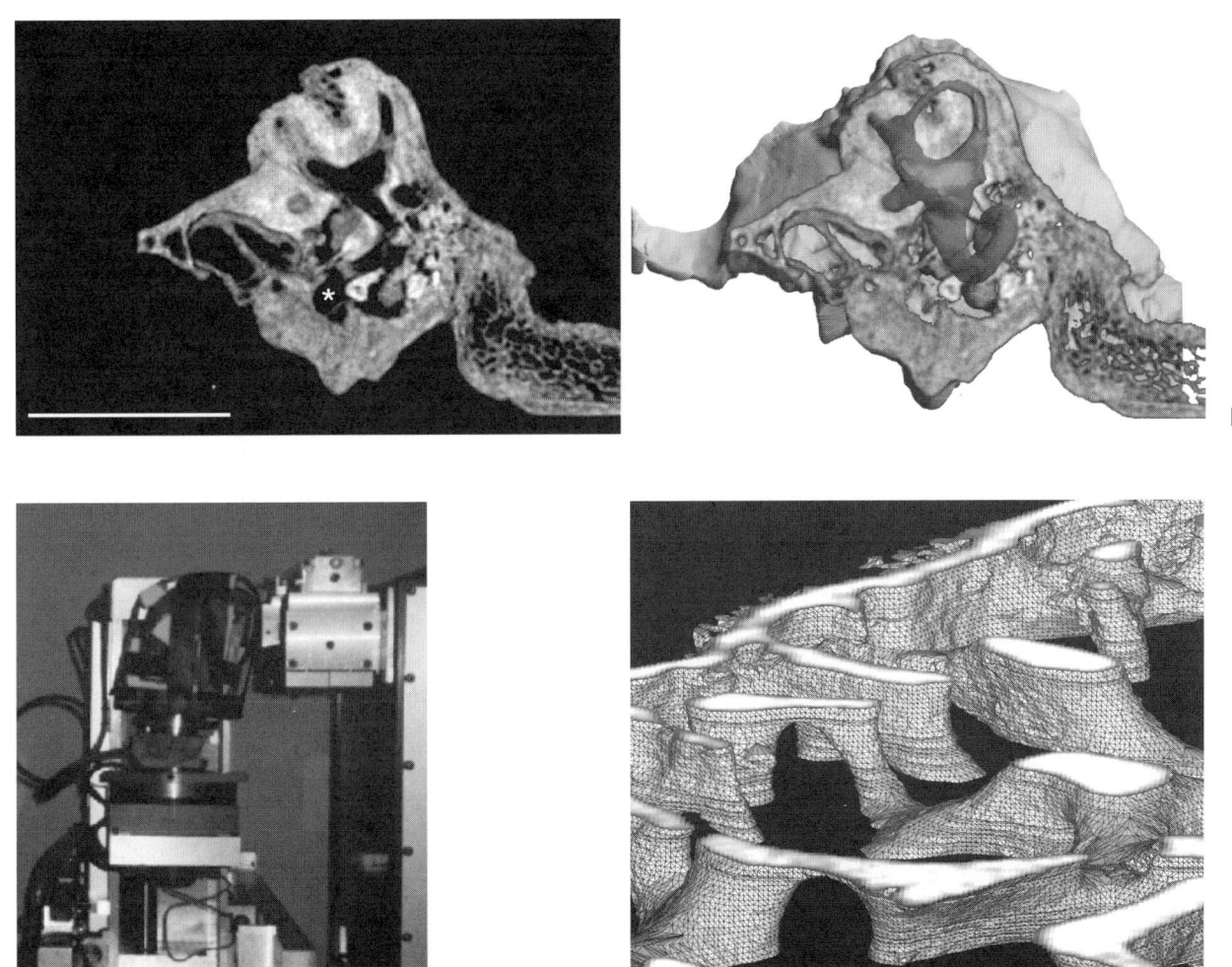

**FIGURE 3-19** Industrial tomography and microtomography, **A, B:** Cross-sectional image and three-dimensional reconstruction of the Le Moustier 1 Neanderthal left inner ear region. Note the dense object that represents the anvil shown in Fig. 2-11 (scale bar is 10 mm). **C:** The same fossil fragment mounted in an industrial scanner. **D:** Microtomography. Three-dimensional reconstruction of the trabecular structure in a terminal phalanx of *Macaca mulatta* (scale bar is 0.1 mm).

sis of geologic samples (Flannery et al., 1987). Recent studies demonstrate the potential of μCT in comparative analysis of trabecular bone architecture from a biomechanical perspective (Borah et al., 2000, 2001; Fajardo and Müller, 2001). The miniaturized tomographic setting of μCT requires extremely focused X-ray sources, or shorter-waved gamma-ray sources. Because X-ray diffraction and dispersal are substantial in this small level of scale, only small objects can be analyzed.

A further step in sophistication is *synchrotron tomography*. Electromagnetic radiation produced by a synchrotron has higher energy (thus shorter wavelength) than X rays and is typically monochromatic; whereas an X-ray source produces a continuum of wavelengths, a synchrotron produces radiation within an extremely narrow band of wavelengths. Short-waved monochromatic synchrotron radiation yields cross-sectional images with high spatial and contrast resolution, but the technique has serious trade-offs due to scattering.

### 3.3.8 Three-Dimensional Data Acquisition with a Medical Scanner

Given the extended list of parameters, trade-offs, and techniques to consider, the reader may wonder how to select parameters and balance them for optimal results. The first and foremost issue to address is whether CT scanning is the appropriate technique for data acquisition within the given bioscientific task. Historically, CT was developed for the analysis of cranial structures, particularly the diagnosis of brain diseases. If we recall that X-ray absorption is a function of the average atomic number of the elements constituting a given material or tissue, it is obvious that contrasting calcium-rich skeletal tissue and surrounding soft tissue is resolved easily. Fine-tuning a CT device and using special image reconstruction algorithms permits resolution of much subtler density differences (e.g., soft tissues within internal organs).

Although CT-based diagnostic imaging is a science in itself, the aims of this section are to demonstrate uses of medical CT scanners in related fields, such as paleoanthropology and paleontology, and to suggest parameters for acquiring quantitatively reliable volume data (hands-on advice is given in Appendix C). Initial settings during fossil scanning differ from initial settings during classic medical exams in the following respects.

- Fossil specimens generally exhibit a wider range of variation in density than clinical patients. During fossilization, bone tissue undergoes both de- and remineralization. Accordingly, within the same specimen, skeletal structures might have become denser than bone in some areas and less dense than bone in other areas. Furthermore, the skeleton is no longer encased by soft tissue; instead sediments, which are often similar in density to the fossilized bone, or even have higher density, usually cover remaining bone.
- During CT-based analysis of fossils, or recent museum specimens, an interest exists in measurement—the acquisition of metrically reliable data—that is typically equal to an interest in diagnosis—the visual assessment of the imaged structures.
- Parameters, such as X-ray dose and scan time, are less constrained when scanning fossils than when scanning living patients. This is a major advantage in fossil scanning because it facilitates optimization of scanning proto-

col with respect to image quality. Medical CT scanners are adapted to the spatial dimensions of the human body and optimized for the range of attenuation values displayed by human tissues. Spatial resolution, as measured by the minimum separable distance between two points, is approximately 0.4 mm but, with interpolation techniques, determining the location of object boundaries to a tenth of a millimeter is reliable in some instances (see Chapter 4). In practical terms, this signifies that, given a pixel size of 0.2 mm, two white pixels are separated by one black pixel. Zoomed image reconstructions at higher magnifications, that is, with smaller pixel sizes, do not provide additional detail. For a $512 \times 512$ pixel image with a pixel size of 0.2 mm, the field of view thus has a smallest operational diameter of about 10 cm.

A point that needs special attention during fossil CT scanning is how X-ray attenuation values are calibrated. On medical scanners, X-ray attenuation is measured in *Hounsfield units*, HU (Fig. 3-20).[6]

The Hounsfield scale is calibrated to 0 for water and −1000 for vacuum (in practice, air is used as a proxy). The normal range of attenuation values found in a medical CT image can be represented with a pixel depth of 12 bits ($2^{12} = 4096$ units mapped onto a scale ranging from HU = −1000 to HU = 3095). Soft tissues, such as fat and muscles, have HU-values around −100 and +40, respectively. Bone ranges around a typical average value of HU = 800, whereas tooth enamel is slightly higher (above HU = 1000), and titanium implants exceed biological tissue densities with values around HU = 2000. Attenuation values of HU > 3095, known as *overflow*, may occur in dental fillings, metal implants (e.g., hip prostheses), and heavily mineralized fossils. Overflow is treated variably by CT manufacturers. Some devices use an extended Hounsfield scale (and thus more than 12 bits per pixel), whereas others reset the scale at HU = −1000 for overflow values of HU > 3095, and still others fix values of HU > 3095 at HU = 3095 (Fig. 3-20).

## 3.4 MAGNETIC RESONANCE IMAGING (MRI)

Similar to computed tomography, magnetic resonance imaging (MRI) acquires volume data from the human body. However, whereas CT measures *absorption* of X rays by atoms, MRI measures electric signals *induced* by atomic nuclei in a strong magnetic field (Lauterbur, 1973).[7]

---

[6]The attenuation scale is named in honor of Sir Godfrey N. Hounsfield, who, together with A. Cormack, received the 1979 Nobel Prize for Physiology or Medicine when they conceived CT.

[7]P. C. Lauterbur is one of the 2003 Nobel laureates for chemistry.

Hounsfield Units

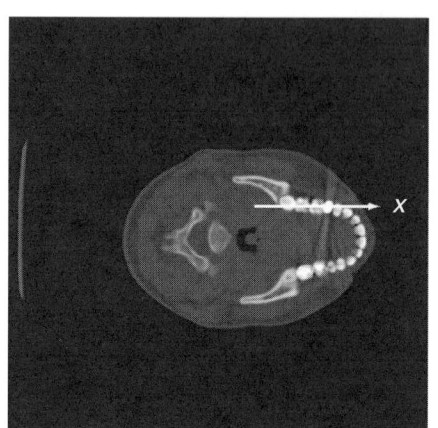
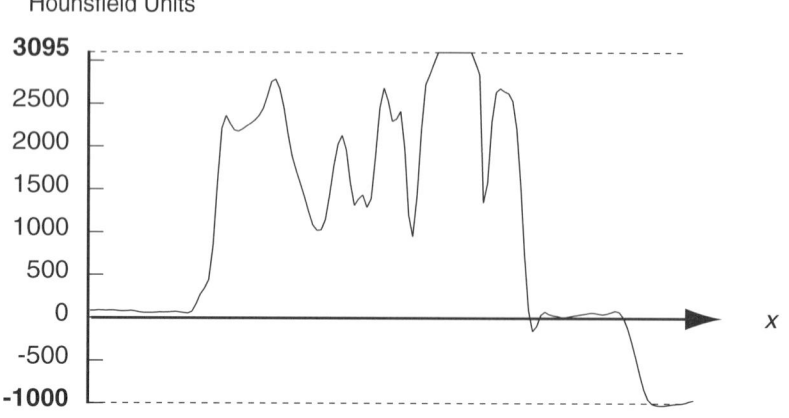

A

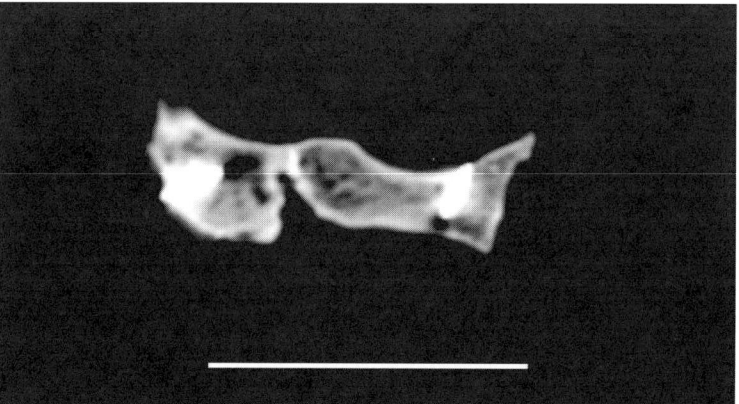

B

**FIGURE 3-20**   Hounsfield scale overflow caused by metal "implants." **A**: Image artifacts caused by dental implants in a patient (note fixation of the HU values). **B**: Basicranial fragment of the Le Moustier 1 Neanderthal fossil. Artifacts were created by two metal rods that were inserted at the beginning of the twentieth century to position the fragment at the skull base (see also Box 3-2).

MRI is based on nuclear magnetic resonance (NMR), which is a spectroscopic technique that physicists and chemists use to analyze interatomic distances with the aim of recovering the three-dimensional structure of molecules. As the name reveals, the resonating entity is the atomic nucleus, or more precisely, its spin, which is similar to a small magnetic dipole with north and south ends. Spin can be imagined as the "sense of rotation" of a particle around its magnetic north-south axis. Eliciting and receiving an MRI signal occurs in three stages:

On a macroscopic scale and under everyday conditions, the net signal of large populations of atomic nuclei tends to be zeroed out statistically; the key is to bring individual atomic nuclei to synchrony, such that a strong bulk signal can be received. This is achieved by lining up spins of atomic nuclei along the principal direction of a strong external magnetic field. Best suited for this task are the atomic nuclei of hydrogen (H), because of their high relative sensitivity and high natural abundance.

Spins are deflected from their principal direction with an electromagnetic pulse in the radio frequency (RF) domain, which corresponds to the resonance frequency of H atoms. Like toy spins, deflected nuclear spins exhibit a precession movement; their main axis "rotates" around the vertical axis at a given frequency, called the resonance frequency. In a receiver coil located in the vicinity of the object under investigation, precession induces a small electric current, in the same way a dynamo, or turbine, generates electric current in a power station. This current is the signal that is measured during MRI.

Spins tend to return to the principal direction imposed by the external magnetic field. This process is called *relaxation*. As an effect of relaxation, the electric signal fades out. This is why the entire procedure of excitation and signal detection is usually repeated over and over again.

For each voxel of the object under investigation, the MR signal must be received and decoded. The ingenious idea behind MR as an imaging technique, as first proposed by Lauterbur (1973), is to modify the external magnetic field such that the signal received from a given voxel conveys information about its location in space. As in computer tomography, image reconstruction proceeds via backprojection, because signals from various voxels are superimposed. How this is achieved is explained in more detail in Figure 3-21 and in Box 3-3.

Compared to CT, MRI is more technically demanding but also less invasive (the energy of radio signals is about one millionth of the energy of X rays and does not damage tissues). However, strong magnetic fields required to elicit a signal in the radio frequency domain must be created with a current sent through superconducting coils that are cooled to near-zero temperatures, usually with liquid helium. These fields are in a tesla range that is 30,000 times stronger than the magnetic field of the Earth. Consequently, extreme caution is necessary to avoid potentially harmful displacement of ferromagnetic objects, either outside or inside the human body. Benefits of MRI technique include independence from ionizing radiation and the ability to differentiate between various soft tissues.

MR imaging is rapidly evolving as a research field. In the past decade, several new developments, such as real-time imaging of the heart, 3D analysis of blood flow patterns, and analysis of activity patterns in the working brain (functional MRI, fMRI) have been realized. Why is MRI pertinent for virtual reconstruction? With increasing spatial resolution and accuracy in volume data sets, MRI could become an invaluable noninvasive tool for the acquisition of 3D anatomic data

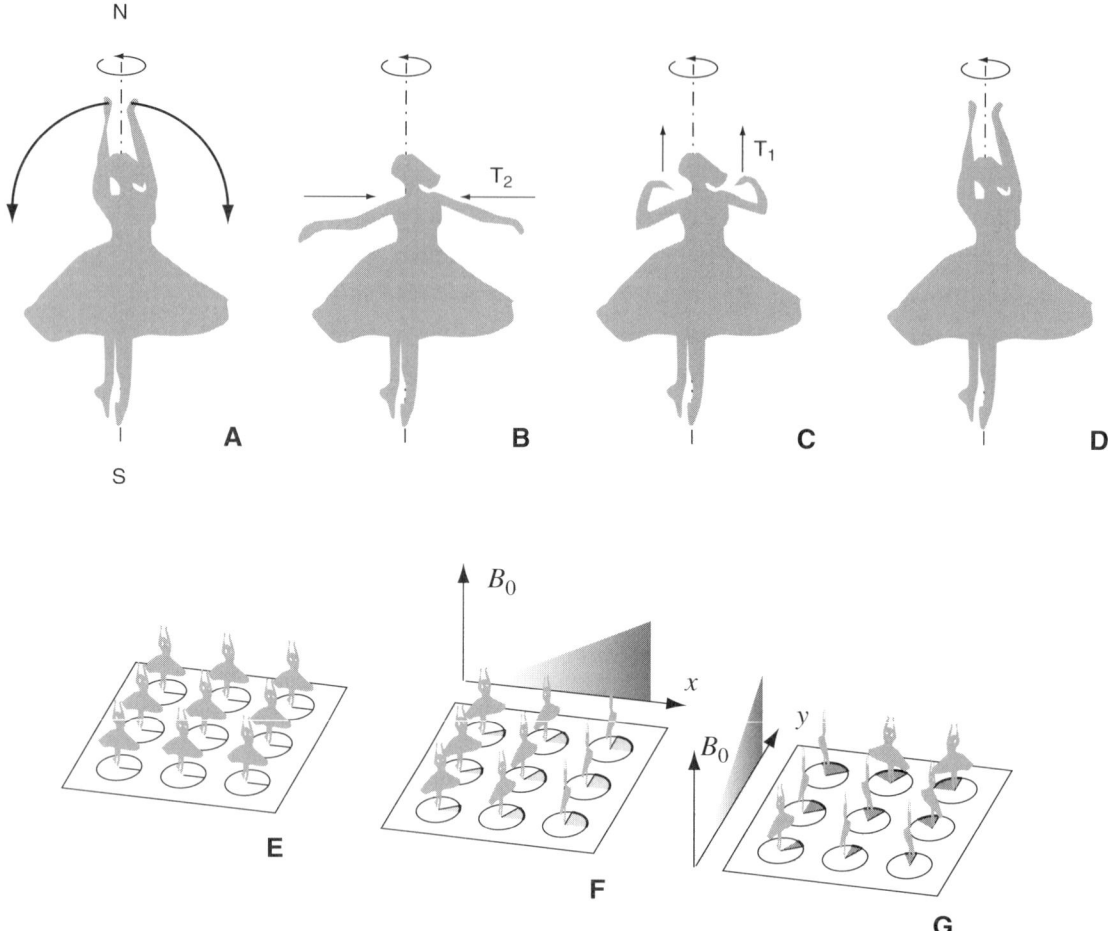

**FIGURE 3-21**  Magnetic resonance imaging. **A**: The spin of an atomic nucleus under the influence of a strong external magnetic field (N-S) behaves like a ballerina spinning around her body axis. Application of a radio frequency (RF) pulse causes the spin to be deflected; metaphorically, this corresponds to downward rotation of the arms of the ballerina. **B–D**: As soon as the RF pulse ceases, the spin recovers and assumes its original alignment. Metaphorically, inward and upward movement of the arms of the ballerina correspond to $T_1$ and $T_2$ signals induced in an external receiver coil. **E** and **F**: Encoding positional information in an MRI slice. Spin directions of all voxels (represented as dancers of a ballet company) are initially aligned (**E**). Application of a magnetic field gradient ($B_0$) in $x$ direction (**F**) and then in $y$ direction (**G**) results in spatial modulation of the spinning frequencies. The differences in rotational speed (frequency) and differential "advancement" around the spinning axis (phase) permit decoding of the ($x,y$) coordinates of the signal received from each voxel.

from "standard" (nonclinical) samples, for example, longitudinal growth studies and quantitative analyses of anatomic variation in three dimensions. Such data could standardize pathological data and growth curves or, in forensic applications, derive average relations between soft and hard tissues.

At first thought, fossils seem unlikely candidates for MRI because protons tend to be immobilized by the surrounding material matrix. However, recent studies demonstrate that million-year-old specimens produce surprisingly strong proton

BOX 3-3

## SPINS AND MRI

In quantum physics, "spin" is a property of matter that is as fundamental as mass or electrical charge properties. Isolated electrons, protons, and neutrons have a spin of magnitudes 1/2. If elementary particles aggregate, they usually align their spins in antiparallel directions, +1/2 and –1/2. The famous Pauli principle states that two electrons of identical energy must differ in spin states. Likewise, atomic nuclei, which are aggregations of protons and neutrons, have zero net spins or small multiples of ±1/2. Hydrogen atoms (H) are ideal elements because they represent isolated protons (spin = 1/2). Moreover, protons are abundant in our body, as an effect of the high proportion of water ($H_2O$) and the ubiquity of C–H bonds in organic molecules. If the human body is exposed to a strong external magnetic field, its protons behave like little magnets that align spins along the magnetic field lines (in practice, it is only a fraction of all protons). Each proton may assume two alternate states—parallel (+) or antiparallel (–) to the external field—corresponding to different energy levels. Let us denote the energy difference as $E$. Transition between low- and high-energy states is brought about by absorption or emission of a photon (a light quantum) of energy $E$. Because photons are electromagnetic waves, we can calculate the frequency $\nu$, using Planck's constant $h$

$$E = h\nu$$

The magnitude of the energy difference $E$ depends on the external magnetic field $B$, and on the physical properties $\gamma$ of the atomic nuclei, such that

$$E = h\gamma B$$

For H-atoms, $\gamma = 42.58$ MHz/T (megahertz per tesla). MRI scanners use magnetic fields $B$ in the range of 0.3 to 3 T (in current research scanners up to 8 T), which typically means that $\nu$ is in the range of 15 to 127 MHz, which corresponds to the frequency of a radio signal. Let us now consider one voxel inside the human body. Because this voxel contains large numbers of atomic nuclei, we no longer observe single spin-flipping events, but must rely on statistical averages. For example, the proportion of nuclei exhibiting antiparallel ($N^-$) spins relative to parallel ($N^+$) spins, the so-called *equilibrium magnetization*, depends on ambient temperature and follows a Boltzmann distribution. At room temperature, the $N^+$ population dominates slightly, so net magnetization $N^- - N^+$ is positive.

**BOX 3-3**

## SPINS AND MRI (*Continued*)

How are MRI signals generated? First, a voxel is irradiated with a short radio wave signal (the radio frequency pulse, RF). Metaphorically, the spin then behaves like a ballerina (Fig. 3-21A–D). While she initially spins with her arms extended *above* her head (i.e., along the body axis), the RF pulse lets her spread them sideways (Fig. 3-21A and B). As soon as the radio energy supply ceases, the voxel returns to its equilibrium state. Metaphorically, the ballerina moves her arms toward the body and upward until the initial state is reestablished (these two "movements" are separated conceptually but occur simultaneously). The signal corresponding to "inward movement" measures net magnetization along field lines of the magnetic field, that is, in the north-south direction; it fades away with an exponential time constant $T_2$. The signal corresponding to "upward movement" of the arms measures how deflected atomic spins tend to reestablish upright (N-S) directions; this occurs with an exponential time constant $T_1$.

MR imaging fundamentally reduces to the recovery of tissue-specific data from a voxel at a given position $(x,y,z)$. The signal emitted by each voxel must contain information about the density of H-atoms and chemical composition within that voxel and information about the position of the voxel within the body. Resonance properties of H-atoms are influenced by the chemical environment of the atoms (primarily due to surrounding electrons, which modify the external magnetic field). This creates tissue-specific values of $T_1$ and $T_2$ (and many further parameters not discussed here), such that, within a given time, different tissues will emit signals at different intensities. A second requirement is fulfilled by imprinting spatial information onto the external magnetic field and the radio pulse, and subsequently decoding this information while receiving the resonance signal (Fig. 3-21E–G).

As we are dealing with the recovery of frequency information in space, it is appropriate to compare the object under MRI investigation with a symphony orchestra. Imagine that each instrument positioned at $(x,y,z)$ in the orchestra represents one voxel $(x,y,z)$. Metaphorically, the imaging task in MRI is to locate each instrument within the orchestra and then assess the specific contribution of it in the total sound of the orchestra, while the director interpreting the score assumes the role of the RF pulse (our sense of hearing is highly efficient in extracting spatial information from a superposition of sounds).

signals (consult the Fraunhofer Institute's NMR website: `www.nmr.fhg.de`). The physical basis of these observations requires further, detailed investigation, but it seems plausible that fossil MRI will become a significant method of data acquisition. For example, X ray-free investigation of fossil material would be less harmful to any remaining DNA in the fossil.

## 3.5 SURFACE SCANNERS

Many imaging tasks require detailed information of surface structures and textures instead of volume data. For example, photography captures surfaces and textures quite well, but it is limited to acquiring two-dimensional projections of three-dimensional objects. Since the invention of photography, various trigonometric techniques have been devised for acquiring true three-dimensional data from a series of photographic images. Computerized image processing techniques, as well as the advent of laser technology, have revolutionized surface data acquisition because they facilitate exhaustive sampling of three-dimensional vertex coordinates from surface points. Their respective technical implementations are known as photogrammetry and surface (or range) scanning.

Surface scanners are true three-dimensional digital cameras that create depth maps. These scanners produce pixel depths that represent actual spatial depth values, that is, distances $z$ between points $(x,y)$ in the image plane and corresponding surface points of the object. Modern surface scanners typically combine depth data and photographic digital images from the same point of view, thus yielding three-dimensional textured surface data.

Like 2D cameras, 3D scanners capture light that is reflected from the surface of an object. However, 3D scanners use a light source to project patterns of known dimensions onto the object so that depth cues are retrieved from the reflected signal. Two methods are most common.

In *photogrammetry*, a light/dark grid or spot pattern is projected onto an object. Photographs from different viewpoints enable reconstruction of the 3D surface from "deformation" of the grid. Because the $x$-$y$ geometry of the pattern is known, $z$ coordinates can be established by geometric triangulation. The second method, *laser range scanning*, involves a very narrow laser beam that scans a surface along virtual gridlines spread over the object. A plethora of techniques are used to sense depth information from the reflected laser light, as listed below.

- The *interferometric* technique exploits superposition between outgoing and incoming light waves to infer distance information from phase shifts.
- Medium- to long-range scanners use the principle of *radar*; they measure the time difference between an outgoing laser pulse and an incoming "echo."

• Most desktop scanners that are suitable for bioscientific tasks, such as scanning human faces and fossil specimens, use optical *triangulation* (Fig. 3-22A).

Surface data acquisition is typically less demanding technically than volume scanning with CT or MRI. However, before quantitatively reliable surface data can be obtained, several issues must be resolved. An underlying problem in photogrammetric and laser techniques is the projective nature of the depth map: Basically, a 2D gridlike structure is "wrapped" onto a 3D object structure that has unknown geometric complexity. As shown in Figure 3-22D, only those surface

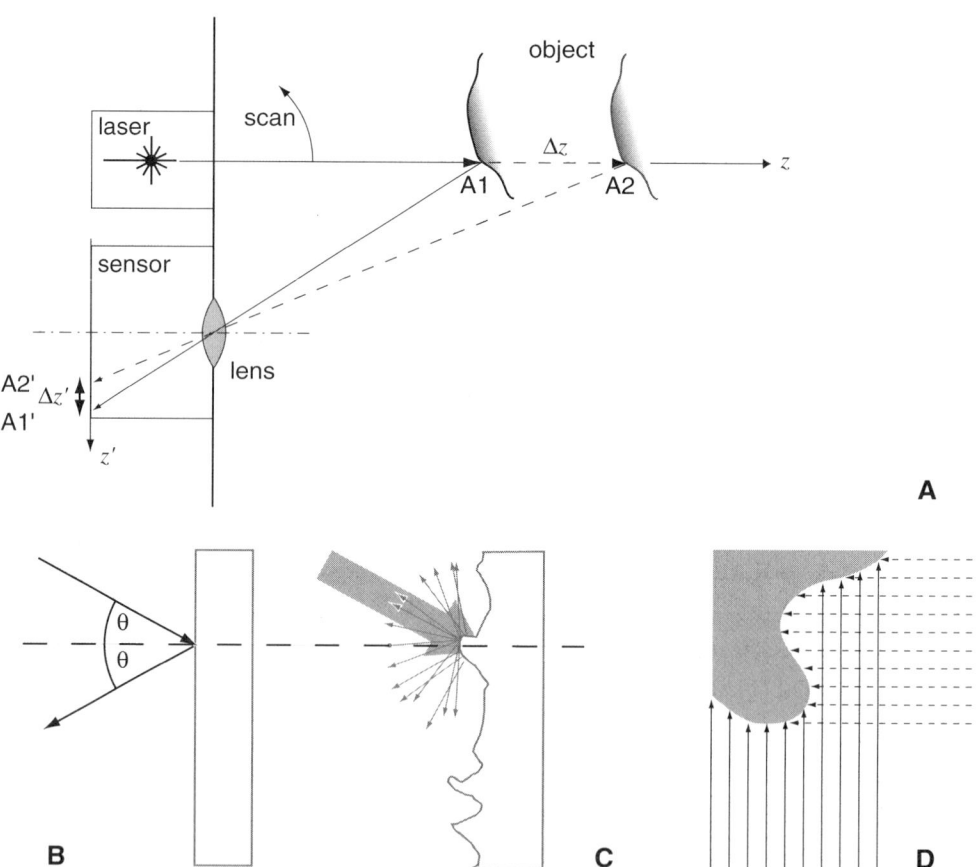

**FIGURE 3-22** Laser surface scanner. **A**: Principle of distance triangulation. A laser beam sent out along axis $z$ hits an object surface at A1. Diffuse reflection generates a signal that is received by the optical sensor system along $z'$. Distance is measured by the relative displacement of the signal, $\Delta z'$, as shown by displacement of the object to position A2. Surface scanning involves sweeping the laser over an object. **B**: During direct reflection, the angle of incidence $\theta$ is equal to the angle of reflection. **C**: Diffuse reflection arises through submicroscopic irregularities of the object surface and the effects of light diffraction. **D**: Shadowing effect. Unidirectional scanning cannot collect data from hidden regions of an object surface. However, this problem is resolved by combining data that are obtained from different views.

regions that face the data acquisition device can be probed. In these techniques, hidden areas appear as "holes" in the reconstructed surface. Moreover, because accuracy of the depth signal depends on the angle of incidence of the laser, regions nearly parallel to the line of vision are resolved poorly. These problems are solved by combining depth maps that are acquired from different viewpoints of the object. To this end, an object should be mounted on a computer-controlled turntable that provides a consistent geometric framework. Most scanners offer software tools to "stitch" together partial surface data and simultaneously eliminate redundancies and inconsistencies.

Because surface scanning relies on diffuse reflection, a second problem arises (Fig. 3-22C). Diffuse reflection is the part of incident light that is diffracted at the object surface and reflected in all spatial directions (note that specular reflection occurs in a single direction that is determined by the angle of incidence). Unfortunately, the specular signal is stronger than the diffuse signal, such that glossy and shiny surfaces must be treated with talcum powder to reduce specular reflection and attain significant diffuse reflection. An opposite problem arises in dark surfaces where diffuse reflection is too insignificant to permit sampling of depth information. Again, talcum powder may help overcome this difficulty.

## 3.6  3D DIGITIZERS

In applications primarily concerned with anatomic landmark positions, it is often straightforward to acquire these data directly, rather than deriving them from volume or surface data. In analogy to two-dimensional digitizing panels that record the position of a digitizing pen, three-dimensional digitizers use a pointer to mark and record vertex positions in space. We may distinguish among three principal types, mechanical lever-arm devices, 3D sensing devices, and optic devices. In mechanical digitizing devices, a hand-guided pointer is fixed to a lever arm construction. Joints connecting the arms have one or two degrees of freedom of rotation, such that the pointer can conveniently be positioned in space without being obstructed by the object under investigation. The spatial position of the pointer is evaluated with trigonometry, using arm lengths and angles of rotation measured in each joint. 3D sensing devices track the spatial position of a sensor in a magnetic field. These devices use a "3D pen" that can be moved freely in space, which represents a considerable advantage during measuring manipulations. However, the magnetic field generated by the device gets easily distorted by metallic objects in its vicinity, so that, overall, 3D sensors are less precise than mechanical digitizers. Typically, therefore, they are used in virtual reality applications, where recording absolute positions in space is less relevant than tracking the relative movement of a user's hand. Finally, optic devices represent a touch-free alternative to measure spatial coordinates on small objects. In the ReflexMicroscope™, for example, the user observes the object under investigation through a binocular microscope and records landmark locations by positioning a light point under stereoscopic vision.

During 3D digitizing, a similar general problem arises as during surface scanning: Data from object regions inaccessible to the device in one session must be recorded in a subsequent session, while the object is reoriented. Most digitizers provide special-purpose software to deal with this problem. For example, it is possible to define an object-centered system of measuring coordinates by specifying three landmark locations on the object. In the reoriented specimen, reference to the same landmark locations permits recovery of the initial system of measuring coordinates.

## 3.7 FURTHER READING

Biomedical imaging, notably CT and MRI, is covered by a series of excellent textbooks, on both introductory and expert levels. Classics of tomography (CT, MRI, and other techniques) are Kak and Slaney (1988), Kalender (2000), and Seeram (2001). Herman (1980) provides an understanding of CT from the image reconstruction perspective. A very useful introduction into multislice CT is given by Cody et al. (1998).

An excellent introductory textbook on MRI is Prince (2003); Lang (2000) is an encyclopedic but highly instructive compendium. General textbooks on biomedical imaging are Webb (2003) and Sutton (2002); Mudry et al.'s volume (2003) comprises a collection of contributions treating all aspects of biomedical imaging.

Historically interesting articles on the beginnings of CT and MRI are Hounsfield (1973) and Lauterbur (1973). Acharya et al. (1995) give a terse tutorial on biomedical imaging techniques within a clinical environment. Pioneering studies outside the clinical area, notably of fossil and skeletal material, are Jungers and Minns (1979), Conroy and Vannier (1984), Wind (1984), Wind and Zonneveld (1985), and Zonneveld and Wind (1985).

MicroCT has become widely used in paleontological and paleoanthropological applications, but also in medical diagnostics (Rüegsegger et al., 1996) and geology (Flannery et al., 1987). Here we mention some papers that are directly related to issues of data acquisition and quantization, but there is a continuity to papers dealing with issues of data segmentation and visualization, and paleodiagnostics (see also Further Reading in Chapter 6).

Müller et al. (1985) consider technical settings for microCT. The papers of Gantt et al. (2002), and Fajardo et al. (2002) are important for anthropologists, as they indicate new directions for measuring classic anthropological parameters such as trabecular anisotropy and dental enamel thickness. Borah et al. (2000, 2001), and Cooper et al. (2003) provide interesting insights into issues of microCT data processing and biomechanical analysis. It should be noted that microCT has also been applied to the study of plant microstructures (Stuppy et al., 2003).

# IMAGE DATA PROCESSING

**4**

Even the simple act that we call "seeing a person that we know" is, in
part, an intellectual act.
> —Marcel Proust, *Du côté de chez Swann* (*Swann's Way*)

## 4.1 RECOVERING OBJECTS FROM IMAGES

In Chapter 2, we provided a scheme for hierarchically ordering biomedical data.
We indicated the general path of data processing steps that is necessary for deriving surfaces from volume data, contours from image data, and anatomic landmarks from surface data. The basic concept involved identifying objects within a
data set by identifying their boundary structures, which correspond to abrupt
intensity changes in image data. In Chapter 3, we discussed various strategies for
acquiring volume, surface, and point data from biomedical objects. Volume data
acquisition with medical imaging devices such as CT and MRI received considerable attention because these methods yield the most comprehensive data sets,
from which any additional type of data can be derived. Although laser scanners
and 3D digitizers yield immediate access to surface geometries and landmark
locations, these data are only implicit in CT and MRI data volumes.

Here, we explore how to make such information explicit. The principal tasks
in image data processing are *object recognition and reconstruction*. Our visual system
accomplishes similar tasks, so it is useful to make analogies between how computerized and neuronal methods recover object information from image data.

Recall that we have mentioned parallels between a photographic camera, as
a data acquisition device, and our eyes. Each acquires data from objects in the
physical world and temporarily stores them as images on the retina or on a charge-
coupled device (CCD). For a computer, an image is a set of intensities $I$ at coordinate positions $(x,y,z)$, as is the case in our visual system, where a retinal image
is a two-dimensional pattern of electrostatic discharges of sensory cells. However,

*Virtual Reconstruction: A Primer in Computer-Assisted Paleontology and Biomedicine.*
By Christoph P. E. Zollikofer and Marcia S. Ponce de León.
Copyright © 2005 John Wiley & Sons, Inc.

whereas our brain "sees" objects in a retinal image virtually immediately, computers do not operate as quickly. Visual data processing is so fast and efficient that we do not realize the long sequence of computational steps that recognizing objects requires. Experimentally, this process can be slowed to experience how the visual system extracts object information from an image in a step-by-step procedure (Fig. 4-1).

The fundamental idea of biomedical *image data processing* is translating tasks executed by our visual system into tasks that are executable on a computer (Fig. 4-2). This occurs in the following subtasks.

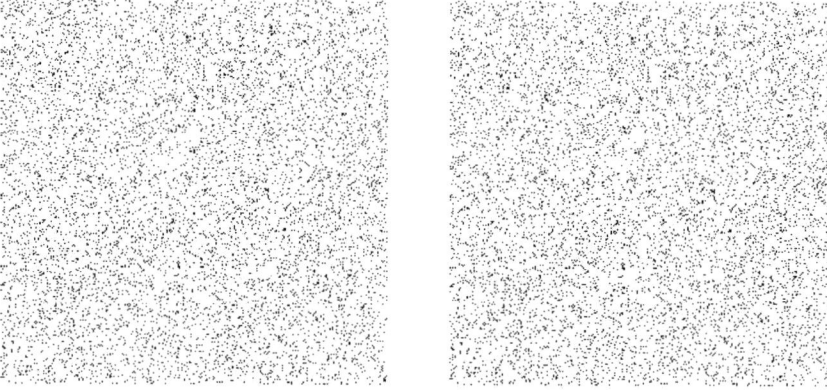

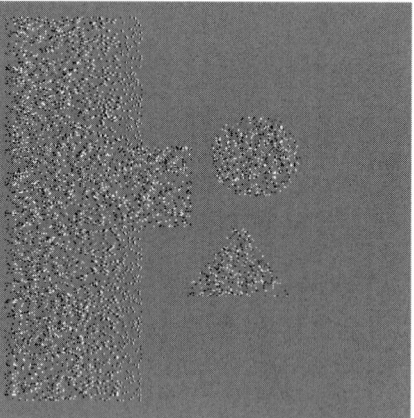

**FIGURE 4-1**  Image data processing and analysis in the visual system. Our visual system extracts object data contained implicitly in these stereo-pair images through a stepwise process (the image below is a disparity map representing the difference between left and right images). In the first phase, referred to as image data *processing*, the spatial disparity between the 8300 black pixels in each image is calculated to create a depth map; each pixel is associated with a distance from the observer. During the second phase, referred to as image data *analysis*, the depth map is interpreted in terms of objects in space; a triangle, a disk, and a square floating above or below ground. These shapes are perceived as clearly delimited, white objects with black speckles, although no material delimitation exists in the image. Depth information in the inclined background on left side of the image occurs independently of recognizing specific shapes.

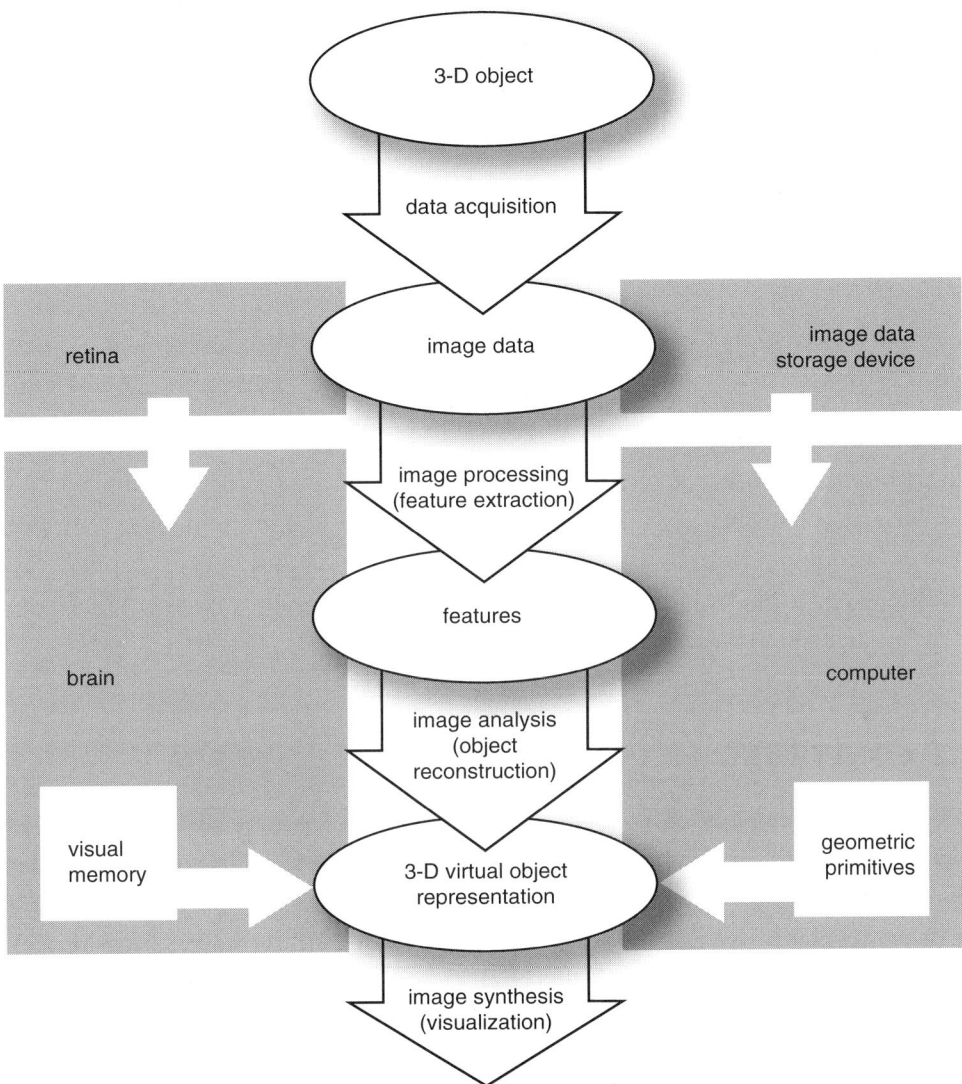

**FIGURE 4-2** Computer-based image data acquisition, processing, analysis, and synthesis and analogous processes in the visual system.

*Image processing* involves transformation and modification of images. The basic goal is identification of significant structures against a background of less significant information. This uses so-called filter functions, which serve as tools to reduce image noise, extract features such as edges, and segment and classify an image into various subfields.

*Image analysis* procedures transform information in images (original or prefiltered) into quantitative geometric representations of 2D and 3D objects. These procedures (re)construct contours, surfaces, landmark configurations, or other data, as specified in Figure 2-14.

*Image synthesis* comprises procedures that create 2D and 3D object data with geometric primitives. This approach differs from image analysis, where information is derived from existing objects, and relates to computer-aided design (CAD). Nevertheless, methods of image synthesis often improve the efficiency of complex image analysis procedures.

*Image visualization* requires procedures for the (re)construction of a virtual environment with 2D and 3D object data. Chapter 5 outlines these final steps in transforming a real physical object into its virtual counterpart.

How can a specific image data processing task, which exists as a concept in our mind, be performed by a computer? First, the concept should be formulated in mathematical terms. Second, these functions should be transformed into an algorithmic structure, that is, a sequence of tasks and conditional subtasks that resolve a problem step by step. Third, algorithms must be implemented with programming languages (such as Fortran, C, C++, Java) and then compiled into machine-specific instructions. These instructions are the *software*; specific software is called an *application program*. Here, we will focus on the first two steps.

## 4.2 CONVERTING A CT IMAGE INTO A SCREEN IMAGE

Let us start with a practical example of image data processing. Medical CT images have a pixel depth of 12 bits that represent 4096 Hounsfield density values ranging from −1000 to 3095, whereas typical 8-bit computer screens display a range of 256 gray values. How are the 4096 density values coded in the image data file best transformed into 256 gray values on the computer screen? The function that translates CT input into display output uses the concept of a *window* (Fig. 4-3). The output "sees" the input through a window, which has a *width* (the range of input values visualized) and a *level* (the middle input value).[1] These parameters permit adjustment of contrast and brightness of the screen image. Output values that fall outside the range of values in the window are standardized to minimum (0) or maximum (255) intensities.

Figure 4-3 depicts various window width and level settings and the effects their selection has on a CT image. Most CT consoles provide predefined settings for optimally rendering tissue-specific Hounsfield density and contrast values. Figure 4-3D shows an interesting special case, where the window width is zero. The resulting *threshold function* generates a binary image that differentiates regions below and above the window level. Threshold functions are important data segmentation tools because they partition image data into distinct sets.

---

[1] The window function is similar to the response function of technical and biological light detectors discussed in Appendix A.

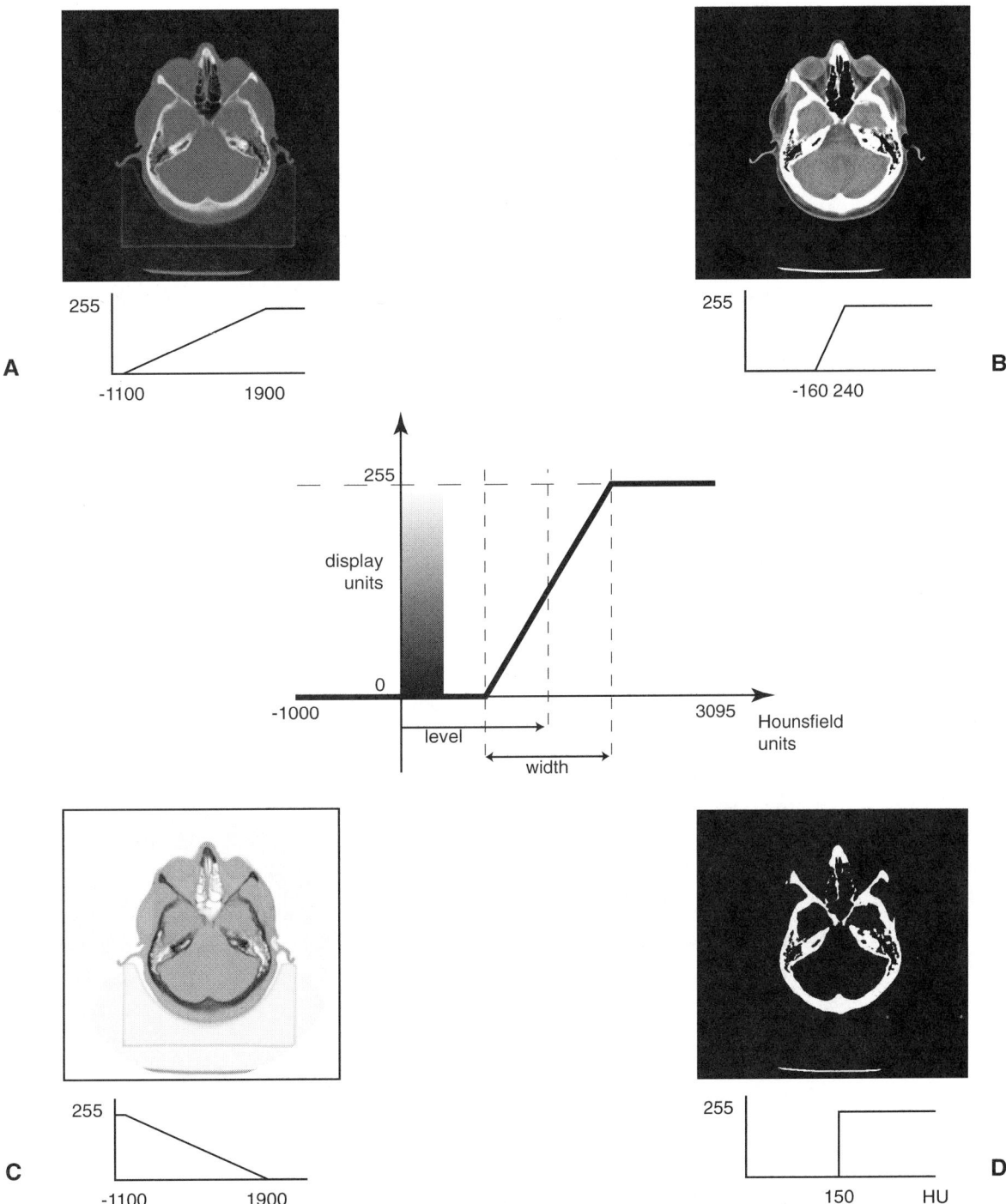

**FIGURE 4-3** Image data processing with window functions. The graph in the center of the image is a manifestation of the general function used to visualize a CT image (pixel depth for Hounsfield values is 12 bit) on a computer monitor (pixel depth for grey-level values is 8 bit). Parameter settings that are illustrated in the inset graphs have several consequences: **A** enhances bone structures (note "underexposure" of soft tissue); **B** enhances soft tissue structures (note "overexposure" of bone tissue); **C** produces a negative image; **D** is a threshold function that creates a binarized image of the bone structures.

The DataFormat Applet on the Web Companion provides an implementation of the window function.

## 4.3 FILTERING IMAGES

### 4.3.1 Coffee and Kernels

Let us generalize the procedure of applying a mathematical function to image data by proposing the following definition: A mathematical function that is applied to every pixel in an image (or voxel in a data volume) is a *filter*, or *convolution* function. The specific function that convolves an image is also called a *kernel*. In the previous example, we used a kernel of dimension zero because the windowing function sent output to the same pixel location from which it received input. Most filter kernels, however, gather input data from within a local neighborhood surrounding a pixel, then process the input, and send the result to the corresponding pixel in the output image (Fig. 4-4). This is analogous to coffee brewing (Fig. 4-4A), where the coffee filter (kernel) receives input from many coffee grounds (input pixel neighborhood), then extracts the relevant substances (information), and funnels the coffee (output) into one cup (output pixel). However, image filters differ from real-life coffee filters in that adjacent pixels (cups) gather information (coffee) from overlapping image regions. In Figure 4-4A and -B, this fact is shown by the virtual overlapping of filters funneling coffee into adjacent cups, and by the overlapping input regions of kernels funneling output to adjacent pixels, respectively.

Figure 4-4C shows in detail the structure of a two-dimensional filter kernel and how it generates output for one pixel. This kernel has the shape of a square containing $N \times N$ elements ($N = 3$ in our example). Each element is characterized by a weighting factor, $w_{i,j}$. Hence, a $3 \times 3$ kernel has the following general structure:

$$
\begin{array}{|c|c|c|}
\hline
w_{-1,1} & w_{0,1} & w_{1,1} \\
\hline
w_{-1,0} & w_{0,0} & w_{1,0} \\
\hline
w_{-1,-1} & w_{0,-1} & w_{1,-1} \\
\hline
\end{array} . \tag{4-1}
$$

It uses input data $I$ from a $3 \times 3$ neighborhood around pixel $(x,y)$ in the input image:

$$
\begin{array}{|c|c|c|}
\hline
I(x-1, y+1) & I(x, y+1) & I(x+1, y+1) \\
\hline
I(x-1, y) & I(x, y) & I(x+1, y) \\
\hline
I(x-1, y-1) & I(x, y-1) & I(x+1, y-1) \\
\hline
\end{array} . \tag{4-2}
$$

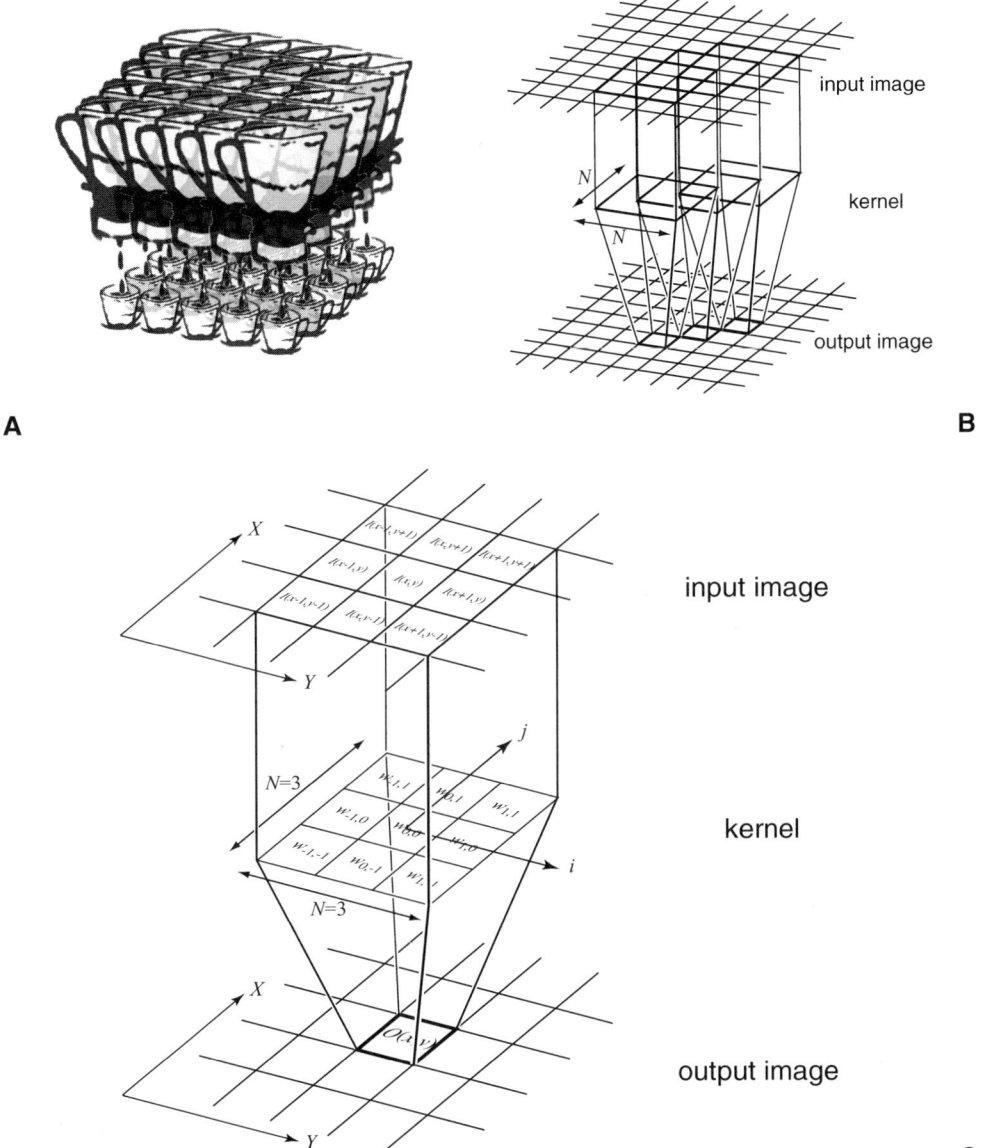

**FIGURE 4-4** Filtering an image. **A**: The coffee filter metaphor. **B**: An image filter uses a kernel that processes data from an $N \times N$ input region and sends the result to one output pixel. This procedure is applied to all image pixels. **C**: Details of the convolution procedure. To evaluate output at pixel position $(x,y)$, input values $I(x + i, y + j)$ are multiplied with corresponding weights $w_{ij}$, then summed.

These input values are multiplied with corresponding weights $w_{i,j}$ and then summed up. The result, $O(x,y)$, is sent to the pixel at position $(x,y)$ in the output image. This procedure looks as follows:

$$
\begin{aligned}
O(x,y) = \; & I(x-1,y+1) \cdot w_{-1,1} + I(x,y+1) \cdot w_{0,1} + I(x+1,y+1) \cdot w_{1,1} \\
& + I(x-1,y) \cdot w_{-1,0} + I(x,y) \cdot w_{0,0} + I(x+1,y) \cdot w_{1,0} \\
& + I(x-1,y-1) \cdot w_{-1,-1} + I(x,y-1) \cdot w_{0,-1} + I(x+1,y-1) \cdot w_{1,-1} \quad (4\text{-}3)
\end{aligned}
$$

Generalizing from a $3 \times 3$ to an $N \times N$ kernel, we may write the following convolution formula:

$$
O(x,y) = \sum_{i,j=-(N-1)/2}^{(N-1)/2} \left( w_{i,j} \cdot I(x+i,y+j) \right) \quad (4\text{-}4)
$$

The terms in this equation signify the following: $O(x,y)$ is the output value at pixel position $(x,y)$; $I(x+i,y+j)$ is the input value at position $(x+i,y+j)$, which is multiplied by the corresponding kernel weight, $w_{i,j}$; $\Sigma$ is a symbol of summation; the terms below and above $\Sigma$ signify that weighted input values are added up over the entire $N \times N$ neighborhood, that is, for all possible combinations of indices $i$ and $j$, which both range from $-(N-1)/2$ to $(N-1)/2$.

The simplest possible implementation of an $N \times N$ kernel is the *mean filter* (this is also the closest computational analog to a coffee filter). Its kernel averages input values over an entire $N \times N$ pixel neighborhood. Accordingly, all $N \times N$ kernel elements have the same weight $w = 1/N^2$, and the kernel function evaluates the arithmetic mean of all input values:

$$
O(x,y) = \sum_{i,j=-(N-1)/2}^{(N-1)/2} \left( w \cdot I(x+i,y+j) \right); \quad w = 1/N^2 . \quad (4\text{-}5)
$$

Accordingly, a $3 \times 3$ mean filter kernel has the following structure:

| | | |
|---|---|---|
| 1/9 | 1/9 | 1/9 |
| 1/9 | 1/9 | 1/9 |
| 1/9 | 1/9 | 1/9 |

$(4\text{-}6)$

The effects of mean filter kernels of different size $N$ are shown in Figure 4-5.

Most convolution kernels exhibit a spatially structured pattern of weights. Let us consider the following example:

| $\frac{-1}{9\pi^2}$ | 0 | $\frac{-1}{\pi^2}$ | 0 | 0.25 | 0 | $\frac{-1}{\pi^2}$ | 0 | $\frac{-1}{9\pi^2}$ |
|---|---|---|---|---|---|---|---|---|

$(4\text{-}7)$

This one-dimensional kernel of size $N = 9$ is applied to pixel arrays (i.e., one-dimensional images). Equation 4-7 is an implementation of the ramp filter, which

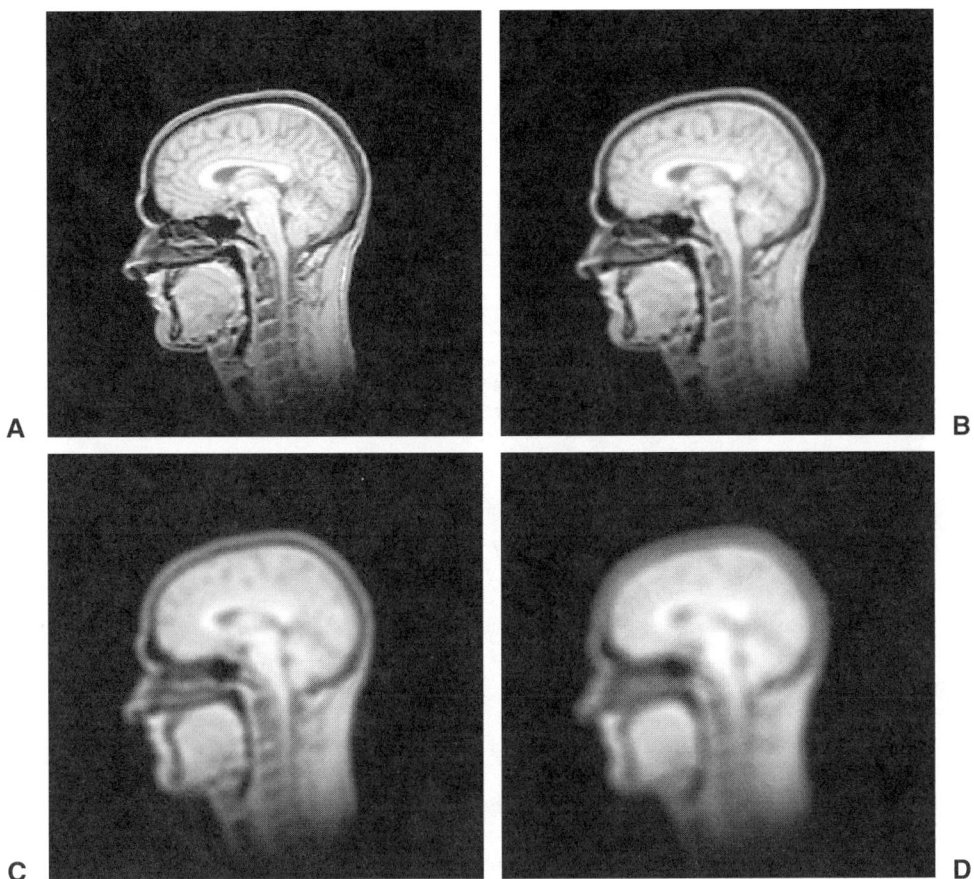

**FIGURE 4-5** Applying a mean filter to an image. The original image (**A**) is filtered with kernels of size $3 \times 3$, $5 \times 5$, and $7 \times 7$ (**B–D**). As kernel size increases, noise and image details are smoothed gradually.

performs filtered backprojection during CT image reconstruction (see Fig. 3-11). As we learned in Section 3.3.3, this filter linearly amplifies higher image frequencies, resulting in edge enhancement. Figure 4-6 shows its structure in more detail, as well as its effects on a single line of a sinogram (the effects on the entire sinogram are visualized in Fig. 3-11).

As a two-dimensional example, let us consider the spatial structure of a $3 \times 3$ *Laplace kernel*, which is routinely used for the detection of boundary structures in an image:

$$
\begin{array}{|c|c|c|}
\hline
-1 & -2 & -1 \\
\hline
-2 & 12 & -2 \\
\hline
-1 & -2 & -1 \\
\hline
\end{array}
\qquad (4\text{-}8)
$$

An in-depth discussion of this important filter follows below. Using the Filter Applet on the Web Companion, you can observe its effects on various input

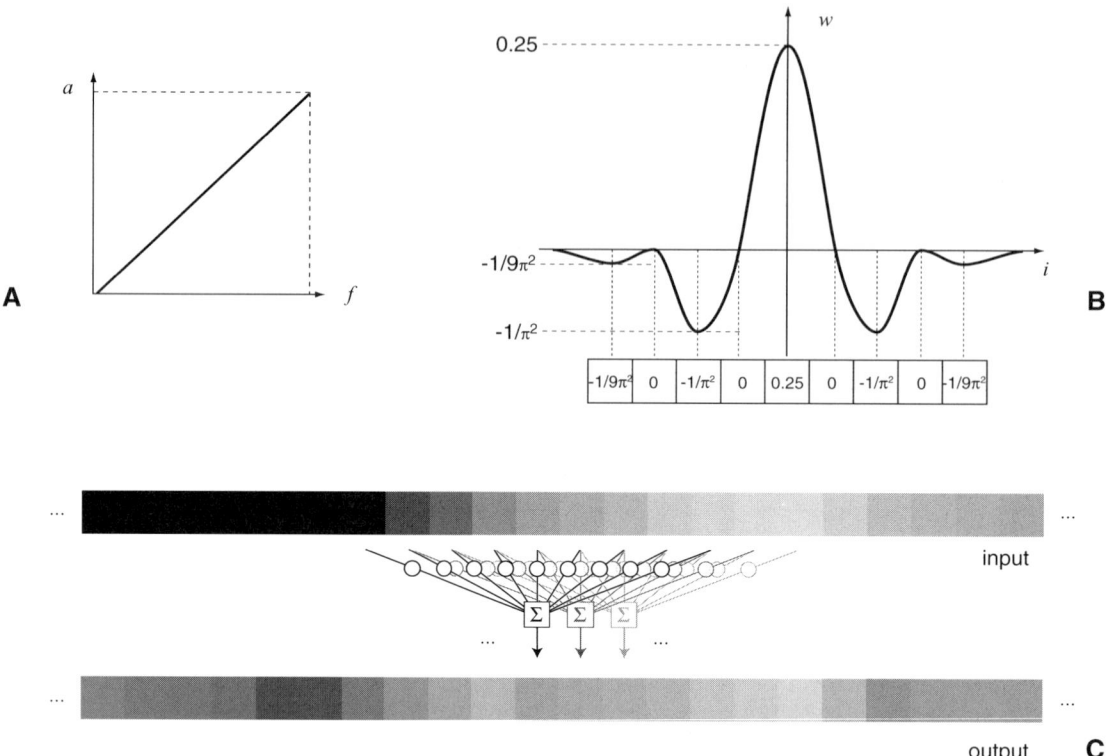

**FIGURE 4-6**  Applying the ramp filter to an CT absorption profile. **A**: The structure of the ramp filter in the frequency domain (see Section 4.3.2). Spatial frequencies $f$ are amplified linearly with factor $a$. **B**: The ramp filter can equivalently be expressed as a spatial kernel function, expressing weights $w_i$ as a function of spatial position $i$. Computer implementations use a discrete (i.e., pixelized) version, which is shown just below the curve graph. **C**: Application of the filter kernel to an attenuation profile. Note that the kernel processes data from 9 input pixels to generate 1 output pixel.

images. This applet provides various predefined filter kernels, and also encourages experimenting with user-specified kernels.

### 4.3.2 Convolution and Fourier Analysis

Up to this point, we implicitly assumed that the size $N$ of a kernel is considerably smaller than the image size, $L$. Using a slightly different approach, we assume a virtual kernel, whose size equals the size of an image to which it is applied. Accordingly, an $N \times N$ kernel can be thought of as a virtual $L \times L$ kernel, in which weights $w_{i,j}$ outside the $N \times N$ neighborhood have values close to zero. In this scenario, every pixel in the input image influences every pixel in the output image. The use of filter functions, which act on a global scale rather than a local scale, is an important concept underlying *Fourier-based filtering*.

In the modulation transfer function (MTF; see Box 3-1), an image represents a superposition of various sine wave functions with different spatial frequencies (high frequencies correspond to short wavelengths, and vice versa). Because images are one-, two-, or three-dimensional data structures, these waves have one, two, or three dimensions, respectively. The function that converts an image from the spatial domain into the sine wave domain is the *Fourier transform*. With a Fourier transform, an entire image is the input, and the output is the decomposition of the image into a spectrum of spatial sine waves with characteristic amplitudes and phase shifts. It is often said that the Fourier transform converts an image from the *spatial domain* into the *frequency domain*.[2]

Fourier analysis provides an elegant alternate method to convolution of an image with a spatially structured filter kernel. Like images, filter kernels can be transformed into the Fourier domain, so the entire filtering procedure can be performed with wave functions. This occurs in the *convolution theorem*, which states that convolution of an image with a filter kernel in the spatial domain corresponds to *multiplication* of the Fourier transforms of the image and the kernel (Fig. 4-7).

Multiplication in the Fourier domain is amply used in CT during filtered backprojection, where the Fourier transform of the attenuation profiles is premultiplied with the ramp filter (Fig. 4-6A), as an alternative to the application of the spatial filter kernel (Fig. 4-6B).

From the perspective of Fourier transforms, reconsider the mean filter introduced above (Fig. 4-5). Because "blurring" an image involves removing high spatial frequencies, the abrupt border of the mean filter kernel in the spatial domain (the rim of the coffee filter; see Fig. 4-4) introduces oscillations in the frequency domain (Fig. 4-8A). As an alternative, the *low-pass filter* suppresses all spatial frequencies above a given cutoff value. In this filter, however, the abrupt border in the frequency domain introduces oscillations in the spatial domain (Fig. 4-8B). The ideal compromise between mean and low-pass filters is the *Gaussian filter* (Fig. 4-8C). In the spatial and frequency domains of the Gaussian filter, the weighting function follows the Gaussian distribution. Because the frequency distribution of a random variable around its mean value is expressed, the Gaussian filter is most appropriate for removing random (Gaussian) noise from an image.

### 4.3.3 Statistical Filters

The averaging effect of mean, low-pass, and Gaussian filters not only eliminates image noise, but simultaneously smoothes relevant small-scale features in the image. How can the smoothing of these features be compensated? Once more, let us change the perspective. Filters act as operators that collect statistical informa-

---

[2] The reader may recall that sine functions are periodic and have no spatial limits. Accordingly, a Fourier transform is only possible for an image without spatial limits. The solution for this is to create a periodic image by repeating it ad infinitum in *x* and *y* directions.

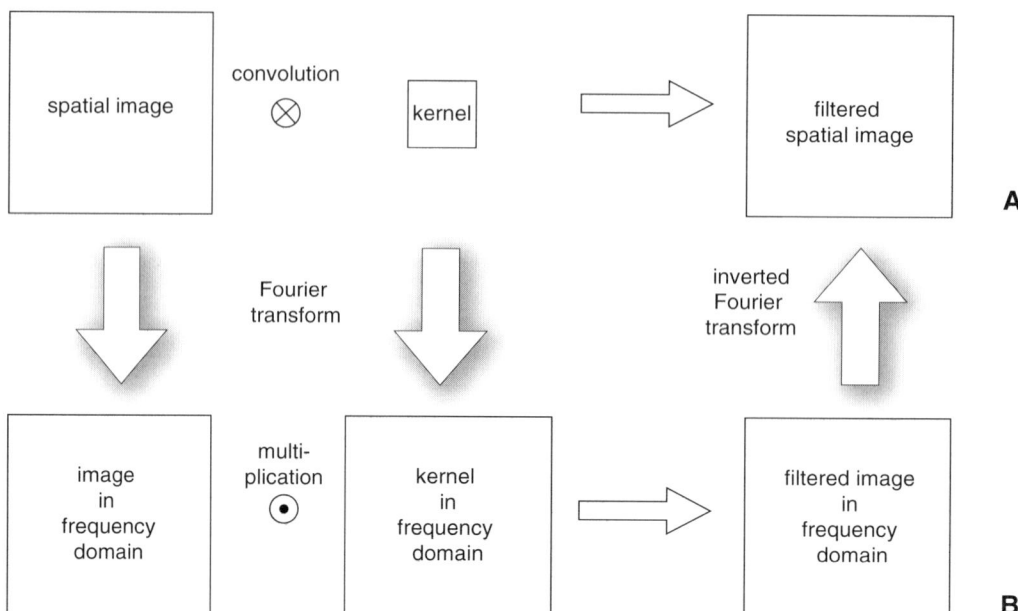

**FIGURE 4-7**  The convolution theorem. Convolution of an image with a filter kernel in the spatial domain (**A**) is equivalent to multiplication of the image with a kernel in the frequency domain (**B**). The Fourier transform is used to convert images (and kernels) from the spatial into the frequency domain; the inverted Fourier transform reconverts images from the frequency into the spatial domain.

tion from an image. The mean filter, for example, computes the *arithmetic mean* of input values within an $N \times N$ pixel neighborhood (see Eq. 4-3). However, other statistical properties also evaluate average values. For example, the *median* is the middle value in the frequency distribution of all measurements in a sample (50% of the measurements are smaller than this value, and 50% are larger than this value). Because the median emphasizes ordering of measurement values rather than magnitude of measurement values, the median value is calculated independent of outliers in the sample. A median filter is highly efficient in eliminating so-called impulse noise ("salt-and-pepper" noise) from an image (Fig. 4-9).

The median filter is one example of a wide variety of statistical filters, or histogram filters, that analyze the frequency distribution, or frequency histogram, of an image and calculate corresponding statistical characteristics.

### 4.3.4 Edge Detection Filters

As mentioned in Chapter 2, the first step in biomedical data processing is identification of object boundary structures, such as contours and surfaces, that can be used to reconstruct 2D and 3D graphical object representations. Boundary structures are local discontinuities in image intensity, the so-called *edges*. The principal

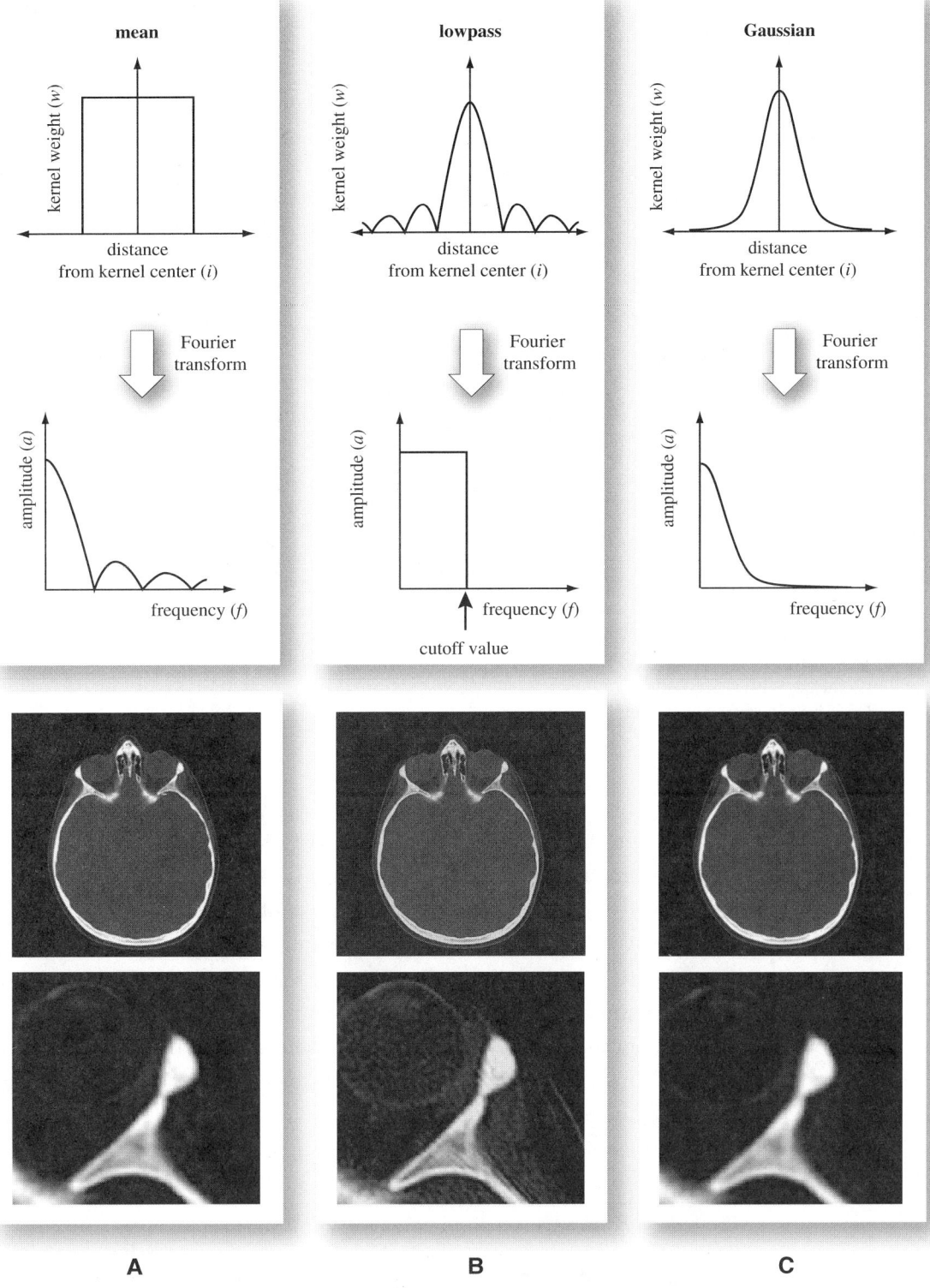

**FIGURE 4-8** Removing high-frequency portions from an image, with a mean filter (**A**), a low-pass filter (**B**), and a Gaussian filter (**C**). The top graphs show each filter kernel in its spatial domain (kernel weights $w$ vs. distance $i$ from center of kernel) and in its frequency domain (amplitude $a$ vs. frequency $f$). Note that the Fourier transform of a Gaussian function is also a Gaussian function. The bottom graphs show the effects of each filter on a CT image (overview and detail region).

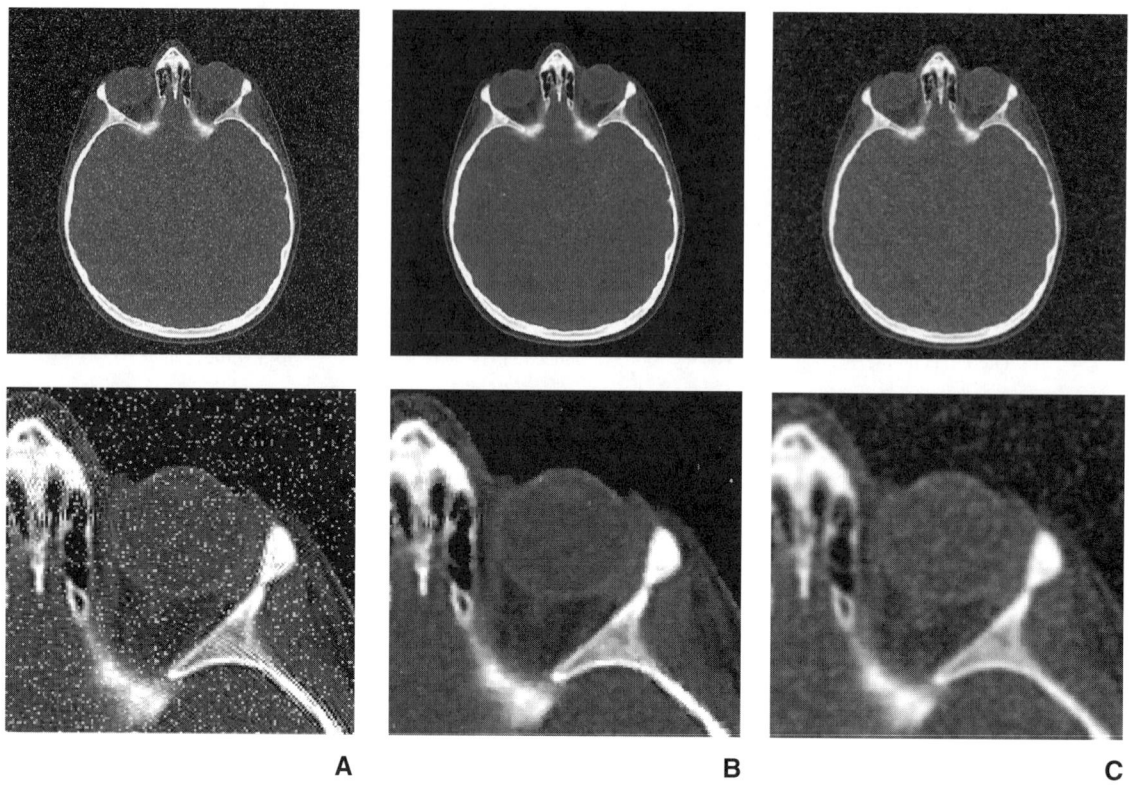

**A**                                    **B**                                    **C**

**FIGURE 4-9** Suppressing impulse noise with a median filter. "Salt-and-pepper" noise in image **A** is removed efficiently with a $3 \times 3$ median filter (**B**). Applying a Gaussian filter (**C**) disperses this type of noise throughout the image rather than removing it.

task is to detect discontinuities independent of absolute image intensity. Discontinuity, in a mathematical sense, is "change in change." This concept is illustrated in Figure 4-10. Image areas exhibiting constant intensity, as well as areas where the intensity changes gradually, do not contain boundaries. Rather, we perceive discontinuity, which is indicative of an object boundary, only where the image intensity changes fairly rapidly, that is, within a short distance.

How can the notion of "fairly rapid change in intensity" be described quantitatively? In an image, rate of change in intensity is measured by the first spatial derivative, the so-called intensity *gradient* (Fig. 4-10A,B).[3] Whereas a constant gradient represents gradual changes in image intensity, positive or negative peaks in

---

[3] The term derivative comes from infinitesimal calculus and denotes a function that operates on continuous intensity distributions. Strictly speaking, derivatives are calculated only for images with infinitesimally small pixels. In real images, however, pixels have a finite size and a discrete location, and they represent discrete intensities. In image processing, the equivalent of a *derivative* is a *discrete difference*. Nevertheless, the term derivative indicates sufficiently the mathematical concept behind a discrete method of data processing.

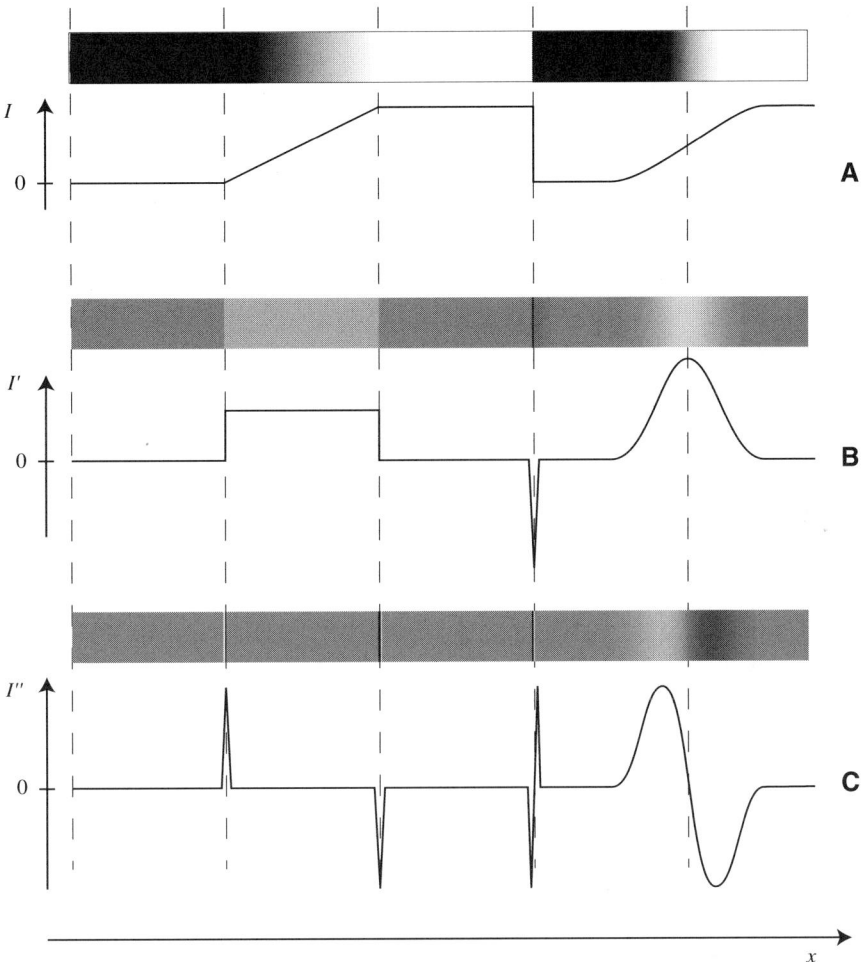

**FIGURE 4-10** Edge detection and spatial derivatives in one dimension. **A** is an image exhibiting changes in intensity $I$ along the $x$-axis and the underlying intensity function $I = f(x)$. **B** and **C** are the first and second spatial derivatives, respectively. A boundary is perceived where the second derivative exhibits a zero crossing. The zero crossing location is independent of $I$ values in the original image or values of the first derivative. Absolute image intensities and absolute gradient values do not influence boundary detection.

the gradient represent more abrupt changes. Maximal local change in image intensity occurs at the tip of the peak.

The second spatial derivative measures rate of change of the intensity gradient (Fig. 4-10C). At gradient peaks, the second derivative switches signs and exhibits a so-called zero crossing. This is the exact location of the discontinuity in the original image.

To visually demonstrate first and second derivatives of a two-dimensional image, we plot pixel intensities $I(x,y)$ into the third dimension (Fig. 4-11A). As in a real landscape, slopes in the resulting "imagescape" differ in steepness and orientation, which are measured with the length and orientation of the *gradient vector*,

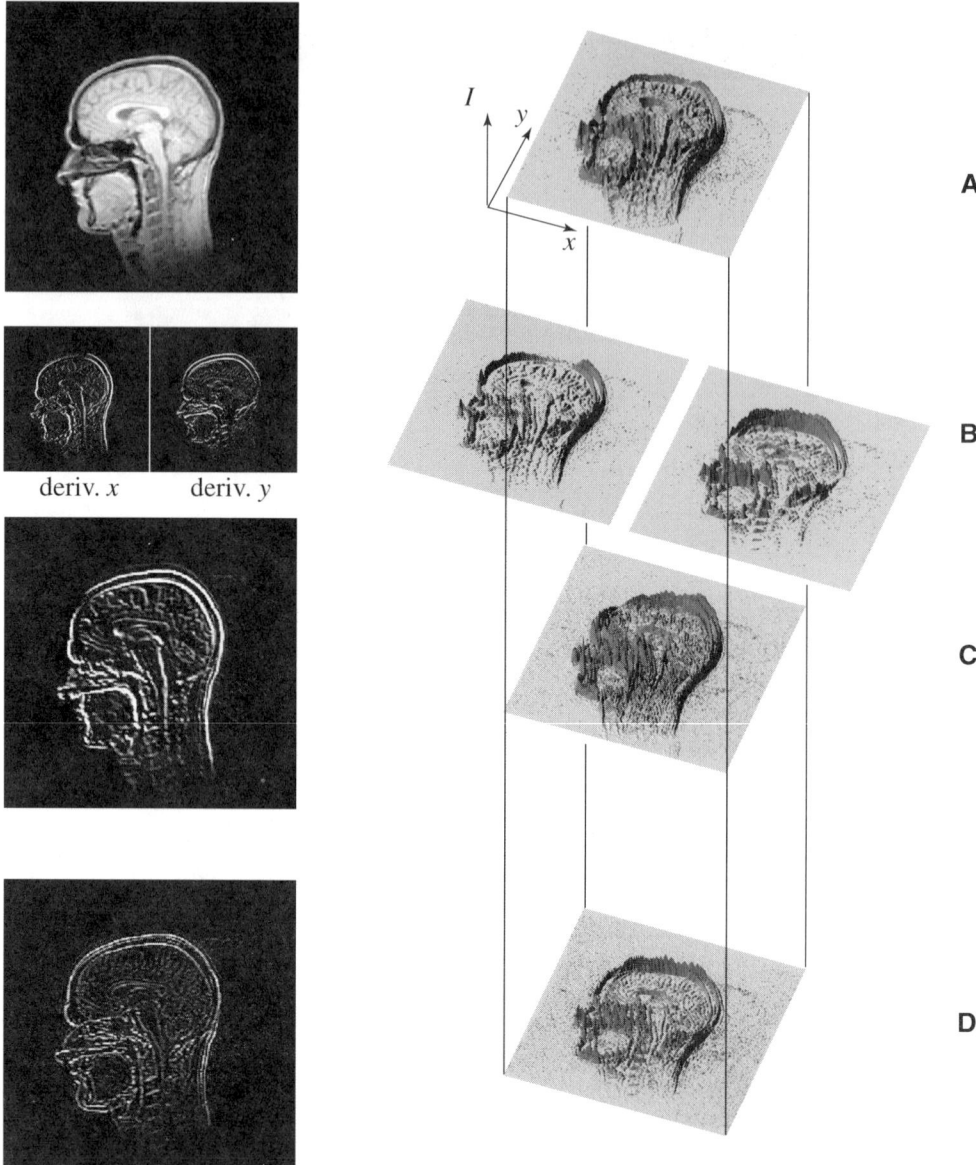

deriv. *x*          deriv. *y*

**FIGURE 4-11**   Edge detection and spatial derivatives in two dimensions. **A** is the original image represented as an "imagescape," in which intensities are plotted into the third dimension. **B** depicts first spatial derivatives in *x* and *y* directions. **C** is the magnitude of the gradient field. **D** is the second spatial derivative, which is evaluated with a Laplace filtering procedure.

respectively. The following thought experiment may help in understanding the concept of a gradient vector: When an imaginary sphere is placed at a given pixel, the gradient vector indicates its initial rolling speed and direction. Gradient vectors are calculated as follows: evaluate the first spatial derivatives in *x* and *y* directions (Fig. 4-11B; so-called partial derivatives), then add the resulting $x$ and $y$ vector components.[4]

Comparing an original image with its first spatial derivative shows boundary structures where the gradient is strong, that is, where intensity changes abruptly. Furthermore, the actual course of a boundary structure is always perpendicular to the direction of the gradient field. One method for determining the location of edges with this information is to calculate the magnitude (i.e., length) of gradient vectors. This operation transforms all slopes of the original image into elevated regions (Fig. 4-11C). Specifically, boundaries become sharp crests, with heights that are positively correlated with relevance. The second spatial derivative of an image is evaluated with the so-called Laplace operator (Figure 4-11D; we already encountered a $3 \times 3$ kernel implementation of the Laplace operator in Eq. 4.4).[5]

In Figure 4-11D, the Laplace operator detects boundaries at many more locations than actually "seen." This occurs because gradients in a real image fluctuate at various amplitudes and spatial scales. Whereas the Laplace filter finds zero-crossings throughout the image, our visual system usually eliminates small-scale fluctuations. This filtering property of our visual system is difficult to replicate on a computer. Although mean filters and Gaussian filters reduce local fluctuations and contrast by *integrating* over a local pixel neighborhood (Figs. 4-5 and 4-8), Laplace filters do exactly the opposite—enhance local fluctuations and contrast—by *differentiating* over the local pixel neighborhood. Accordingly, edge detection efficiency can be improved with a two-step procedure. First, local fluctuations are eliminated with a Gaussian filter of appropriate kernel size. Second, the Laplace filter is applied subsequently to detect significant zero crossings. Results from this so-called Laplacian-of-Gaussian filtering are shown in Figure 4-12. Zero crossings in this image now indicate relevant boundary structures (see also Box 4-1).

## 4.4 EXTRACTING ISOSURFACES

### 4.4.1 Determining Boundaries in CT Images

The general aim of edge-detection methods, in practice, is the characterization of object boundaries independent of absolute image intensities. However, there is one important special case of edge detection—extraction of boundary structures

---

[4] Derivatives in $x$ and $y$ directions are called partial, because they represent independent directional components of the total derivative.

[5] Named after Pierre Simon Laplace (1749–1827). For historical reasons, the term *Laplace operator* is more common than *Laplace function*. The term operator often is used synonymously with the term function.

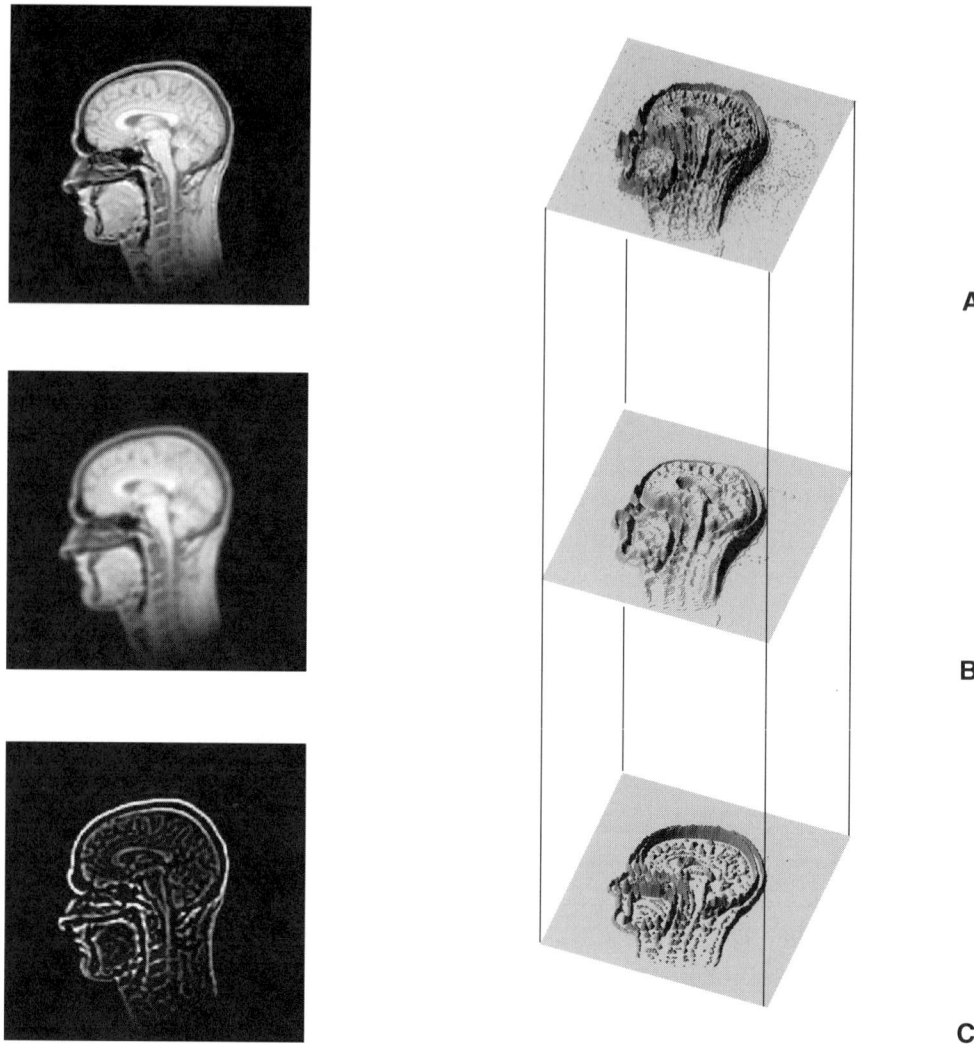

**FIGURE 4-12** Combining boundary detection with noise elimination. The original image (**A**) is filtered with a Gaussian kernel. **B** is the resulting noise-reduced image. Application of the Laplace filter improves edge detection (**C**; compare this image to Fig. 4-11D).

from CT images. In a CT image, each pixel represents the X-ray density of an object at that position. Because image intensity values $I(x,y)$ directly express material properties, an image can be segmented into different areas by characteristic average densities. Segmentation can be accomplished with a window function and threshold filters (Fig. 4-16).

Let us examine in detail how bone tissue is extracted from CT image data. Skeletal parts exhibit a higher density than surrounding soft tissue in living spec-

## LAPLACE AND THE VISUAL SYSTEM

How does our visual system detect boundaries? In the retina, the layer of sensory neurons stores an input image as a pattern of electrical charge differences, where each neuron represents a pixel. Subsequently, the layer of horizontal cells acts as a filter that convolves the retinal image and sends output to the following layer of bipolar cells.

As shown in the schematic drawing (Fig. 4-13), bipolar cells (B) receive inhibitory (negative-valued) input from neighboring horizontal cells (H) and excitatory input from sensory cells (rods, R). A combination of central excitation and lateral inhibition is equivalent to positive and negative weights in the center and periphery of a Laplacian kernel. Accordingly, our retina computes an analog to the second spatial derivative (Fig. 4-14).

Contrast enhancement at edges can be experienced in the "Mach band" optical illusion.[6] As an effect of neuronal lateral inhibition, the intensity within each band appears to change from left to right (Fig. 4-15A). The Hermann–Hering grid[7] in Figure 4-15B demonstrates contrast enhancement by lateral inhibition depending on the spatial scale. "Whiteness" in the grid is a function of spatial vicinity of dark squares. Accordingly, contrast enhancement is stronger along grid lines than at grid crossings because the distance between neighboring black squares is smaller in the former. This creates artifactual dark blobs at grid crossings, but because lateral inhibition is virtually absent in the central region of the retina blobs disappear when inspected (see Fig. 2-5).

imens, or relative to air in museum specimens.[8] We use a threshold function established at 150 Hounsfield units (Fig. 4-16C) to partition the object in the image into bone ($\geq$150 HU) and nonbone (<150 HU). In the binary image, edges occur between bone and nonbone pixels. The exact position of borders can be determined even more precisely (e.g., at the subpixel level) if we consider the physics of the CT data acquisition device. Each X-ray detector within the detector array

---

[6] First described by the physicist, sensory physiologist, and philosopher Ernst Mach (1838–1916).

[7] Named after physiologists Ludimar Hermann (1838–1914) and Ewald Hering (1824–1918).

[8] In fossils, the situation is more complex. Portions of a fossil structure may be embedded in sediment exhibiting density values that can be very similar to those of fossilized bone. Furthermore, fossilized skeletal parts can exhibit density fluctuations created by differential mineralization.

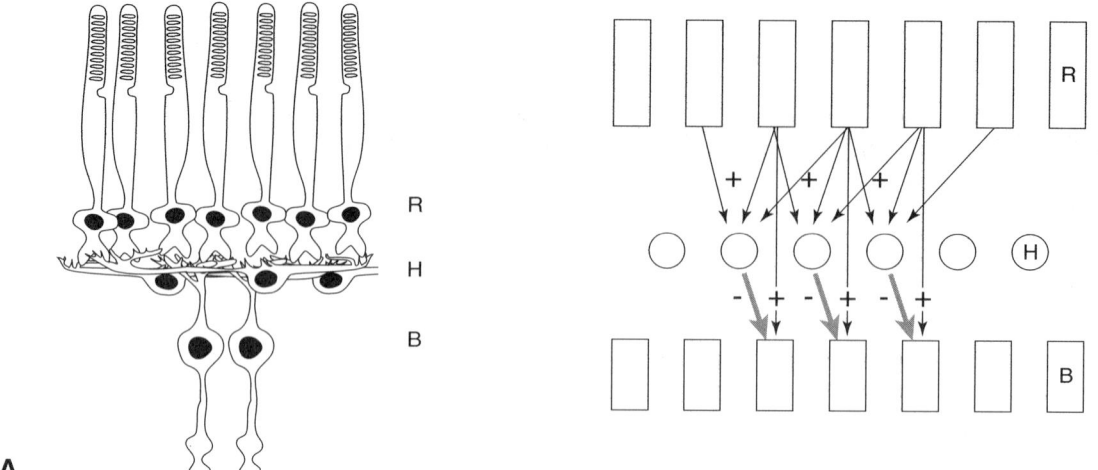

**FIGURE 4-13** Anatomical **(A)** and computational **(B)** organization of the neuronal layers of the retina. Light receptor cells (R) convert light into electric signals, which are sent as positive and negative signals to horizontal cells (H) and bipolar cells (B), respectively. Also see Figure 2-5.

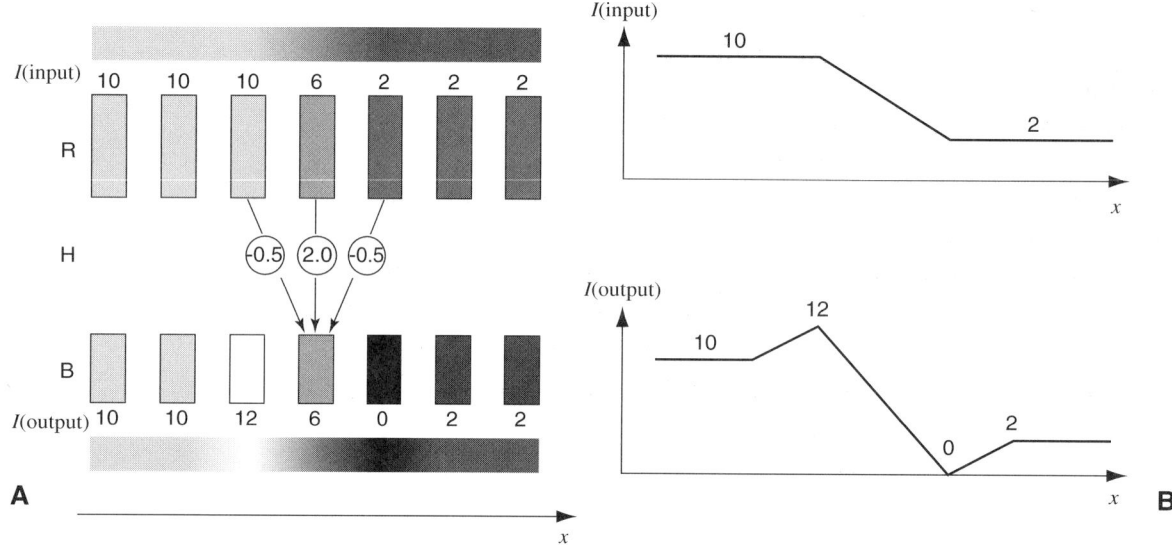

**FIGURE 4-14** The logics of edge enhancement through lateral inhibition in the retina. **A** shows in a model calculation how a ramp-shaped input intensity distribution is processed through neuronal layers to yield edge enhancement in the output (R, H, B: receptors, horizontal and bipolar cells). **B**: graphs of the input and output intensity distributions.

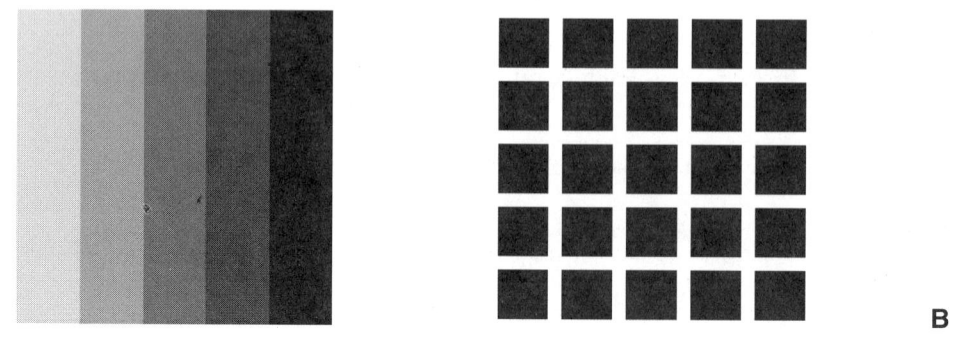

**FIGURE 4-15** Optical illusions illustrating effects of lateral inhibition. **A**: Mach bands. **B**: Hermann-Hering grid.

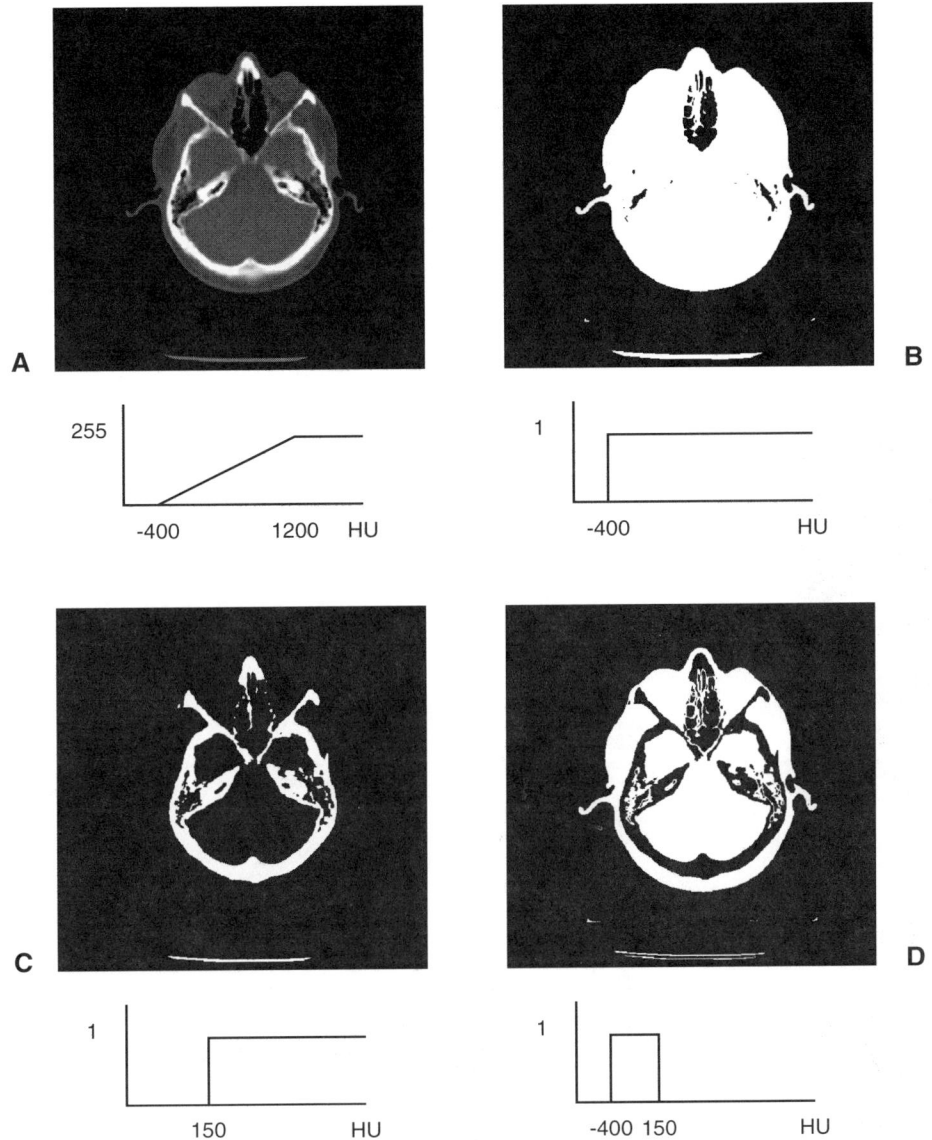

**FIGURE 4-16** Segmentation of a CT image according to local Hounsfield density. **A** is the original image displayed with a wide window setting (see also Fig. 4-3). **B** is the image displayed at the soft tissue threshold, −400 HU. Dark areas inside the skull indicate the nasal cavity and mastoid air cells in the temporal bone. **C** is the image displayed at the bone threshold, 150 HU. **D** is the soft tissue window minus the bone window, ranging from −400 to 150 HU.

of a CT gantry (Fig. 3-14) acts as a Gaussian filter (see Appendix A) that smoothes the stairstep-shaped drop in X-ray density associated with a border, as shown in Figure 4-17A. Nevertheless, Gaussian smoothing preserves the original image geometry without distortion, such that the actual location of a physical boundary

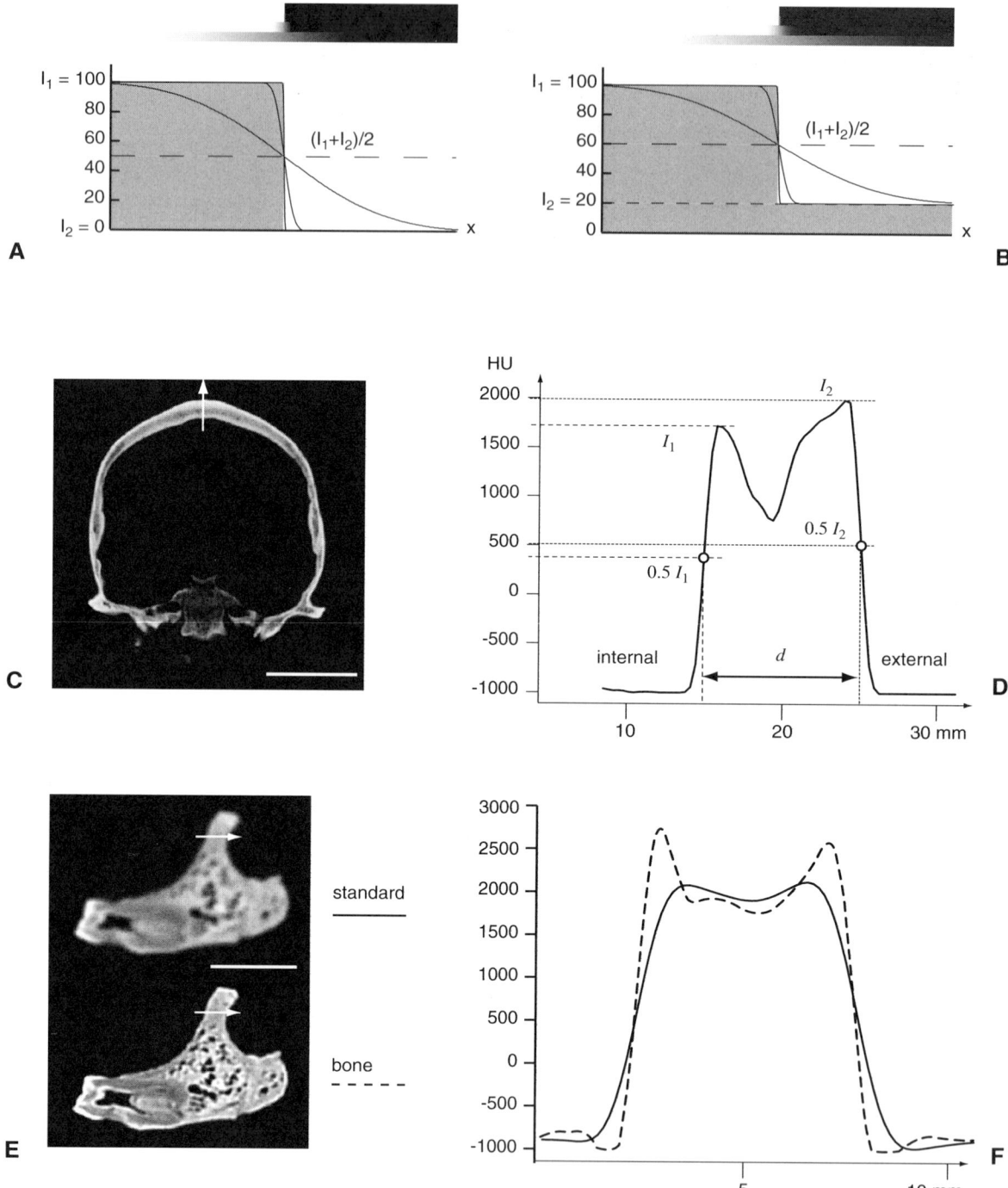

**FIGURE 4-17** Density-based measurement of linear distances in CT images. In **A** and **B**, a sharp drop in Hounsfield density (gray area) is mapped into a smooth decay during CT imaging. Exact location of the edge occurs at 50%-intensity between the two levels, $I_1$ and $I_2$, and is independent of the amount of smoothing. **C** and **D**: Density-based measurement of bone thickness equals $d$ (modern human skull; scale bar is 5 cm). **E** and **F**: Image reconstruction with a bone filter yields density functions that no longer represent true Hounsfield values and, consequently, should not be used to determine edge locations (Sakajia Neanderthal maxillary fragment; scale bar is 2 cm).

equates with the position where measured image intensity drops to 50% of the object density.[9]

Generalizing this argument to two objects with Hounsfield densities $I_1$ and $I_2$, we may state that the edge is located midway between $I_1$ and $I_2$ (Fig. 4-17B). Sampling linear distance measurements from a CT image is a direct application of this method. Figure 4-17C is a coronal cross section through a modern human skull. The density distribution along a midsagittal section contains two small plateaus at 1540 HU on the inner table of the cranial vault and 1995 HU on the outer table, with values dropping to –1000 HU for air inside and outside the skull.[10] Thickness of the cranial vault equals the distance between internal and external 50%-density points.

One important practical consideration is that the method of density-dependent edge location is only reliable when CT images are reconstructed with the standard ramp filter (see Fig. 3-12 and Fig. 4-6). Although reconstructing a projection data set with a bone filter kernel yields a detailed rendering of fine structure of the bone, Hounsfield values in the image no longer reflect true physical object densities, as can be verified in the corresponding density profile (Fig. 4-17E, F).

## 4.4.2 From Edges to Isocontours and Isosurfaces

Here we examine how methods of thresholding a density-dependent boundary detection also generate isocontours and isosurfaces, that is, boundary structures located at the same single density value throughout an image. In Chapter 2, free-form objects of any geometric complexity are described as ordered sets of simplexes—two-dimensional contours by sets of line segments and three-dimensional surfaces by sets of triangles. Accordingly, we first locate all simplexes in an image data set, and then we determine their orientation to differentiate between inner and outer sides of an object and organize the simplexes into ordered sets. Line segment simplexes and triangle simplexes can be located and extracted from binarized data sets with *morphologic operators* (Höhne and Hanson, 1992; Soille, 2002). These behave as filter functions that search a local pixel neighborhood for specific spatial patterns (so-called morphologies) and output those that are identified as simplexes.

Applying the threshold function yields a binary image that distinguishes between locations inside (white) and outside (black) the object (Fig. 4-16). To

---

[9]The 50%-intensity point of a Gaussian distribution occurs at the location with the steepest slope (the point of inflexion). This corresponds to a zero crossing in the second derivative, which is consistent with our earlier definition of an edge point.

[10]The reason why the outer table appears denser than the inner table is that this dry skull was scanned with a scanner calibrated for patients. Less beam hardening than expected results in higher reconstructed Hounsfield density.

extract line segments, we use the *marching-squares* algorithm (also known as "marching-ants"; see Fig. 4-18), which operates on the smallest possible two-dimensional neighborhood, that is, a square measuring $2 \times 2$ pixels. In a binarized image, the four pixels have $2^4 = 16$ potentially different black/white patterns. Each possible pattern relates to a unique configuration of oriented line segments that subdivide the neighborhood into areas inside and outside the object. Scanning an entire binarized image with the marching-square algorithm yields a collection of line segments that form closed contours when ordered. Some contours will have positive (clockwise) orientations, whereas others will have negative (counter-clockwise) orientations (Fig. 4-18C). The sign represents external and internal delimitations of the objects, respectively. In the final step, density-dependent boundary localization determines the exact position of each line segment between neighboring pixels.

In three dimensions, the situation is relatively more complex but still equivalent to the situation in two dimensions (Fig. 4-19). The smallest possible voxel neighborhood contains $2 \times 2 \times 2 = 8$ elements, resulting in $2^8 = 256$ possible black/white patterns. The *marching-cube* algorithm–the extension of the marching-square algorithm to the third dimension–ascribes to each pattern a set of boundary triangles whose orientation in space determines the inside and outside of a reconstructed object (Lorensen and Cline, 1987). Arranging triangles into surfaces is accomplished via shared edges or vertices.

The marching-square and marching-cube algorithms have several properties that are inherent strengths, but simultaneously act as limitations. Because they

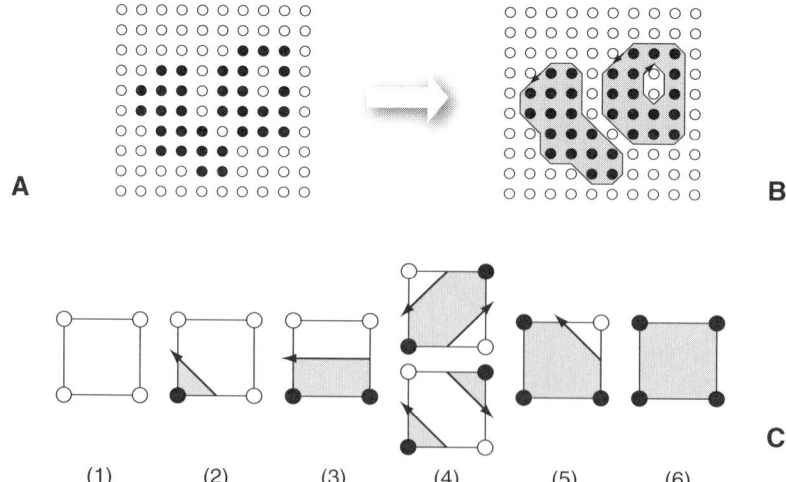

**FIGURE 4-18**   The marching-squares (MS) contour reconstruction algorithm. **A:** A binarized image. Black and white pixels indicate filled and empty areas (inside and outside an object, respectively). **B:** Contours generated with the MS algorithm. **C:** The MS algorithm is a morphologic filter that yields directed contour segments for each of the six basic patterns of filled/empty pixels in a $2 \times 2$ square (note ambiguity in case 4).

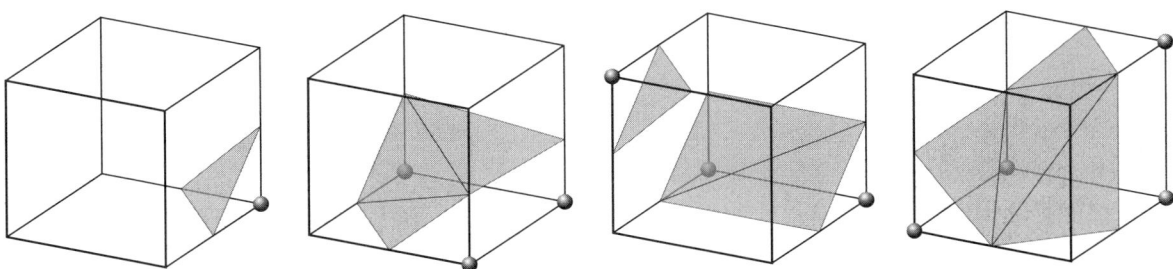

**FIGURE 4-19** The marching-cubes (MC) surface reconstruction algorithm operates as a morphologic filter on binarized voxel data sets. All possible combinations of filled/empty pixels in a $2 \times 2 \times 2$ cube can be classified into 14 different cases (Lorensen and Cline, 1987), 4 of which are illustrated here. In each case, the surface separating inside from outside of the object is specified by a set of triangular tiles.

operate by closest-neighbor relationships, these algorithms emphasize local structure of the boundaries, irrespective of large-scale features of an object. This divide-and-conquer approach facilitates constructing free-form objects of variable complexity from a relatively limited number of elementary boundary topologies (16 and 256 in two and three dimensions, respectively). On the other hand, limiting topologies to small-scale neighborhood relationships will introduce stairstep-like surface artifacts where the object surface curves smoothly; overall, this effect also inflates the number of contour segments or triangles. A notable issue concerning the marching-cube algorithm is that more than 256 topologically correct solutions exist for the 256 binary patterns (we also encountered a topological ambiguity in the marching-squares algorithm; see Fig. 4-18C). To tackle these problems, a wide variety of solutions have been proposed to fix any potentially resulting ambiguities in affected free-form objects, and to decimate the number of surface elements by merging small neighboring elements into larger ones (Montani et al., 1994; Delibasis et al., 2001; Lee and Lin, 2001; Rajon and Bolch, 2003).

As an alternative to the marching-cube algorithm, which operates on nearest pixel neighborhoods, *contour-based triangulation* establishes surface segments on a larger scale, by connecting contours in adjacent CT slices with triangle strips (Boissonnat, 1988; Ekoule et al., 1991; Wallin, 1991; Meyers et al.., 1992). As seen in Figure 4-20, this approach has its own challenges. The task of stitching together $M$ contours in layer $z$ with $N$ contours in layer $z + 1$ in a topologically reasonable way is a complex activity.

## 4.5 INTERACTIVE SEGMENTATION

A fundamental difficulty in reconstructing objects from image data is that object information in the data set is usually incomplete, such that supplementary

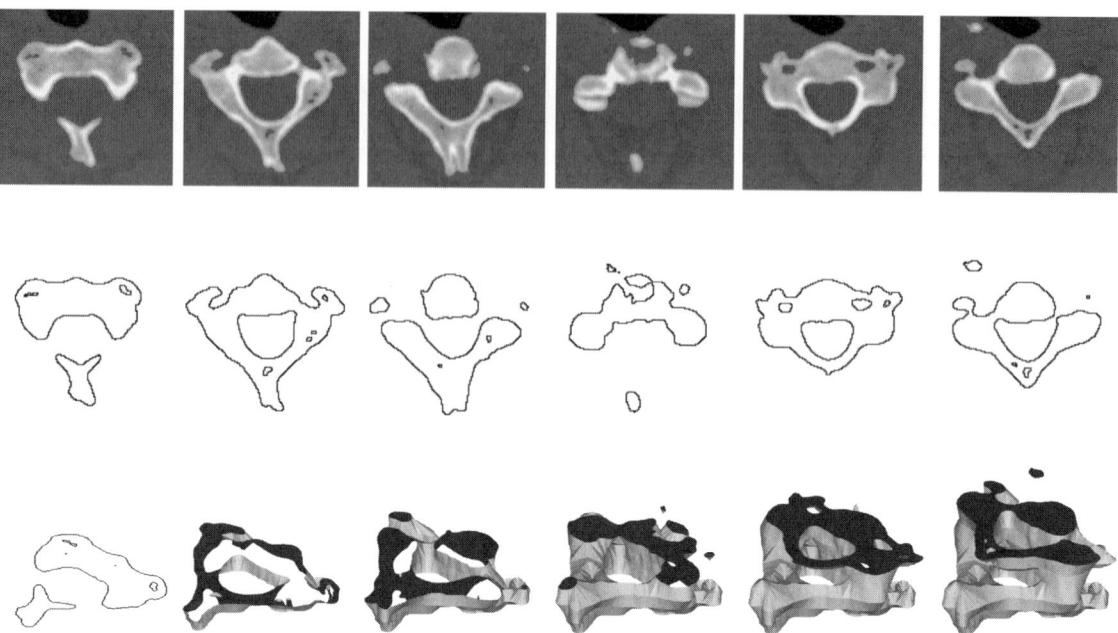

**FIGURE 4-20** Contour-based triangulation. Contours extracted from adjacent cross-sectional images are connected by nearest-neighbor triangulation.

information is necessary to reconstruct the "original" object geometry. The afore-mentioned examples demonstrated that filters are more effective and surface reconstruction methods more accurate when incorporating expert knowledge about imaging modalities and expected object properties into the filter design. For example, bone structures can be extracted easily from patient or fossil CT data if the expected Hounsfield density range of bony tissue, or the Gaussian intensity profile at tissue boundaries, is provided. Once these parameters have been deter-mined, bone extraction becomes an automated task because the computer pro-ceeds through image stacks, performing segmentation and triangulation without further user instruction.

An important group of morphologic operators that requires knowledge-based initial user interaction is represented by *erosion-and-dilation* procedures (Fig. 4-21) (Soille, 2002). Let us consider the automated extraction of the brain surface from a patient MRI. In MRI images, the brain's gray matter differs only slightly in signal intensity from the surrounding meningeal tissue, such that automated threshold-ing always results in some connectivity between these structures. The solution to this problem consists in removal of connective structures of a given "thickness" before isosurfaces are extracted. Imagine the original isovolume as being exposed to erosional forces that wear away a one-pixel-thick hull during one time step. Obviously, during the first iteration, two-pixel-wide structures are eroded completely, during the second, four-pixel-wide structures disappear, and so on.

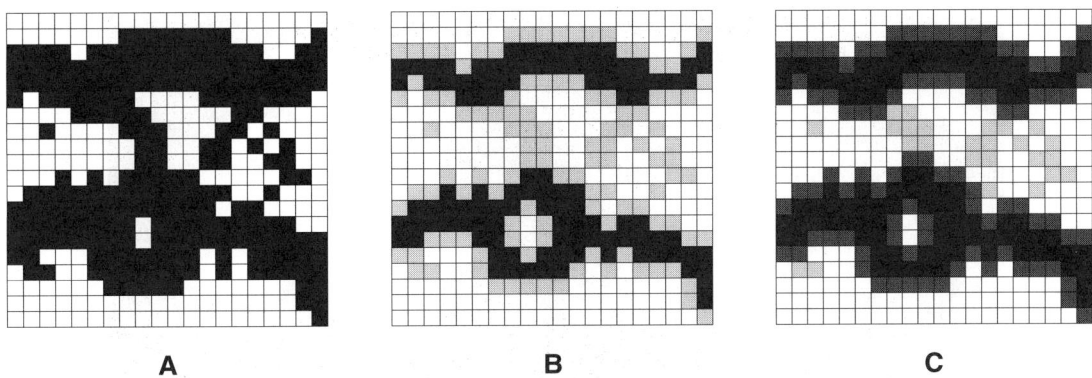

**FIGURE 4-21** Erode-and-dilate algorithm, applied to a small patch (20 × 20 pixels) of an MRI image. **A:** A binarized image (black/white pixels denote inside/outside of an object). **B:** Erosion of a one-pixel-wide boundary around the object (light gray). **C:** Dilation of the eroded object through apposition of a one-pixel wide boundary (dark gray). As an effect, all original structures less wide than three pixels are deleted, notably connections between "objects" in the upper and lower half of the image.

The important point is that structures wider than a user-defined minimum width shrink but are conserved. Because this shrinking process follows reversible rules, it can be undone by application of the corresponding amount of dilation, that is, the iterative apposition of pixels. Overall, the original object structure is reconstituted, with the exception of the connective structures. Figure 4-21 shows an example with one single erosion-dilation iteration.

In many instances, automated segmentation and surface reconstruction yields unsatisfactory results despite initial user interaction. As shown in Figure 4-22, bone density may fall below expected values, or surrounding structures may exhibit similar density to bone, such that automated segmentation fails to isolate bone structures. Either scenario would create spurious results during surface reconstruction.

When automated procedures yield inconsistent results, user interaction is required to provide additional object details. An often-used method is *region-growing* (or seed-filling) (Haralick and Shapiro, 1985). The user plants a "seed" pixel within a structure of interest, which is then allowed to "grow" into neighboring pixels according to some connectivity criterion (e.g., similar density).[11]

Another interactive strategy involves using *active contours*, or *deformable models* (or templates) (Kass et al., 1988; Terzopoulos et al., 1988; Caselles et al., 1993; Chung and Ho, 2000; Harders et al., 2002; Malladi and Sethian, 2002). These are geometric primitives, for example, a circle in two dimensions or a sphere in three dimensions, that are user-positioned at appropriate locations in the image and then morphed to potential object boundary structures. This approach uses

---

[11] Readers using image processing software will be familiar with the "magic wand" tool as an instance of region-growing.

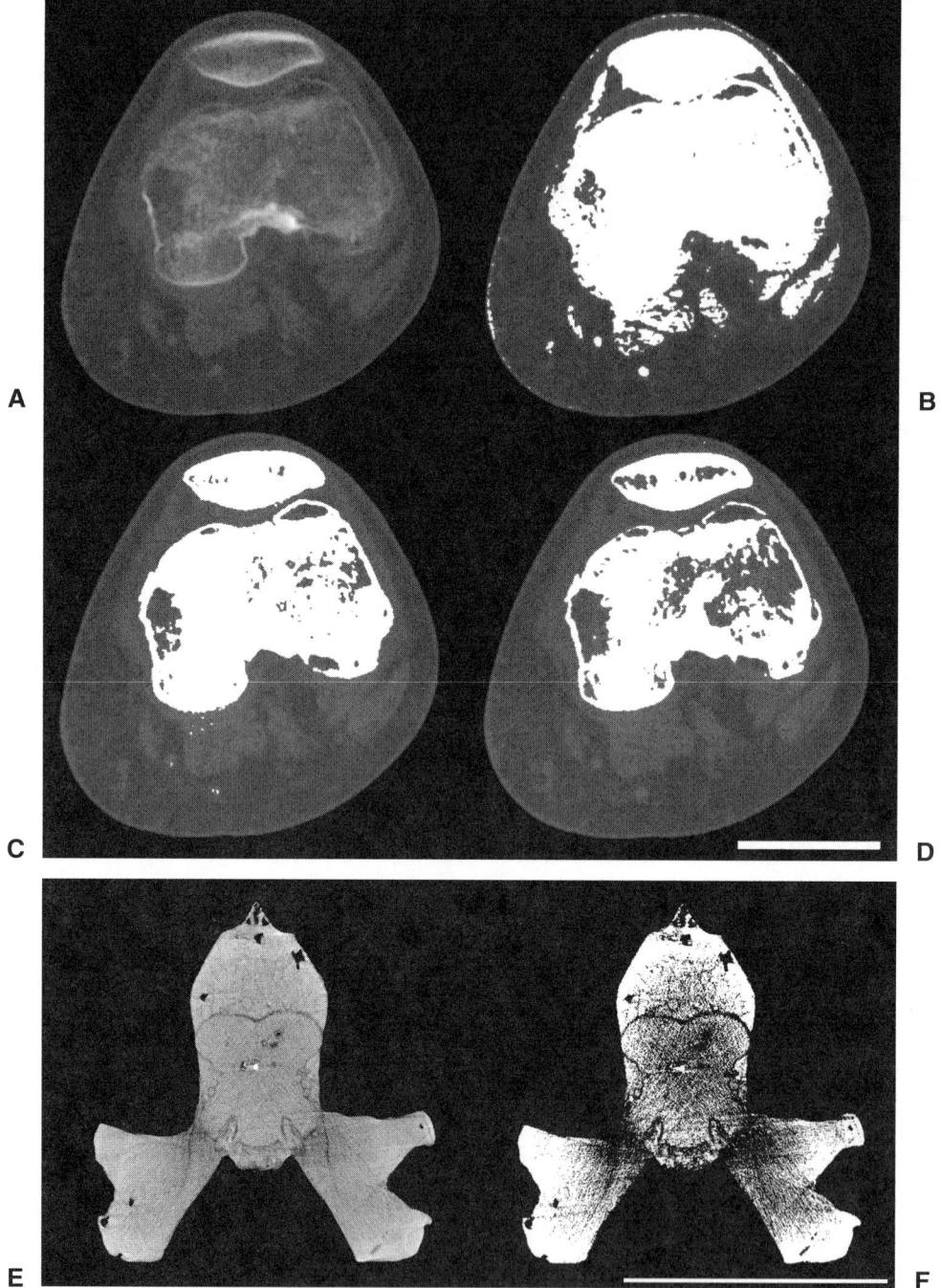

**FIGURE 4-22**   Limits of threshold-based bone extraction from CT images. **A–D** depict a cross section through a human knee that is reconstructed with a standard (ramp) filter and displayed with parameters of window center = 250 HU and window width = 1500 HU (**A**) and thresholded at 50 HU (**B**), 100 HU (**C**) and 150 HU (**D**). Density of cancellous bone within the patient's femoral head and patella occupies the same range as surrounding soft tissue. This similarity is primarily due to partial volume averaging (see Figs. 3-14, 3-15). This type of imaging artifact is a consequence of trabecular structures not necessarily having lower densities than compact bone structures, but having dimensions below the resolution of the CT image. **E** and **F** are coronal sections through the skull of a pterosaur (industrial CT scan) and reveal a sediment-filled braincase. The X-ray density of bone parts is similar to the X-ray density of the sediment, such that manual segmentation procedures are required to define the endocranial cavity. Scale bars are 5 cm.

elements of image synthesis to perform image analysis. For example, a predefined geometric form serves as a guide to the form of an unknown object in the image data set.

How is a deformable template morphed to object structures in the image? In practical terms, this occurs similarly to shrink-wrapping a vacuum bag around food items, or inflating a balloon to fill a cavity. The shared critical feature is that the object to which the template is fitted can be specified relatively loosely (Fig. 4-23)—its boundary structures may be known only partially—but the object can still be recovered with high efficiency by imposing constraints on the process of template deformation. For example, the energy-minimum principle, which guarantees that the template behaves like a rubber band wrapped around some object, holds that the template adjusts its form until it attains a state of minimum internal energy. In addition, topology preservation is a useful constraint. Topology deals with the geometric properties of object surfaces. Two objects have the same topology if they can transform into each other by simple rubber sheet deformation, excluding the use of scissors and glue (i.e., no holes are created or filled; the

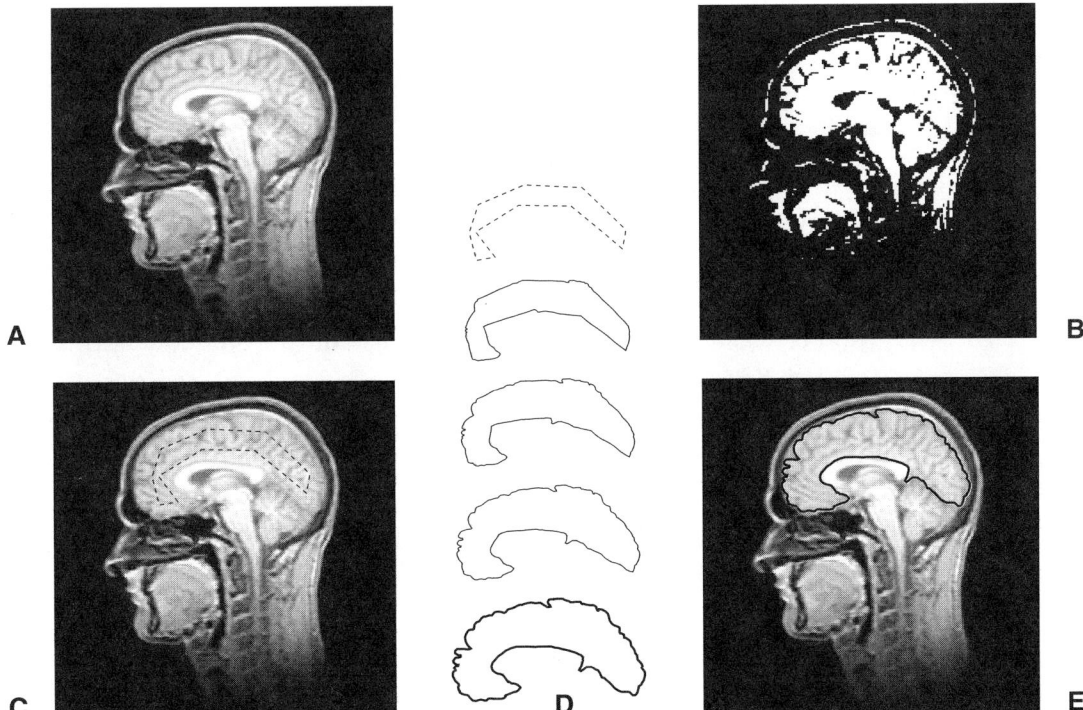

**FIGURE 4-23** Deformable models. On this midsagittal MRI scan of a human head, the boundaries of the cortical area of the brain are easy to perceive (**A**) but difficult to extract with thresholding procedures (**B**). A template contour line placed manually inside the region of interest (**C**, dashed line) is expanded, refined, and deformed according to local image properties; the process of adaptation (**D**) stops when the deformable template attains a substantial drop in image intensity (**E**) or when it tends to be stretched beyond some maximum amount of deformation energy.

coffee cups in Fig. 4-4A have the same surface topology as a doughnut). For example, the brain surface is highly convoluted and difficult to extract from an MRI scan. However, its topology resembles that of a sphere enough that this surface can be extracted with a spherical deformable template.

The idea of deformable models leads us to still another strategy to model the surface structures of a biomedical object. Up to this point, we have considered *discontinuous* surface structures consisting of a set of simplexes, such as line segments or triangles (see Fig. 2-11). Using the concept of a spline function (Fig. 2-13), it is also possible to define deformable models as *continuous* surfaces and to adapt them to surface structures in two- or three-dimensional biomedical image data (Meinguet, 1984; Lounsberg et al., 1992). Furthermore, with a physics-based approach, growing surfaces can be used as deformable models (Chan and Vese, 2002).

Deformable models are useful tools to segment standard data sets when density-based edge extraction methods do not work (Harders et al., 2002), and various extensions of the original concept have been proposed to adapt not only the shape but also the topology of deformable models to image properties (McInerney and Terzopoulos, 2000).

Many image data sets, especially from paleontology, are individualized and contain structures that cannot be identified readily with standard filters or templates. In these situations, manual image data segmentation methods are helpful. By classifying image pixels using the computer mouse or graphic pen as an eraser, pencil, paint bucket, electronic knife, or chisel, an object may be segmented manually. Interactive segmentation uses the concept of image layers. Rather than modifying an original image, virtual overlay sheets are used because they can store additional pixel-based information. This information can be used to classify image regions (e.g., according to tissue specificity) during subsequent processing steps (see Fig. 6-3A).

In summary, because object reconstruction from image data can be an ill-posed problem (i.e., critical information is incomplete), experience is required to provide additional insights for recovering object structures. As we have seen, many paths lead to Rome, that is, from two-dimensional or three-dimensional image data to the free-form representation of a biomedical object (Fig. 4-24).

## 4.6 FURTHER READING

Methods and techniques of image data processing are treated in a large number of textbooks, such that the sources mentioned here represent a small and subjective choice. Gonzalez and Woods (2002), Pratt (2001), and Lim (1991) treat all aspects of digital image processing. Fourier analysis is introduced in Morrison

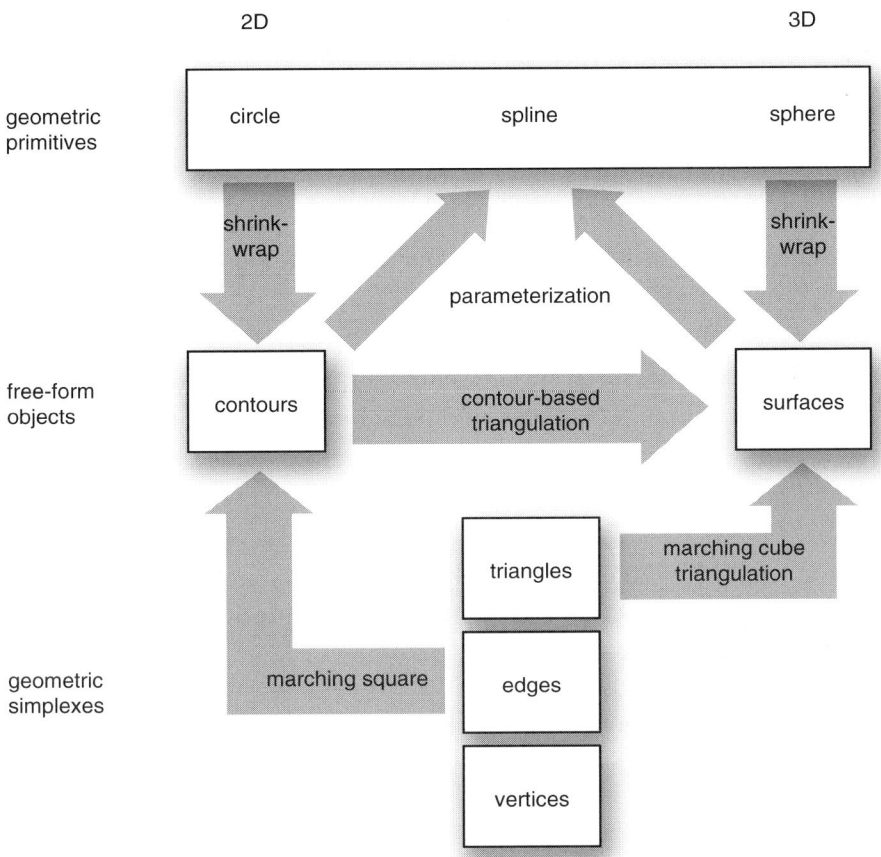

**FIGURE 4-24** A summary of the various paths of image data processing that lead from two- and three-dimensional data to reconstructed free-form objects. Methods of image analysis are used to transform vertices, edges, and triangles into two-dimensional polygonal contours and three-dimensional triangulated surfaces. These discontinuous contours and surfaces can be further transformed into continuous spline-based contours and surfaces. With methods of image synthesis, geometric primitives (discontinuous or continuous) are shrink-wrapped to recover objects from image data.

(1994) and Gray and Goodman (1995). Soille (2002) provides an important textbook on morphologic operators, and Singh et al. (1988) and Metaxas (1997) treat the recovery of biomedical objects from images using deformable models. McInerney and Terzopoulos (1996), and Jain et al. (1998) give valuable reviews on this increasingly important topic. Aspects of computational geometry are covered by de Berg et al. (2002).

*Biomedical imaging* comprises all aspects of image data from acquisition over processing to visualization and interaction, such that there is some overlap between references cited in Chapter 3, here, and in Chapter 5. A suite of papers presents an overview of the first decade of three-dimensional medical imaging (Hemmy et al., 1994; Zonneveld, 1994; Zonneveld and Fukuta, 1994). Although we now live toward the end of the second decade, these papers give a valuable overview of concepts, applications, and research topics that are still central to biomedical imaging. A recent review and interesting outlook is given by Sinha et al. (2002). Further reviews are by Pham et al. (2000), Robb (1999), and ter Haar Romeny (1998). The annual number of publications in the field of biomedical image processing and visualization is literally growing exponentially, and its importance in a clinical environment can be gauged by simply enumerating biomedical imaging journals.

One of the great challenges of medical image data processing is the geometric characterization of the brain surface, notably from MRI data. Difficulties arise from various sides, among others the relatively low contrast of this structure relative to surrounding tissues, the complexities of a corrugated surface structure, individual variation, and functional plasticity (Van Essen et al., 1998; Thompson et al., 2000).

While CT and MRI data acquisition becomes faster, biomedical imaging ventures into the fourth dimension—time, which adds additional complexity to issues of semiautomated and automated segmentation, for example, of the working heart (Clarysse et al., 1997; Sarti et al., 2002).

# VISUALIZATION AND INTERACTION

**5**

I turned, and perceived an old woman calling after me. "Young man!
Beware, you have lost your shadow!"
—ADELBERT VON CHAMISSO, *THE MAN WITHOUT A SHADOW* (1813)

## 5.1 VISUALIZING DATA IN TWO AND MORE DIMENSIONS

Adelbert von Chamisso, author of a *Grammar and Vocabulary of the Hawaiian Language*, naturalist on the *Rurik* during a three-year voyage around the world, and curator of the Royal Botanical Gardens in Berlin, is probably best known for writing the novel *The Man Without a Shadow*, in which a young and naive Schlemihl, dazzled by the promise of a self-replenishing purse, hands over his shadow to the devil, only to realize his mistake and attempt to recover it throughout the rest of the novel. The consequences of this uneven trade are paralleled by those of another barter: In Offenbach's opera *Tales of Hoffmann*, shadowless Schlemihl meets mirror image-depleted Hoffmann (a real-life colleague of Chamisso) in a duel . . . .

As we will see in this chapter, Schlemihl's and Hoffmann's loss of visual attributes are less relevant in scientific visualization than in romantic novels. Earlier in this book, we used the example of Zeuxis' illusionistic painting of a still life to demonstrate that scientific visualization differs from painting in its ability to render a scene interactively and thus to enhance perceptual equivalence. Virtual reality and its principal aim, photorealistic representation of a virtual scenery, are replaced by *enhanced reality* (Bowskill and Downie, 1995). Some features of physical reality (e.g., gravitational effects and specific optical effects like hard shadows and specular reflections) can be omitted, whereas other effects with no real-life optical counterpart (e.g., the strain distribution in a bone under loading conditions or the pattern of shape transformation during development of the human skull) can be added to the scenery to enhance relevant aspects of the data set.

Up to this point, we have examined various ways of acquiring three-dimensional data from physical objects by transforming them into two- and three-dimensional images and deriving geometric representations from these image data, such as free-form surface descriptions. The subsequent objective within the framework of virtual reconstruction is the creation of a virtual world in which virtual objects can be rendered visually with sufficient perceptual equivalence to permit interactive manipulation with computer tools.

Visualization, as a general concept of computer graphics, encompasses three main fields:

*Data visualization.* The visualization of empirical data of any class in the form of graphs such as *xy*-plots, bar and pie charts, contour plots, and histograms. We deal with this type of visualization in Chapter 8, where we study various concepts of making multidimensional data visually comprehensible.

*Surface-based object visualization.* Surface-based visualization acts on geometric descriptions of three-dimensional objects. As we have seen in Chapter 4, surface structures of biomedical objects can be obtained from 2D/3D image data by triangulation and/or by approximation through discrete or continuous deformable template surfaces. For two reasons, object surfaces provide an ideal basis for fast, interactive rendering of a virtual environment. First, the representation of an object as a surface structure is more compact than the original data volume from which it was derived. Second, most of the lighting and shading effects that make objects visually discernible are generated at surfaces because light is most affected by abrupt changes in physical properties at these locations (recall the corresponding definition of a surface in Chapter 4). Accordingly, the calculation of lighting effects can be restricted to surfaces.

*Image-based object visualization.* As we have stated in Chapter 4, image visualization is the final step in image data processing. In Figure 4-3, for example, we visualized a 2D image with a windowing function whose parameters were adjusted to highlight structures of interest in the depicted cross-sectional CT slice through a patient's head. Here, we will extend image visualization methods into the third dimension. *Volume visualization* is especially useful when object boundaries within the data volume are difficult to define, that is, amorphous. For instance, when an object exhibits no sharp drop in object density, such as in a cloud, or when the object boundary itself is exceedingly complex, such as the structure of bronchial ramifications in a lung or trabecular structures in cancellous bone, volume visualization can be advantageous.

Here, we will focus on the last two categories, surface- and volume-based visualization of 3D data from physical objects.

## 5.2 INTERACTION WITH VIRTUAL WORLDS

We may discriminate between two quite different approaches of interacting with objects in a virtual world under visual control (Fig. 5-1). In the first, which we name the *walk-* or *fly-through paradigm*, users explore objects that are larger than themselves. This approach is most familiar in architectural design and the gaming industry, where the observer/player moves through virtual buildings and landscapes and is confronted with virtual persons. This approach has been adopted more recently in medicine; for example, in virtual endoscopic diagnostics, the observer steers a virtual probe through blood vessels, airways, and the digestive system.

The second approach may be termed *CAD paradigm*. In a computer-aided design (CAD) setting, users who are typically larger than the objects of interest manipulate objects with virtual instruments. As a rule, in this approach, the observer remains stationary at a virtual desktop where he or she examines, positions, and alters virtual objects within a relatively confined region of virtual space.

Virtual reconstruction predominantly employs the second of these two approaches. Whether reconstructing fossil specimens from fragments, sampling morphometric data, planning a surgical intervention, or designing prosthetic parts, the observer is manually active, but otherwise stationary, while interacting with the virtual objects.

**A**                                                                                          **B**

**FIGURE 5-1** The fly-through (**A**) and CAD (**B**) paradigms of visualization and interaction. **A**: Users move through a scenery with objects larger than themselves (the ghost of Hamlet's father floating through the castle of Helsingör). **B**: Users are stationary and manipulate objects smaller than themselves (Hamlet contemplating a skull).

## 5.3 THE GRAPHICS RENDERING PIPELINE

*Graphics rendering* is a technical term for computer-based visualization of virtual objects. The sequence of tasks performed during rendering is called the *rendering pipeline* and is similar to the sequence of tasks a photographer follows while setting up a scene and taking the picture.

1. Define a scenery where you want to place objects.
2. Select a set of objects exhibiting various physical properties, such as different forms, colors, and textures.
3. Position and orient the objects in the scenery.
4. Install one or more light sources.
5. Place and aim the camera and then shoot the picture (i.e., expose the film).
6. Create a hard copy with the desired scale and format.

Translated into the rendering pipeline, this sequence becomes:

1. Define a world coordinate system.
2. Select a set of *virtual objects* and define their optical material properties (*material definition*), as well as their surface textures (*texture binding*).
3. Apply a transformation to each object, specifying its position (*translation*) and orientation (*rotation*) relative to the world coordinate system. This step is known as the *model transform*.
4. Apply step 3 to light sources and calculate their effects on the objects (*lighting* and *shading*).
5. Place and aim the camera (or cameras, if stereovision is intended). This is the *viewing transform*.
6. Project the scenery, as seen through the virtual camera, onto the computer screen (*viewport transform*).

Let us examine exactly how each of these steps is carried out, first during surface-based rendering and then during volume rendering.

## 5.4 SETTING UP A VIRTUAL ENVIRONMENT

The first step, defining a world coordinate system, enables the user to position all objects, light sources, and cameras with respect to a common spatial (and temporal) system of reference, the world coordinates. Up to this point, the objects to be positioned in this framework only exist as abstract geometric surface descriptions. To make them visible, we need to attach some material (optical) properties to object surfaces, which determine how incident light rays interact with the surfaces. Additionally, we need to define geometric and optical properties of the light sources themselves. And finally, to give the virtual objects a quasi-realistic visual appearance, we need to compute the propagation of a large number of individual light rays through the scenery, in accordance with the laws of optics.

### 5.4.1 Object Materials, Lighting, and Shading

Figure 5-2 shows how an incident light ray interacts with an object surface. As we know from laser scanners (Fig. 3-22), one portion of the light ray is reflected, that is, it bounces off the surface as a mirror image of the incident ray, while another part is scattered as a consequence of the microscopic roughness of the surface. These phenomena are named *specular and diffuse reflection*, respectively. As a third type of interaction, a light ray may be refracted at the object surface and deviated from its original path at some angle. During its course through the object, light is absorbed according to Lambert and Beer's law (see Box 3-2). Furthermore, object color patterns occur because the wavelengths of incident light are absorbed at different proportions. And finally, objects might not only receive light, but also emit it, that is, "glow" at different wavelengths.

With all of these parameters to consider, rendering is a computationally intensive task (see Box 5-1). The principal aim here, however, is to visualize objects in

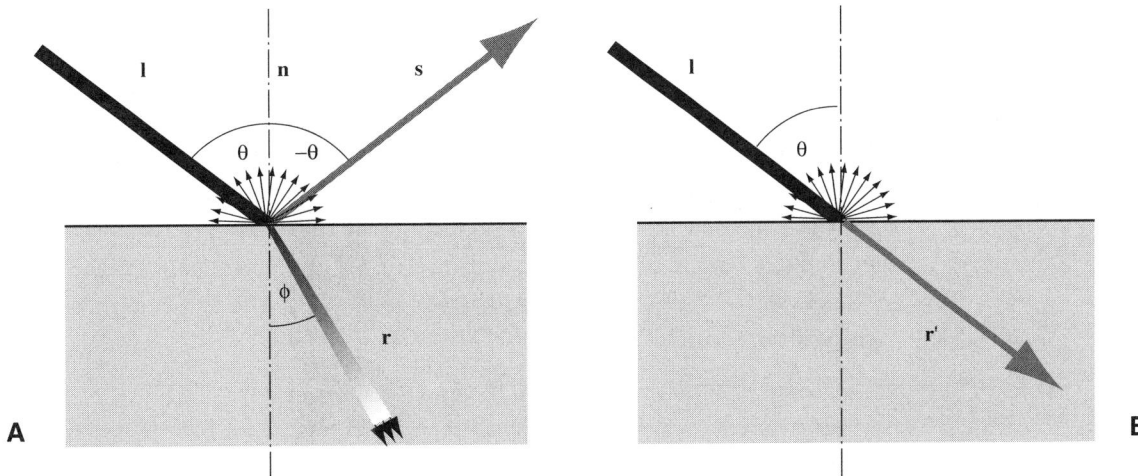

**FIGURE 5-2** Interaction between an object and an incident light ray. **A**: The major physical parameters specifying how an incident light ray **l** interacts with an object surface. Light ray **l** hits the surface at an angle $\theta$ relative to the surface normal **n**. One part (**s**) undergoes specular reflection at an angle of $-\theta$, another is diffusely reflected around the point of incidence, following Lambert's law (see Fig. 5-4A). Following Snell's law, the refracted portion **r** travels through the medium at an angle of $\phi$. Note that $\phi$ varies with wavelength, such that **r** is divided into a fan. Furthermore, **r** is absorbed according to Lambert and Beer's law (see Chapter 3). Object color results from differential wavelength absorption. **B**: The lighting model used in computer graphics for fast rendering abstracts from physical reality; in a first approximation, only diffuse reflection and transparency without refraction are considered.

## BOX 5-1

### PHOTOREALISM AND RENDERING EFFICIENCY

Rendering photorealistic images is a computationally intensive task. It is necessary to evaluate the effects of multiple light sources on each of the objects in the scene and to consider how light rays being modified by one object interact with another object. On each surface, a single "parent" light ray may split into various "child" rays through reflection, scattering/diffraction, or refraction, any of which themselves generate "grandchild" rays on neighboring objects, and so on, such that an exponentially increasing number of path ramifications must be followed before rays reach the virtual camera optics.

Fortunately, things can be simplified by a fundamental law of optics that states that the direction of light rays can be reversed without affecting the properties of an optical system. *Ray tracing* is a rendering method that utilizes this law (Fig. 5-3). Rays are emitted from the camera and sent through the scenery. These primary rays are reflected and refracted at object surfaces until some degree of ramification into secondary, tertiary, and so on, rays, is attained. Although ray tracing methods may speed up the rendering of photorealistic images, they operate more slowly than real-life illumination, during which, at any given moment in time, billions of light rays are propagated, scattered, reflected, refracted, superimposed, and diffracted and simultaneously interact with each other. In contrast, a computer graphics processor can trace only one ray at a time.

Another means of speeding up computation is to simplify object geometries by attaching *textures* to surfaces. Textures are comparable conceptually to photographic images that are pasted onto, or wrapped around, object surfaces. For free-form objects, textures are useful to render surface details and properties that are beyond the resolution and scope of the data acquisition device. For example, CT scans of a fossil do not resolve the roughness caused by erosion of a bone surface, whereas photography reveals these effects through textural or color differences.

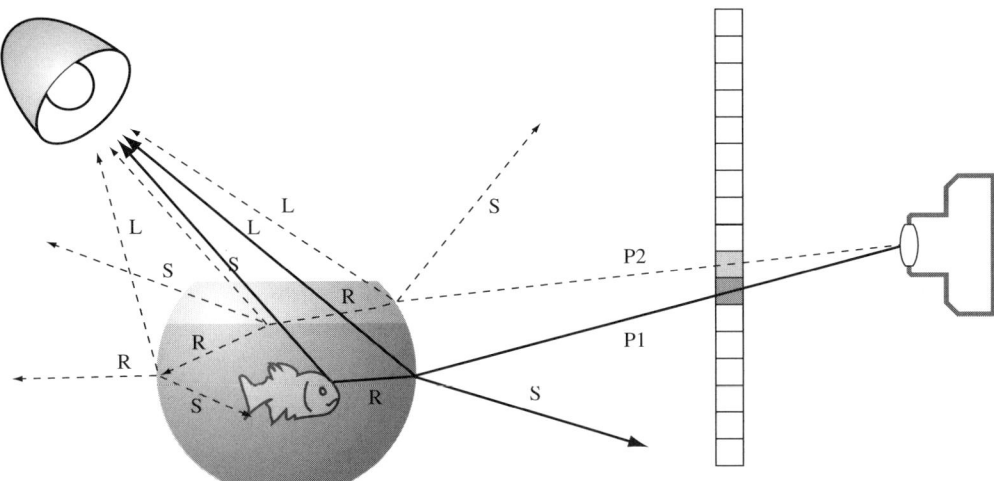

**FIGURE 5-3**    Ray tracing. Two primary rays (P1 and P2) are sent out along lines of sight (i.e., through pixels of the computer screen) and tracked along their ramifications. Ramification occurs through specular reflection (S) and refraction (R) at object surfaces, and by following paths toward light sources (L). The brightness and hue of each screen pixel are evaluated by accumulation of all lighting effects along the tree structure. Typically, ramification is stopped at some level, because computation time increases exponentially.

real time, such that they can be manipulated interactively in virtual space. To achieve real-time rendering, it is necessary to seek a compromise among physical accuracy, computational efficiency, and a visually plausible result of the lighting model. A *lighting model* represents the specifications for light interactions with material and geometric properties of a given object surface. Typically, we restrict such calculations to *local models*, that is, first-order interactions between light sources and each object. This means that we address light coming directly from light sources, whereas second-order interactions, such as mirroring effects caused by indirect light rays, shadows cast by objects occluding the light sources, and refraction, are not considered. Omitting these interactions during virtual reconstruction also creates a practical benefit, as it is easier to work without the distorting effects of refraction or the concealing effects of shadows.

Nevertheless, light traveling through a scene as a consequence of higher-order interactions has a relevant contribution to the overall lighting conditions. This contribution can be modeled by a constant term of nondirectional and omnipresent *ambient light*. The role of ambient light is best illustrated by the scattering effect of the atmosphere, which produces an even distribution of illumination; the absence of this effect, for example in lunar landscapes, creates a striking contrast between lit and shadowed areas.

Standard computer graphics lighting models permit the definition of individual material properties for ambient, diffuse, and specular components of the light reflected at a given virtual surface. Each of these components is specified in terms of RGBA units (see Chapter 2), where the RGB triple determines the color (red, green, and blue components) and A determines the transparency of the object surface.

To complete the lighting model, the influence of object geometry on each of these reflective properties must be specified. This step, called *shading*, requires calculation of how the intensity of reflected light changes as a function of the angle of incidence with the surface and, later down the rendering pipeline, how it changes relative to the position of the observer.

Because ambient light reflection is independent of the object geometry and the position of the observer, it can be added as a constant illumination term to each surface. To calculate diffuse reflection, *Lambert's law* is applied, which states that the intensity of the light reflected from a matte surface drops off with the cosine of the angle of incidence, independent of the observer's position (Fig. 5-4A).[1]

Specular reflection creates highlights on glossy object surfaces. The size and brilliance of these highlights are measured by a surface property called *shininess*, whereas their position on the surface depend on the observer's viewing direction (Fig. 5-4B).

Once material properties for virtual objects in the scenery have been defined, light sources are positioned and operated. Directional sources illuminate the scene from a large distance outside the scenery, thus emitting (virtually) parallel light rays. Sun rays, for example, have a well-defined direction, even though the sun itself has no position within the scenery. In contrast, positional light sources, such as point lights (e.g., candles) and spotlights do have a position within the scene and emit spherical and cone-shaped light volumes, respectively. Ambient light, which is defined without position and direction within the scenery, is also modeled.

The effects of a directional light source are easier to calculate, and thus more quickly rendered, than point and spot light effects. Directional light models, in fact, provide sufficient perceptual equivalence for most practical purposes, particularly when the user is concerned with real-time handling of virtual objects, because one tends to illuminate objects under examination evenly in real-life situations. Combined effects of ambient, diffuse, and specular material components and various lighting conditions are shown in Figure 5-5.

How can laws of reflection be applied to visualize free-form objects that have discrete, triangulated surfaces? Basically, for each triangle, the direction perpendicular to its surface (the *surface normal*) is determined, and shading and reflection is calculated for each flat triangle surface according to methods shown in

---

[1] Not to be confounded with Lambert and Beer's law (Box 3-2).

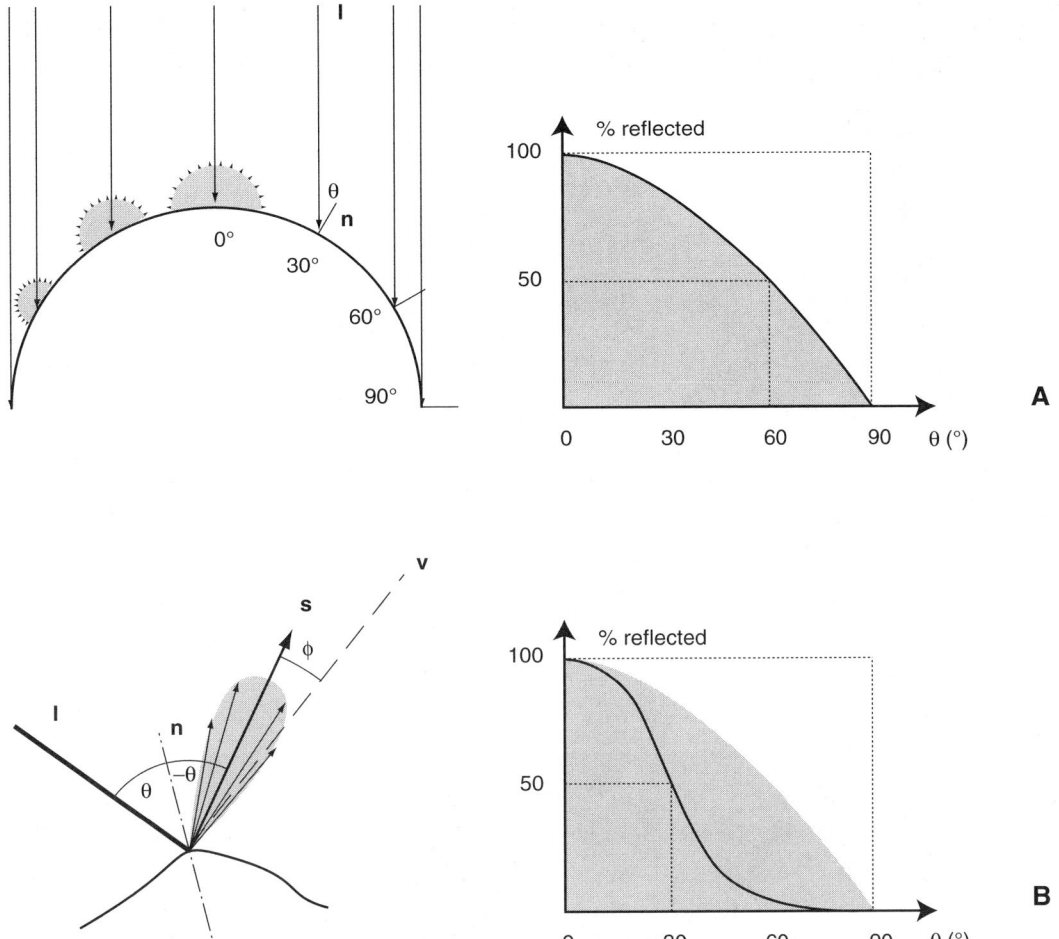

**FIGURE 5-4** Diffuse and specular reflection. **A**: Lambert's law of diffuse reflection states that the total intensity of light reflected off a perfectly matte surface is proportional to the cosine of the angle of incidence $\theta$ of the light ray **l** relative to the local surface normal **n**. **B**: During specular reflection, ray **l** with incidence angle $\theta$ bounces off as ray **s** at an angle of $-\theta$ relative to the surface normal **n**. The intensity of specular reflection is largest along **s** and drops off toward the periphery. The drop-off effect is modeled typically as being proportional to a power function (called shininess) of the cosine of the angle $\varphi$ between **s** and the viewing direction **v** of the observer.

Figure 5-4. This procedure, called *flat shading*, yields good results when the size of the triangles is small compared to overall object size (compare Fig. 5-5F and I) or in technical parts with large flat surfaces and sharp edges. For free-form objects, graphical interpolation between adjacent flat surfaces creates a visual appearance that more closely resembles the actual surface of the original object. This is typically accomplished via a method called *Gouraud shading*, where surface normals are calculated for each vertex of the surface and then interpolated along triangle edges and over triangle surfaces (Fig. 5-5B–E,G,H).

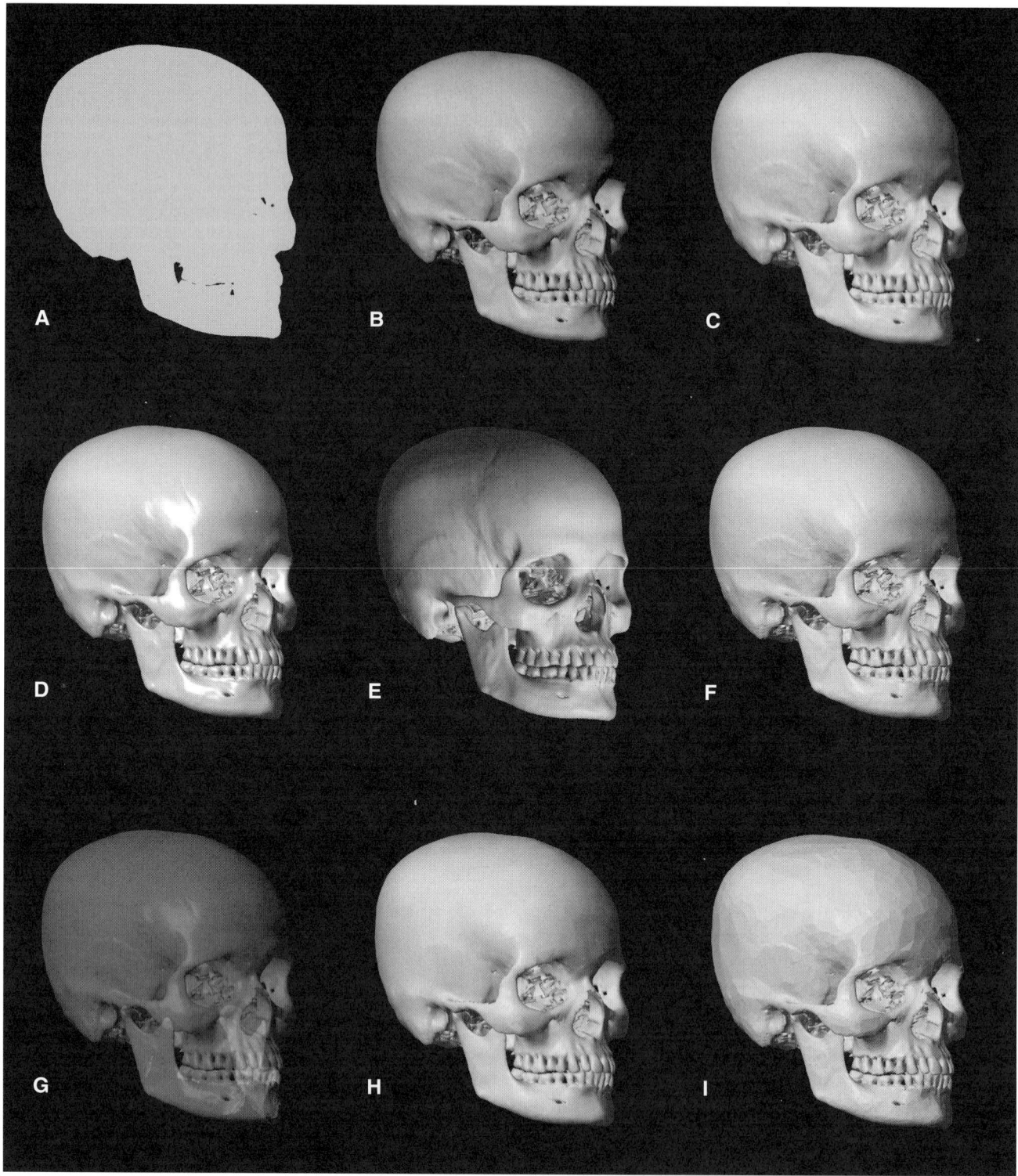

**FIGURE 5-5**   Effects of modeling ambient, diffuse, and specular reflection properties during visualization of a human cranium. **A**: Ambient light only. **B**: Directional light. **C**: Ambient and directional. **D**: Ambient, directional, and specular. **E**: Changing the light position. **B–E**: Gouraud shading. **F**: flat shading of the full data set (529,326 triangles, 2.52 triangles per mm²). **G**: transparency (alpha, Gouraud). **H, I**: Gouraud and flat shading of the reduced data set (198,904 triangles, 0.95 triangles per mm²).

## 5.4.2 Setting Up the Camera

The virtual camera defines which region of the scenery will actually be visualized on the computer screen. In real-world photography, the field of view in camera optics can be imagined as a cone of a given angular diameter (e.g., narrow in telescopes, wide in fish-eye optics). Because of the rectangular shape of the film, however, the viewing cone is trimmed to a pyramid-shaped body that extends, in the third dimension, from the front lens toward infinity (Fig. 5-6A).

In computer graphics, virtual physics is used to model this setting. Although lenses and a focal plane are not required, limits to the viewing volume must be imposed in all three dimensions of space. The resulting box-shaped viewing volume is called a *viewing frustum*, where frustum denotes a cupped quadrilateral pyramid (Fig. 5-6). As we will see in Chapter 6, imposing boundaries to the visual field (so-called *clipping planes*) has great practical significance because this permits the visual dissection of an object at defined planes, as well as visual navigation within the interior of an object through the virtual elimination of parts outside the viewing volume.

The shape of the viewing frustum is of great importance to the ultimate visual appearance of the scene on the two-dimensional computer screen. We may discern between two basic geometries, *perspective* and *orthographic* view. Perspective view can be traced to Renaissance times, where the human subject (in modern terms, the user) was established as the principal point of visual reference. User-centered perspective views dominate everyday life, especially through their application in photography. The main geometric feature of perspective view is the visual convergence of parallel lines and the resulting reduction in size of more distant objects relative to more close objects.

Interestingly, however, we often do not consciously perceive the convergence of parallel lines but intuitively transform convergence into depth or height information. For example, persons at a distance and persons nearby are perceived at similar sizes, although their images on the retina have different absolute dimensions. Computer-based visualization can exploit these perceptual facts in multiple ways. In a technical setting, such as computer-aided design, where virtual objects are constructed according to metric specifications, or during virtual reconstruction, where objects must be positioned with high precision relative to each other, it is more practical to represent the shape of objects without perspective distortion and the size of objects independent of the actual distance from the observer. *Orthographic view* is useful for this purpose because the observer is thought to be (infinitely) far away from the scene, such that parallel lines appear parallel and objects maintain their size independent of relative distances (Fig. 5-7).

As soon as the viewing frustum is defined and the camera is positioned and aimed in space, a virtual picture can be taken. The screen now assumes the

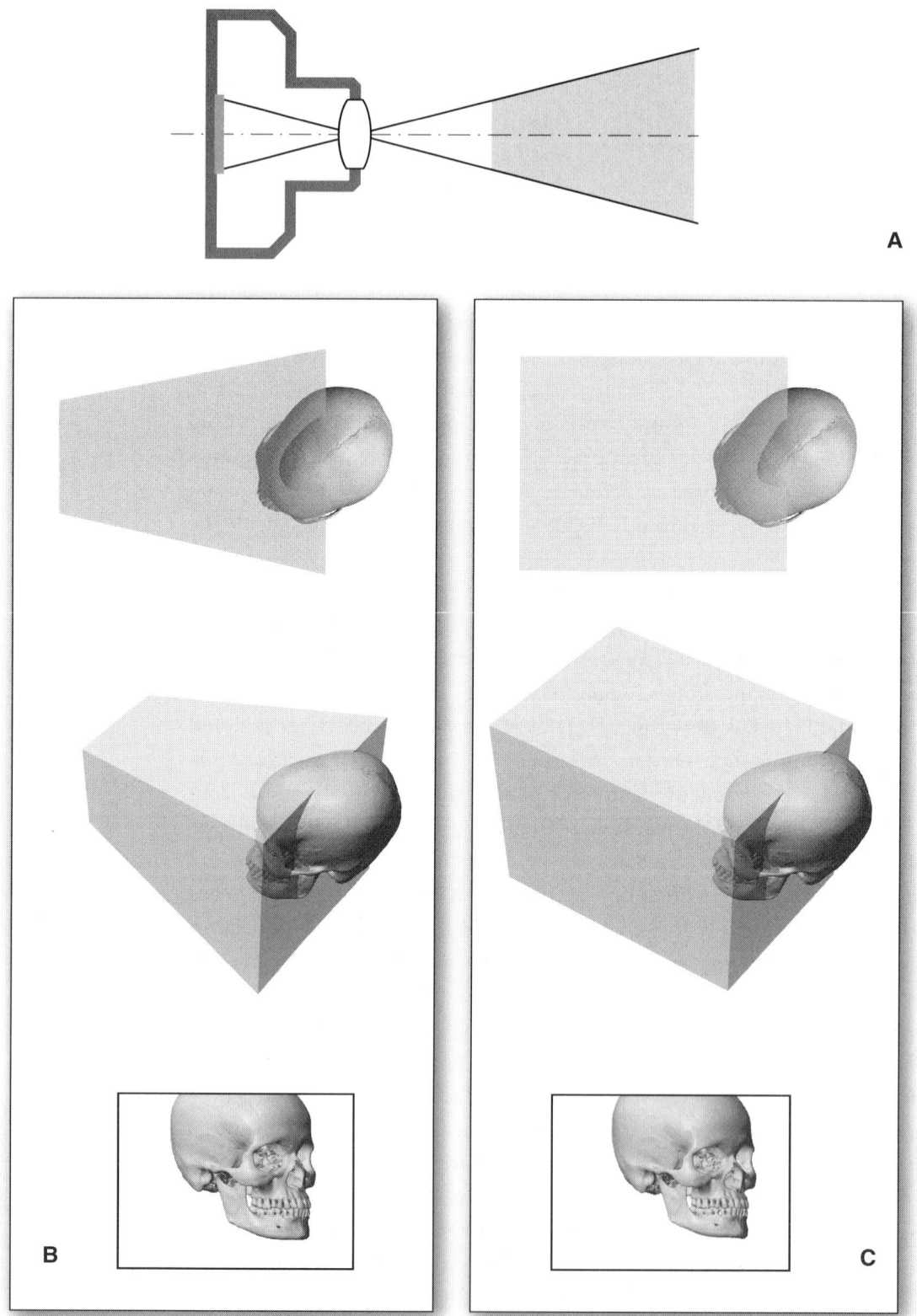

**FIGURE 5-6** The viewing frustum. **A**: The camera requires a lens to produce an inverted image plane (mapped onto film or a CCD) from object space. **B, C**: During computer visualization, the object space is projected onto the front plane of the viewing frustum (**B**: perspective, **C**: orthographic projection). Note that object regions outside the frustum are clipped.

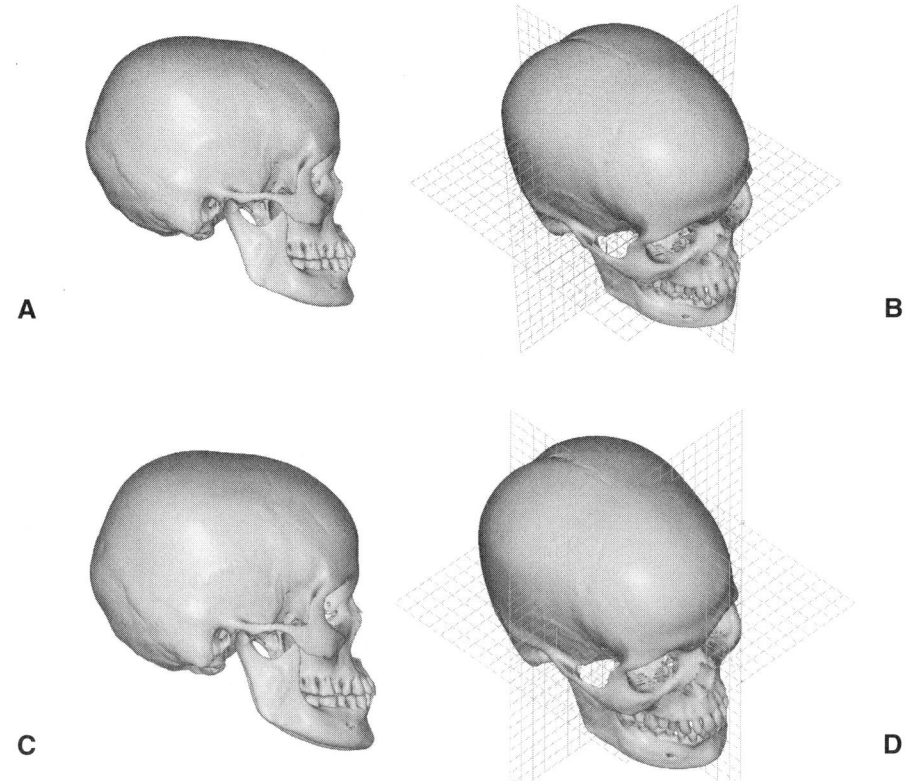

A

B

C

D

**FIGURE 5-7** Perspective (**A, B**) and orthographic (**C, D**) projections of a human skull. Note that slight asymmetry between the left and right mandibular branches is only visible in orthographic view, whereas, in perspective view, the front side of the mandible obscures the rear side.

function of photographic film in a camera, as the entire viewing volume is projected onto it. When we project 3D scenery onto a 2D screen, each screen pixel stands for one line of sight along a specific spatial direction; all objects appearing under this direction are superimposed onto the same pixel (Fig. 5-8). In real life it is evident that, along a given viewing direction, object surfaces closest to the observer obscure surfaces at a farther distance. In computer graphics, the explicit ordering of surfaces along sight lines and the geometric evaluation of their mutual occlusion (*hidden surface removal*) are computationally cumbersome tasks.

An elegant solution to this problem invokes the model shown in Figure 5-8 and the concept of pixel depth (see Fig. 2-4). Rather than using actual object geometries, we focus on the smallest units of visualization, namely, the pixels of the computer screen, and consider the screen as the surface of a voxelized viewing volume rather than a viewing plane. The viewing volume can be imagined as consisting of "filled" and "empty" voxels, corresponding to regions inside and

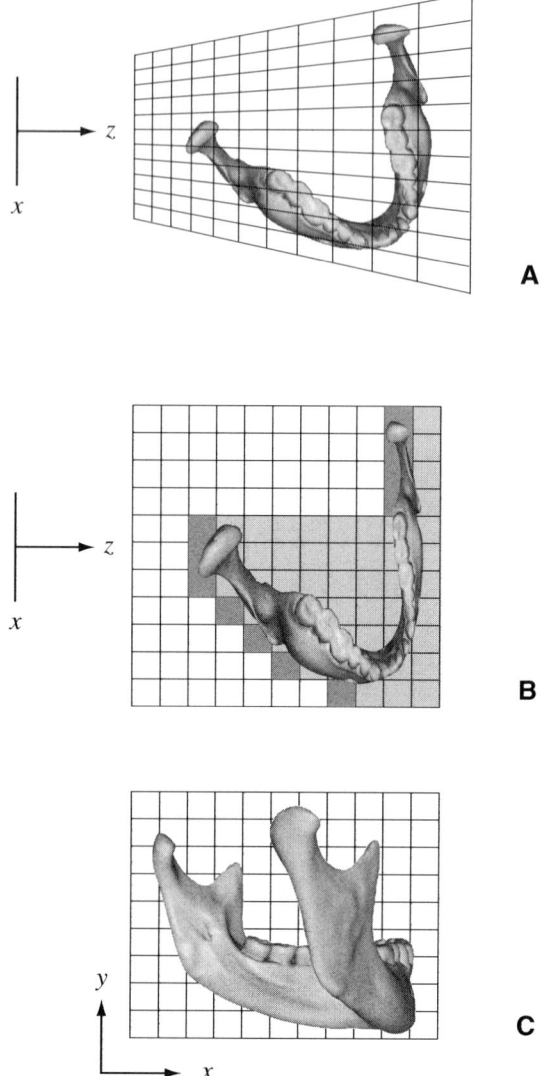

**FIGURE 5-8**  Projecting objects onto the computer screen. **A**: This mandible is situated in a perspective viewing frustum (for sake of clarity, it is shown in only two dimensions); note that the accuracy of depth information is proportional to the distance from the observer, as reflected by the increasing distance between lines parallel to $x$. **B**: The frustum is converted into a standard viewing cube; the object now appears under perspective distortion, while voxels have equal dimensions $x$, $y$, $z$. Voxel positions along the $z$-axis represent distance from the observer. Hidden surfaces can be removed by drawing only the foremost surfaces (dark gray) per voxel row along $z$ ($z$-buffer method). **C**: Screen view of the mandible with hidden surfaces removed.

outside objects. Accordingly, each voxel row starting at pixel position $(x,y)$ on the screen ($z = 0$) and extending along the $z$-axis represents one specific viewing direction, such that pixel depth ($z$) serves as a proxy for the distance of an object point from the viewer. The method is called *z-buffering*. The entire viewing frustum is then "collapsed" along the $z$-axis onto the computer screen. During this process of *projection*, only "filled" voxels closest to the observer (i.e., with the lowest $z$-value at a given $x$-$y$ position) are actually visualized.[2] Extending the concept of filled versus empty voxels to voxels with differential transparency (as represented by the A component of the RGBA color specification), $z$-buffering can also be used to visualize semitransparent objects.

### 5.4.3 Object Manipulation and Interaction

Objects are now placed in space, light sources are switched on, and the camera is ready to take pictures. How is this static setting transformed into a dynamic virtual environment in which objects can be manipulated interactively? First, we need to specify where and how objects and light sources are positioned in space. Subsequently, we must think about our own (i.e., the camera's) movement relative to the objects. These issues are considered in two dimensions, then generalized to the third dimension.

The two-dimensional setting of Figure 5-9A represents a desktop, as seen from above, on which a specimen is moved from position **P** to position **Q** in front of the user. To perform this manipulation, we select a point **r** that is outside the object as the center of rotation, which may represent the elbow joint of a virtual arm performing the movement. In computer graphics, this task is subdivided into several subtasks. First, we define the lower left corner of the table as the origin of the *world coordinate system* $(X, Y)$, which serves as a general positional reference for all objects in the scene. A specimen also has its own system of coordinates $(x, y)$, which is typically the system of the data acquisition device. At the outset, the coordinate system of the specimen is supposed to coincide with the world coordinates (the original position **O** in Fig. 5-9A).

Positioning an object on a plane or in space is a matter of transforming its system of coordinates relative to the general reference system of the world coordinates. Accordingly, the object's new position **P** on the desktop results from translation and rotation relative to the world coordinates. Obviously, there are many possible paths leading from **O** to **P**, and the user decides how rotation and translation are combined to reach **P**. Focusing on beginning and end positions, a straightforward method of transformation from **O** to **P** is given by the following instructions (Fig. 5-9B):

**O** → **O'**: Rotate the object (and its coordinate system) around the origin of the world coordinates by $\alpha$.

**O'** → **P**: Translate the object along vector $\mathbf{v}_1$.

---

[2] Note the close similarities between $z$-buffering and surface scanning: The laser beam of the scanning device stops at the closest surface found in a given direction (see Chapter 3).

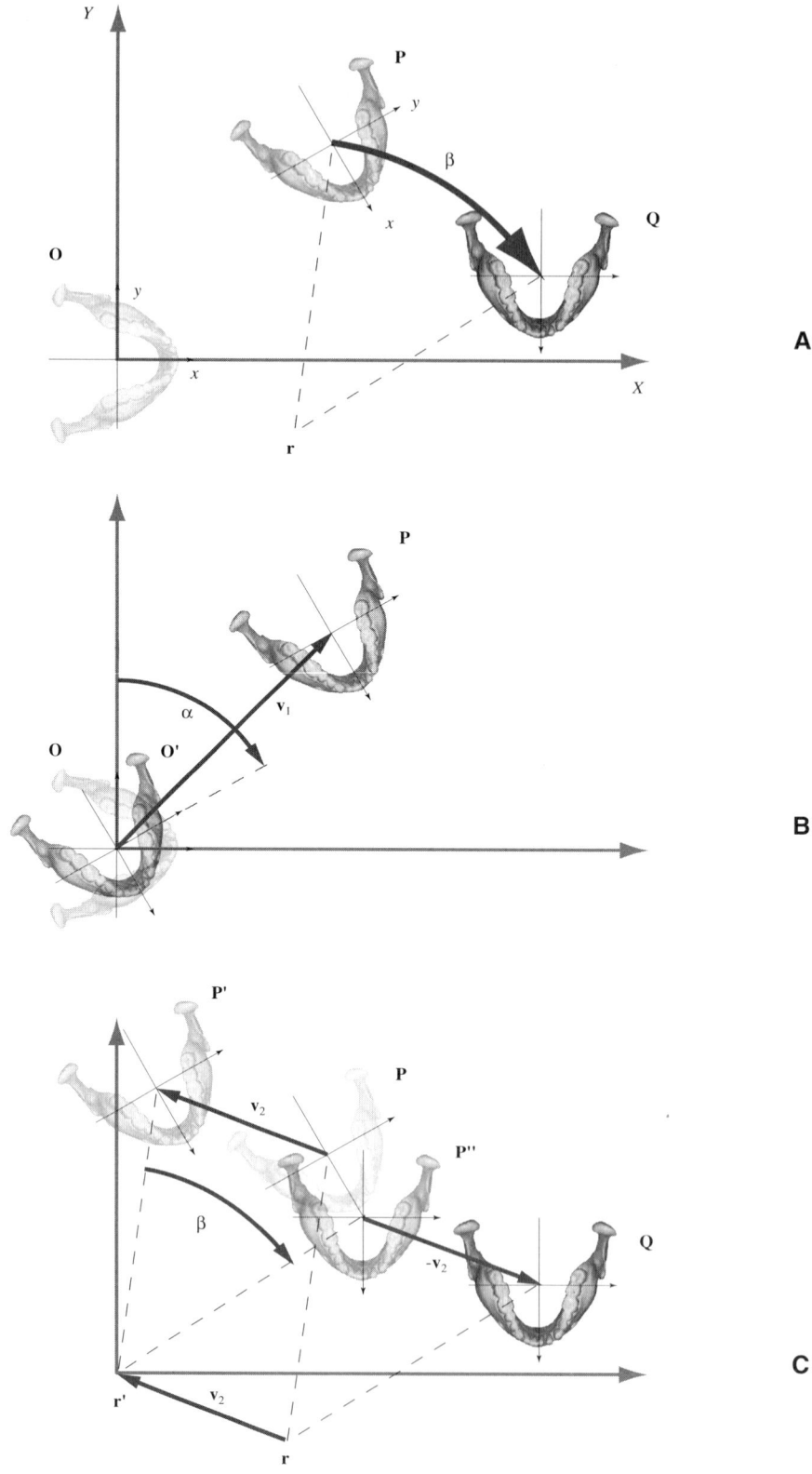

Note that this transform changes the position of the specimen, not its shape. This is called a *rigid transform*.[3] We now move the specimen from **P** to **Q**, again combining translation and rotation.

In computer graphics, the transform from position **P** to **Q** is accomplished as follows (Fig. 5-9C):

**P** → **P'**: Translate the object along vector $\mathbf{v}_2$. This procedure shifts the object such that the center of rotation **r** coincides with the origin of the world coordinates.

**P'** → **P''**: Rotate about the origin by angle β

**P''** → **Q**: Translate back along vector $-\mathbf{v}_2$. This procedure shifts **r'** back to its original position, **r**.

Accordingly, to get from **O** to **Q**, we may simply concatenate the transformations **O** → **P** and **P** → **Q**, resulting in:

$$\mathbf{O} \to \mathbf{O}' \to \mathbf{P} \to \mathbf{P}' \to \mathbf{P}'' \to \mathbf{Q}$$

where each arrow symbolizes a rigid transform.

Figure 5-9 demonstrates that objects in two dimensions have three degrees of freedom of movement—two translational (along axes $X$ and $Y$) and one rotational (about the coordinate origin). Furthermore, we conclude from the above example that transforms can be concatenated to simulate composite changes in *position* (translational component) and *orientation* (rotational component). It is important to realize that the sequence of these operations is critical. Normally, changing the order of operations will change the resulting compound transformation.

In three dimensions, transforms are more complex. Three-dimensional objects have three translational and three rotational degrees of freedom, all of which are independent. Accordingly, more information is required to specify how objects are placed and manipulated in space (Fig. 5-10).

---

[3] The reader may suspect at this point that objects can be deformed, or conversely, object deformations can be corrected with nonrigid transforms (see Chapter 6).

---

**FIGURE 5-9** Positioning and moving objects in a plane. **A**: A mandible on a virtual desktop is located at position **P** and moved into position **Q**. At the outset (**O**), the coordinate system of the specimen $(x, y)$ coincides with the world coordinates $(X, Y)$. **B**: The position **P** is reached by a rotation (α) about the origin of the coordinate systems (**O** → **O'**) followed by a translation along $\mathbf{v}_1$ (**O'** → **P**). **C**: Moving the mandible from **P** to **Q** is performed with three transformations: a shift along $\mathbf{v}_2$ (**P** → **P'**) followed by a rotation (β) about the coordinate origin (**P'** → **P''**) followed by back-translation along $-\mathbf{v}_2$ (**P''** → **Q**).

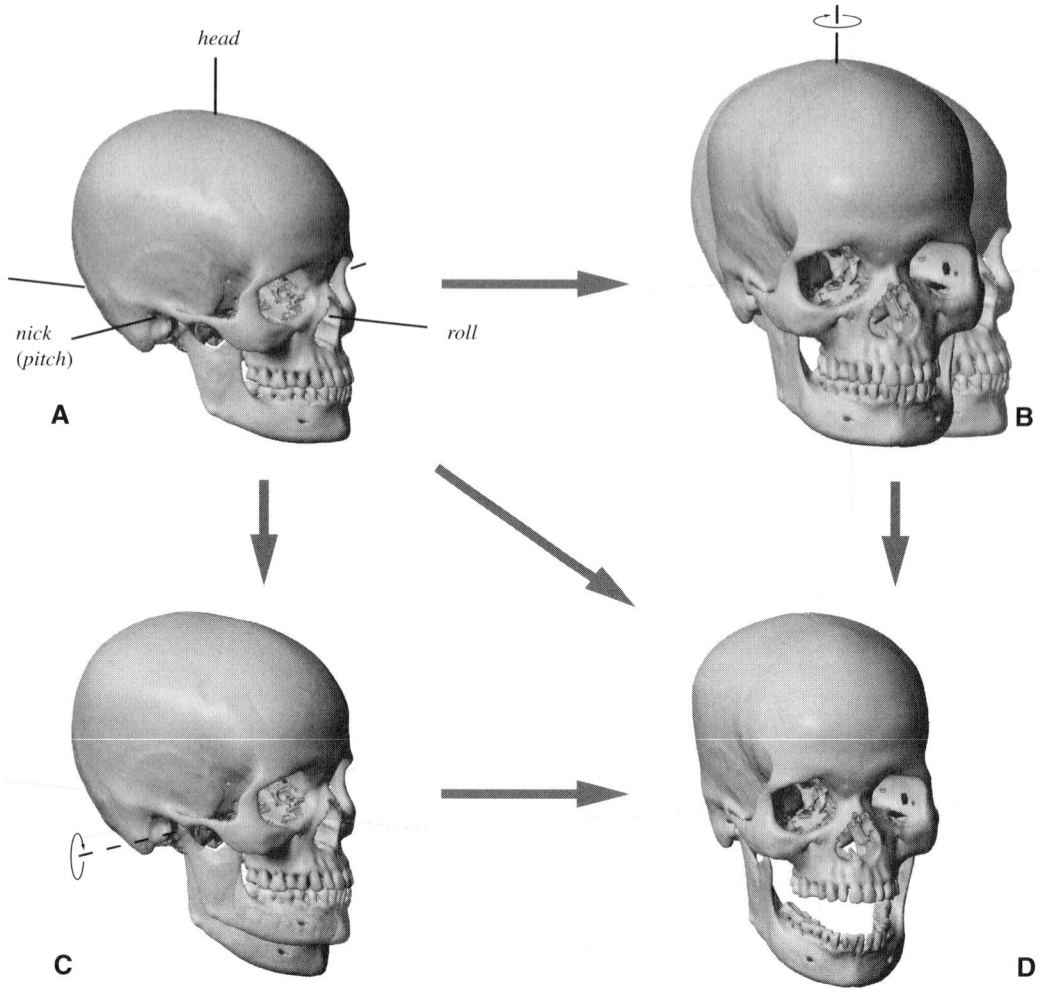

**FIGURE 5-10**   Positioning and moving objects in space. Three natural axes of a skull, permitting head, roll, and nick rotations, as well as translations along each axis (**A**). To rotate the head while opening the mouth (**A** → **D**), the mandible is attached to the skull as a subobject. **A** → **B** → **D** effects a rotation of the craniomandibular complex, followed by mandibular rotation, whereas **A** → **C** → **D** starts with mandibular rotation, followed by rotation of the craniomandibular complex. In both cases, mandibular rotation in space depends on the position and orientation of the skull.

During three-dimensional interaction, it is often convenient to think of the rotation and translation of an object in terms of its own "natural" axes, as well as relative to other objects, rather than relative to the world coordinate system. In the example of Figure 5-10, the skull's movement is best described by rotations about, and translations along, its roll, pitch, and gear (head) axes (see Appendix F.1). Furthermore, the natural range of movement is restricted for the mandible by the temporomandibular joints, which define a primary axis of rotation relative to the head.[4]

---

[4] Actual physiological trajectories of the mandible are more complex, combining rotation and translation.

How can we describe and visualize the combined movement of the head and mandible? Let us consider the specific case in Figure 5-10 to demonstrate a solution. The head is turned toward the observer, and the mandible is lowered relative to the head. As the movement of the mandible depends on the movement of the head, it is sensible to specify positional change of the former relative to the latter, then to specify the movement of the head with the attached mandible relative to the body (not shown here), and finally to specify the movement of the body relative to the world coordinates. During computer visualization, this procedure can be generalized and expressed as a hierarchical tree of transforms (Fig. 5-11).

For each object, therefore, a chain of transformations can be created, beginning with its own transform relative to the object to which it is attached, and followed down to the roots ultimately until the viewport transform is applied to the entire scene. In mathematical terms, each transform can be characterized by a $4 \times 4$ matrix, and concatenated transforms correspond to the multiplication of all matrices down the hierarchy. A more detailed account of these methods is given in Appendix D, as well as on the Web Companion, where the Transformation Applet permits experimenting with matrix-based object manipulation.

Now that the position and orientation of objects in a virtual environment can be specified and changed, we must consider how to accomplish actual interactive

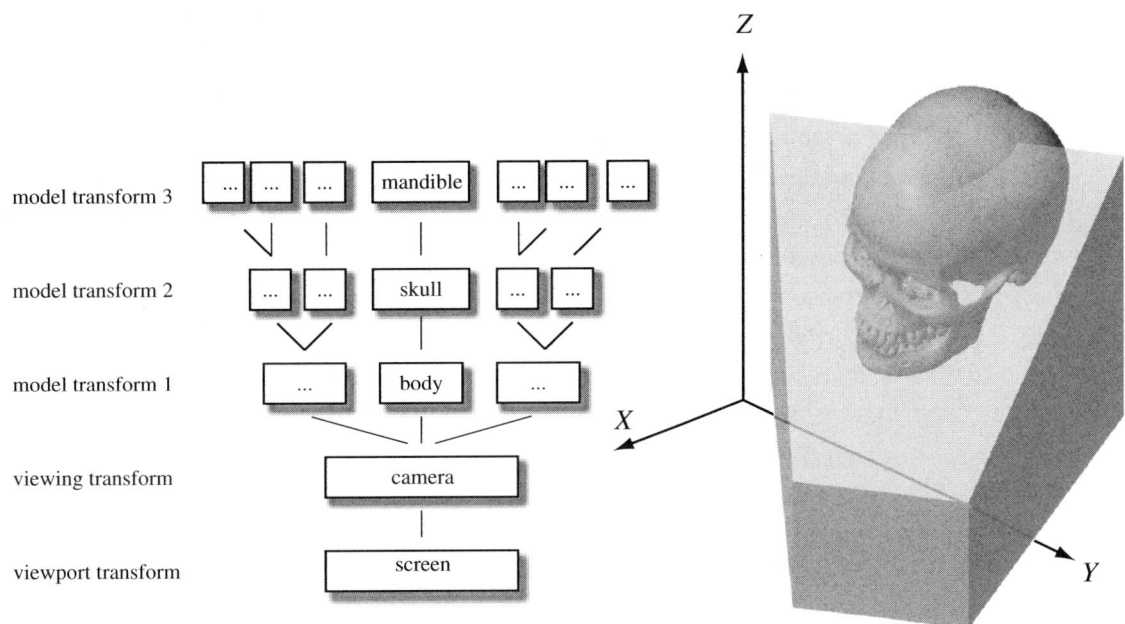

**FIGURE 5-11** The transformation tree. Applying transformations to visualize objects in virtual space (world coordinates *X, Y, Z*) follows a hierarchical scheme. At the root, the viewport transform renders an image on the computer screen. The viewing transform projects the scene onto the near plane of the viewing frustum, and subsequent transforms position object hierarchies in space.

movement of a virtual object on a computer. To do so, two basic requirements must be met (see Chapter 1):

- We need an input device to perform the required manipulation.
- We need sensory—typically visual—feedback to monitor it.

The standard input device in computer graphics applications is the mouse. A mouse yields two-dimensional input, but because the virtual setting is three-dimensional, it is more appropriate to use a three-dimensional (3D) input device, for example, a trackball or a cyberglove. As shown in Box 5-2, however, the additional degree of freedom of movement provided by a 3D input devices generates such intricate handling problems that, for most purposes, the use of a mouse is generally more precise and more practical.

How can 2D mouse manipulations be translated into 3D object manipulations? This is mainly a question of *user interface* design, which specifies the way in which communication among the user, the input device, and the computer is established. One possibility is to let the user explicitly specify this manipulation, for instance, by telling the computer exact numerical values for rotation and translation by typing the respective values into an input field or by moving a value slider. A more intuitive approach is to use the trackball metaphor, which permits the execution of rotations and translations about and along user-specified axes. These procedures are described in Box 5-2.

Visual feedback in a virtual environment is generated by continually rendering the scene while modifying it through user input. As in a movie, the perception of movement in computer graphics is produced through the steady supply of a sequence of static images. However, whereas animations are based on precalculated images, user interaction requires that images be calculated and rendered on the fly. To perceive movement, visual updates of the scene must be supplied in real time, or near-real time, that is, at frame rates of at least 12 images per second (below this rate, object manipulations are no longer perceived satisfactorily as movements).

To summarize, interactive visualization presupposes the conversion of input device data into object transforms, followed by rapid rendering of the altered scene. The iterative routine for interactive visualization is as follows:

- Capture the input device data.
- Convert them into object transforms.
- Update the object positions accordingly.
- Render the scene and display it.

Let us conclude this section on interaction with some consideration of the underlying graphics hardware and software. Most modern graphics workstations

## BOX 5-2

### USING TWO-DIMENSIONAL INPUT TO MANIPULATE OBJECTS IN A THREE-DIMENSIONAL VIRTUAL ENVIRONMENT

A computer mouse provides the computer with data about its position ($x$, $y$ coordinates) on the screen at time $t$. Why has the mouse been so popular historically, and why was it not replaced by three-dimensional tracking devices? First, the simple $xy$ setting permits a variety of interaction paradigms, because direction, speed, and acceleration of mouse movements can easily be evaluated from successive positions and translated into various forms of data manipulation within the virtual environment. Second, some seemingly trivial features of the mouse have contained great practical value. The mouse can be lifted off its pad and repositioned at a new physical location without influencing its actual virtual location. Thus the user can leave temporarily the two-dimensional world of the screen coordinate system and interrupt the data stream until reentering and creating a new frame of reference. Furthermore, when the user releases the mouse, it remains stationary, thanks to friction and gravitational forces. Three-dimensional input devices, such as 3D digitizers (see Chapter 4), data gloves, or other tracking devices do not offer these abilities. A system that tracks coordinates in three dimensions does not permit the possibility of jumping out into the fourth dimension, nor does it allow the device to be left floating in space, which prevents the user from freeing his or her hands and performing other manipulations.

A trackball, which can be imagined as a 2.5-dimensional input device (Fig. 5-12), is one possible solution to these problems. Strain and torque elicited by the user during manual turning, pushing, and pulling of the stationary ball are translated into rotational and/or translational commands. Trackballs offer intuitive navigation capabilities in a virtual environment under the fly-through paradigm but are less suitable for precise manipulation of virtual objects in space under the CAD paradigm because force exerted by the hand rather than hand movement is translated into a virtual movement. Nevertheless, the two-dimensional mouse signal can be converted into intuitive object manipulations in three dimensions via a graphical implementation of the trackball paradigm (Fig. 5-13).

Using a ball that is projected onto the screen, rotational movements about axes $x$ and $y$ are accomplished by dragging the mouse in the center of the field of view, and rotation about $z$ is performed by dragging the mouse in the periphery of the field of view. Object translation in three dimensions is separate from object rotation; typically, a keyboard command, for example, the shift or control key, distinguishes between rotational and translational input.

**FIGURE 5-12**  A trackball serves as an intuitive three-dimensional data manipulation device. The ball can be pushed, pulled, and rotated in any direction; the torque generated during these manipulations is transformed into translational and rotational object movements within the virtual scene. Actions corresponding to "3D mouse clicks" can be performed with a button inserted at the front end of the ball (not visible) while keyboard input expands trackball functionality.

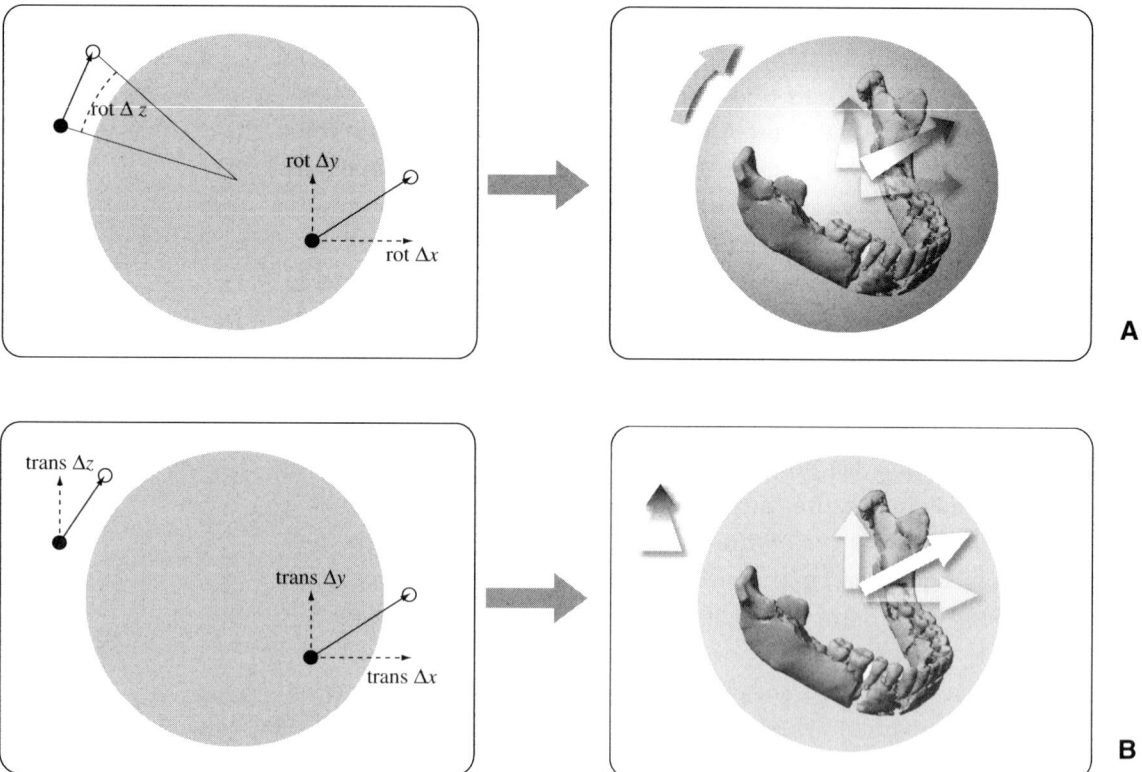

**FIGURE 5-13**  Graphical implementation of the trackball paradigm for rotational (**A**) and translational (**B**) movements. Left graphs show mouse movements; right graphs show corresponding three-dimensional movements. Mouse movements originating inside a given region of interest (gray) are interpreted as rotations about (**A**) or translations along (**B**) axes $x$ and $y$ ($\Delta x$ and $\Delta y$). Mouse movements originating outside the region of interest are interpreted as rotation about (**A**) or translation along (**B**) an axis directed toward the observer ($\Delta z$). Rotation and translation can be discerned by using different mouse buttons or a keyboard command.

and PCs use special-purpose hardware to perform these operations in real time. The manner in which brand-specific graphics hardware carries out commands, such as "rotate about $x$" or "translate along $y$," is no longer of direct concern to the programmer or user. Transforming standard graphics commands into hardware-specific instructions is performed by so-called *graphics libraries*. At present, the *OpenGL standard* (`http://www.opengl.org`) provides a suitable lingua franca for hardware-independent implementation of powerful graphics applications.

Virtual scenes are rendered pixel by pixel. During interactive visualization, however, it is necessary to provide complete virtual snapshots on a screen-by-screen basis, similar to the frames of a movie. This is facilitated through a method called *double buffer* drawing. Graphics computers have two memory areas, *buffers*, that correspond in size to the screen dimensions. The *front buffer* is used to copy a rendered scene onto the screen. While this scene at a given time $t$ is displayed, the next scene (at time $t + 1$) is rendered pixel by pixel into the invisible *back buffer*. After completion, the two buffers are swapped and the whole procedure is repeated, resulting in a continuous update of screen images. The rate at which images are swapped is determined by the rendering speed of the CPU and the refresh rate of the monitor (around 60–90 hertz). Typically, rendering rates are slower than refresh rates.

## 5.5 VOLUME RENDERING

Up to this point, we have dealt with objects positioned in a virtual scenery that was rendered with surface-based visualization techniques. During volume rendering, the task list for visualization that was presented at the beginning of this chapter is shortened considerably because many steps can be collapsed into one and the same logical operation. Volume data derived from CT or MRI can be imagined as a scenery filled by a single, immobile object—the voxel matrix. Voxels, in turn, are subobjects with a standard size and shape, whose positions $(x, y, z)$ relative to the world coordinate system $(X, Y, Z)$ remain fixed. Establishing voxels as stationary objects frees us from the task of performing object transforms. However, this same property restricts manipulations of subobjects within the volume.

The first task in volume visualization is to define material properties for voxels. It is straightforward to use the actual voxel values, which, in the case of CT, represent local X-ray densities. Subsequently, a light source is created and its optical interaction with the voxels is calculated. Because we are looking through a volume, differential transparency is a relevant issue. We are already familiar with this type of visualization that was first demonstrated by Röntgen's original

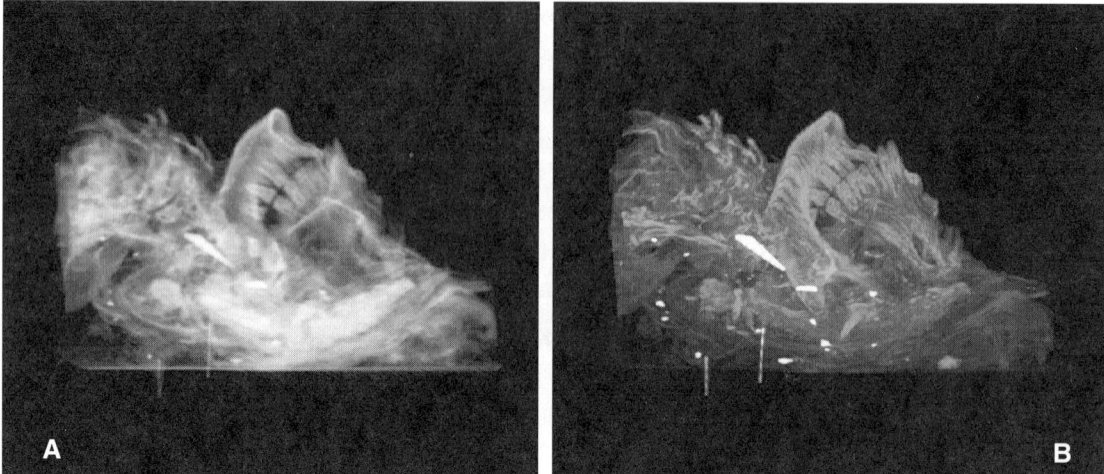

**FIGURE 5-14** Ray casting techniques are used to visualize the internal structures of an ancient bog body from northern Germany. **A:** Virtual light rays sent along the observer's viewing direction sum up local absorbances along their path through the data volume. The resulting image is similar to a classic X-ray projection image. **B:** During maximum-intensity projection (MIP) only the highest voxel intensities along each light ray are sampled. The resulting image thus enhances structures with elevated local density in the data volume, such as the preserved facial skeleton. CT data courtesy of A. Hering (University of Düsseldorf) and M. Fansa (Natural History Museum, Oldenburg).

experimental setup (Fig. 3-5) in which X rays were sent through an object and projected onto photographic film.[5]

Let us start with a basic implementation of volume visualization and consider a CT data volume of dimensions $N \times N \times N$, in which the value of each voxel represents local X-ray density. According to Lambert and Beer's law (see Box 3-2), voxels have individual attenuations. When parallel light rays are sent through one face of the $N \times N \times N$ cube, the resulting image can be evaluated at the opposite face. As stated above, parallel rays imply an observer at an infinite distance. As each pixel of the front face receives a light ray, we calculate light absorption along voxel rows until we reach the back face. This method is called *ray casting* (Fig. 5-14A).

The only optical effect that has been considered up to this point is light absorption by a semitransparent medium along the line of sight of the viewer. To get beyond the resulting radiography-like volume visualization, we may use more complex optical phenomena, such as reflection, refraction, and scattering of the incident light rays, and position light sources off the line of sight. Because each voxel can be treated as a standardized three-dimensional object, lighting and

---

[5] Note the close similarity between X ray-based imaging and volume rendering. In the first case, data are sampled from real object volumes, in the second from virtual object volumes. Likewise, tomographic projection is volume rendering that is restricted to two dimensions (Fig. 3-10).

shading techniques similar to those described for surface rendering apply. Furthermore, in the spirit of enhanced reality, quasi-physical light propagation algorithms can be implemented to provide specific insights into the data volume. An example of visualization based on one of these methods—*maximum intensity projection* (MIP)—is given in Figure 5-14B.

Volume rendering is computationally more demanding than surface rendering. The number of voxels in the visualized volume grows by the third power of the linear dimensions of the visualized image, while the number of surface elements grows by an exponent of 2. One possibility for rendering volumes interactively is to exploit a technique called *texture mapping*. As mentioned in Box 5-2, textures are two-dimensional image data structures that can be "wrapped" onto the surface of three-dimensional object structures. For volume rendering, the three-dimensional data matrix can be represented as a stack of textures attached to consecutive two-dimensional slices. Texture pixels (*texels*) are represented by RGBA values (see Fig. 2-4) that specify the color and transparency of each voxel. On most computer graphics workstations, texture mapping and rendering is performed with dedicated hardware that makes rendering a volume as a stack of textures highly efficient.

To conclude this chapter—and in view of Chapter 6 on virtual reconstruction of fragmentary fossils—we may ask ourselves which of the two techniques of visualizing a virtual environment—surface rendering and volume rendering—is more appropriate for biomedical applications. The answer depends on the actual scientific problem and the interaction paradigm (CAD vs. fly-through) that is most appropriate to solve that problem. During virtual fossil reconstruction and virtual surgery, objects must be assembled, disassembled, modified, and newly created, which requires real-time interaction with, editing of, and visualization of surface-based object descriptions, as exemplified by the CAD paradigm.

Exploration of fossil CT data and image-based medical diagnostics, on the other hand, follow the fly-through paradigm. Entire volume data sets are inspected, for example, during virtual endoscopy, without precomputing explicit object boundary structures and without modifying their spatial structure. Volume data sets typically represent solid objects, such that an enhanced reality approach is used to tune material properties (notably transparency) and lighting parameters to achieve optimum visualization conditions. A further valuable option available with volume rendering is *interactive multiplanar reformatting* (Fig. 5-15); the data volume can be cut along user-defined planes (i.e., at arbitrary angles relative to the orientation of the original CT or MRI cross sections), which greatly facilitates the combined two- and three-dimensional exploration of complex anatomic structures. Nevertheless, because diagnosis and interaction are closely related in biomedicine, there is a trend toward combining surface and volume rendering techniques in the same application.

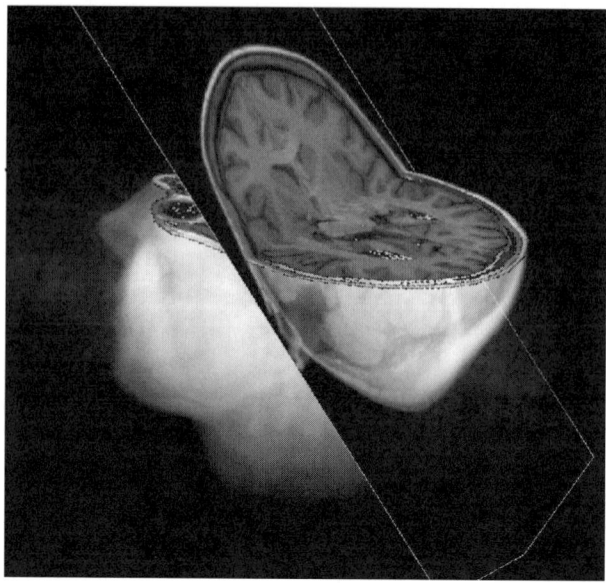

**FIGURE 5-15**   Multiplanar reformatting. This MRI data volume of a patient's head was acquired in sections parallel to the cranial midplane. The resulting data volume can be sectioned interactively along arbitrary planes.

## 5.6 FURTHER READING

Matters of computer graphics, visualization, and virtual reality are well documented in introductory and advanced textbooks. One of the most comprehensive sources for the reader simultaneously interested in principal issues, procedural detail, and mathematical background is Akenine-Möller and Haines (2002). An interesting early paper on methods of biomedical volume rendering is Ney et al. (1990). Valuable references on biomedical visualization/interaction and its applications are Lavallee et al. (2000), Sinha et al. (2002), Adams et al. (1990); Masutani et al. (1998), and Robb (1999 ). Computer graphics programmers will need Foley et al. (1995), and Glassner (1990), and readers interested in geometry and linear (matrix) algebra behind computer graphics will find comprehensive information in Farin and Hansford (1998), de Berg et al. (2002), Eberly and Schneider (2002), and Mortenson (1999). Virtual environments and virtual interaction paradigms are described by Sherman and Craig (2002), Jacko (2003) and in the Human-Computer Interaction Handbook (Jacko and Sears, 2002); see also the articles by Feiner (2002) and Fuchs and Ackerman (1999). Scientific visualization concepts and issues are treated in a valuable edited volume (Brodlie et al., 1992). Applications of the methods described in this chapter to issues of virtual reconstruction are referenced at the end of Chapter 6.

# VIRTUAL FOSSIL RECONSTRUCTION

**6**

Postmodernity is the simultaneity of the destruction of earlier values
and their reconstruction. It is renovation within ruination.
        —Jean Baudrillard (*1929), Cool Memories, ch. 4 (1987).

## 6.1 A BAROQUE PUZZLE

We have now established the full tool kit that permits the preparation and reconstruction of fossils in a virtual environment. But before we delve into the concepts and technicalities of computerized fossil reconstruction, let us once more take a look from postmodernity into antiquity.

At the beginning of the third century, Roman Emperor Septimius Severus initiated a large-scale urban management project dedicated to the establishment of a detailed ground plan documenting all architectural features of Rome. For representational purposes, a monumental version of this ground plan—known as the *Forma Urbis Romae*—was incised onto a large marble wall in a public building. During the steady decline of Rome from the world's metropolis into a medieval village, the plan deteriorated and suffered considerable damage. Most marble slabs were used as building material for new construction projects or simply milled into powder for lime production. More than a thousand years after its creation, Renaissance and Baroque scholars and artists were fascinated by the surviving 10 to 15 percent of the *Forma Urbis Romae* and initiated the first attempts to recover its original form. For example, Giovan Battista Piranesi—who made much of his fame by representing the ruins of Rome in highly evocative copper engravings (*Le Antichità Romane*, 1756)—started his own research into ancient Roman urban structure and dedicated an entire series of engravings to these picturesque but enigmatic fragments (Fig. 6-1).

Ever since, people have attempted to piece together the fragmented and incomplete puzzle of the *Forma Urbis*. A recent approach at digitizing, cataloging,

**FIGURE 6-1**  Fragments of the ancient City Map of Rome (*Forma Urbis Romae,* early third century C.E.); engraving by G. B. Piranesi, 1720–1778 (illustration from Piranesi's *Le Antichità Romane* (1756), courtesy of Alan Wofsy Fine Arts, San Francisco 94126, `www.art.books.com`).

and reconstructing the fragments with computer-assisted procedures is documented at `http://formaurbis.stanford.edu`.

The logics behind the task of reconstructing the map can be stated in one sentence: Recover absolute map coordinates (geographic latitude and longitude) and orientation (N-S direction) of each fragment and place them at their respective positions on the marble wall. During practical reconstruction, however, two major challenges have to be faced. First, many fragments lack architectural landmarks that would permit their orientation and location within a geographic frame of reference. Second, large portions of the original map are lost forever and remain "terra incognita." These problems can be tackled following a top-down approach or a bottom-up approach, or a combination of both. Adhering to the top-down approach, we make use of external sources of information to reconstruct the map. For example, the topography of the ancient city of Rome is relatively well-known from archeological excavations on the spot, from ancient documents containing detailed descriptions of the location of architectural features, as well as from other fragmentary ground plans.

The bottom-up approach does not rely on external information or prior knowledge about how Rome looked. Rather, several smaller reconstruction tasks are performed independently, until a larger picture of the original map emerges. For example, by joining neighboring fragments along shared edges local frames of reference can be established. If direct contacts between individual fragments are not available, continuity between isolated pieces may be established through recognizable patterns such as roads and other large-scale architectural features that permit interpolation between related fragments. Iterating this process leads to formation of locally growing "fragment clusters," which eventually coalesce on a larger scale.

## 6.2 PRINCIPLES OF RECONSTRUCTION

The aims of fossil reconstruction are similar to those of the reconstruction of ancient city maps. In both cases, the proximate goal is the reversal of the detrimental effects of time (diagenetic and archeological time, respectively) back to some moment along the time axis (the death of an individual and a historical key date, respectively), and the ultimate goal is the inference of morphologic/urban patterns and processes *before* that moment (Fig. 6-2; see also Fig. 1-3). In practical

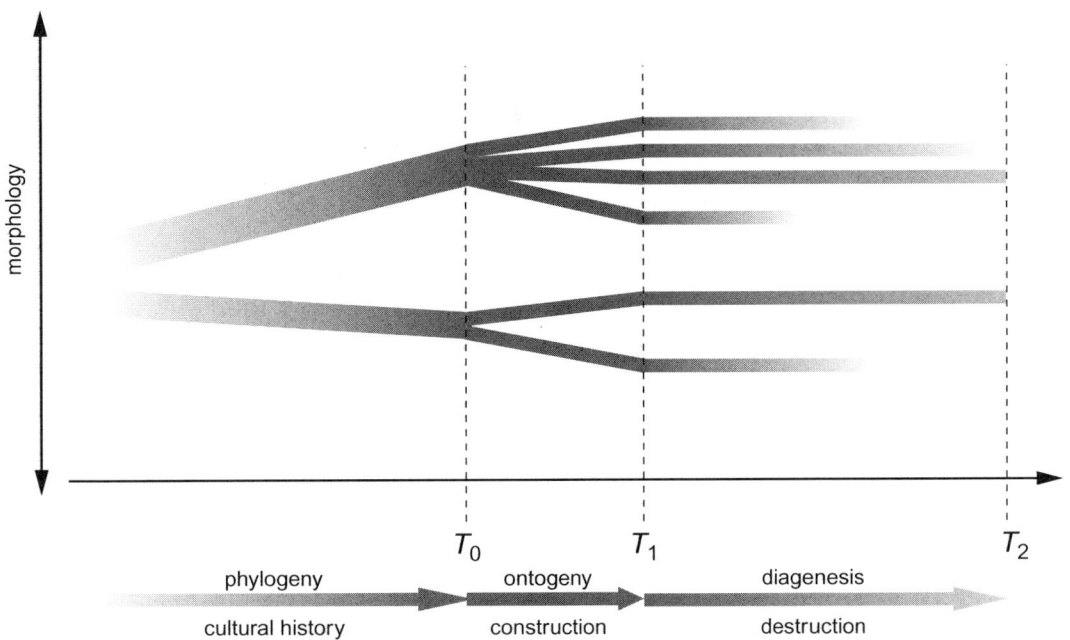

**FIGURE 6-2** The logics of reconstruction. Fossil and/or archeological remains found at time $T_2$ are reconstructed to infer their structure at time $T_1$. This information is used to infer patterns and underlying processes of change from $T_0$ to $T_1$ (during the individual's lifetime and the objects' construction/functioning, respectively) or before $T_0$ (during evolutionary time and in a wider historical context, respectively).

terms, the proximate goal in the Forma Urbis project is to reconstruct the map as it looked around 320 C.E., and the ultimate goal is to better understand the structure, development, and history of Rome as the capital of the expanding Roman empire. In paleontology, the proximate goal is to reconstruct a fossil specimen as a three-dimensional object at the time of the individual's death, and the ultimate goal is to better understand the structure, development, and/or evolutionary history of the extinct group of organisms that it represents. This chapter is about the first step, the reversal of the effects of diagenetic time, and Chapter 7 is dedicated to the second step, reconstruction of patterns of morphologic change during evolution and development.

During fossil reconstruction, similar practical problems arise as during map reconstruction, although at a higher level of complexity. First and foremost, fossils are three-dimensional objects. As a consequence, isolated fragments have six degrees of freedom of motion (three translational, three rotational; see Chapter 5), whereas the pieces of a puzzle only have three (two translational, one rotational). This increases considerably the number of ways in which fragments can be placed and oriented relative to each other. Moreover, some degree of physiological and diagenetic "matching uncertainty" must be taken into account, for example, in skeletal parts associated via anatomic joints, and in fossil fragments joined along eroded fracture lines. With an increasing numbers of fragments, matching uncertainty may lead to error propagation over the reconstructed fossil morphology as a whole. Third, the pieces of the fossil three-dimensional puzzle may be bent out of shape, notably through taphonomic processes. As an effect, reestablishment of preserved anatomic contacts between neighboring fragments might result in overall anatomic distortion and requires specific tools for correction.

Fossil reconstruction differs in yet another way from the reconstruction of the Roman city map. However fragmentary this ground plan may be, we know that it represents the urban structure of the unique city of Rome at the beginning of the third century. In organismic reconstruction, there is no unique reference structure that would tell us what an individual fossil looked like at the time of death. Obviously, the top-down approach mentioned above does not work. For example, during the reconstruction of the fragmentary Neanderthal child skull from Gibraltar (see Box 6-4), the principal aim is to reconstruct the cranial morphology of this specific individual rather than to reconstruct a "Neanderthal"-type morphology. The latter approach would represent an undue mixture of proximate and ultimate goals of reconstruction, as it implies interference with the reconstructor's ideas and preconceptions about Neanderthal morphology. Let us refer to this problem as the *reconstruction bias*.

How do we obtain unbiased reconstructions? During fossil reconstruction, we must primarily rely on intrinsic morphologic clues (the bottom-up approach) and use extrinsic information parsimoniously (the top-down approach). How we weigh intrinsic versus extrinsic information can be stated in *four principles of reconstructive parsimony*.

- The *first principle* consists in relying, as far as possible, on the morphologic information that can be recovered from the fossilized remains and their taphonomic setting. Virtual methods permit nondestructive retrieval of concealed morphologic information and thus play a key role in the exhaustive use of anatomic clues during three-dimensional reconstruction.

- The *second principle* advocates the utilization of generalized anatomic information, if no direct anatomic clues are available from the fossil specimen. This information must represent the inferred shared ancestral (symplesiomorphic) state of a morphologic character rather than a derived (autapomorphic) state that is characteristic for a specific taxon. For example, during reconstruction of a skull (see Fig. 6-6), bilateral symmetry serves as a robust *null hypothesis* guiding the reconstruction, because it can be assumed that symmetry represents a symplesiomorphic condition, irrespective of the taxonomic status attributed to the specimen. On the other hand, making use of characteristic Neanderthal features known from more complete specimens during the reconstruction of an incomplete specimen with inferred Neanderthal affinities would introduce a bias toward a Neanderthal-specific derived morphology.

- The *third principle* pertains to morphologic terra incognita. It may be necessary to interpolate or extrapolate lost information that cannot be recovered from the preserved fossil morphology. For example, we may wish to reconstruct the lost occipital region of a Neanderthal skull (see Fig. 6-6) to obtain an estimate of its endocranial capacity. Typically, a probabilistic approach, combined with geometric techniques, is used to reconstruct a likely average form and provide a range of variation of this estimate. However, we must take account of the fact that application of geometric principles alone— without attributing biological significance—may be misleading. In sum, we must check whether the information used to interpolate missing structures has a taxonomic bias.

- The *fourth principle* regards consistency. The anatomic hypotheses used during reconstruction must be free of contradictions, and multiple independent reconstructions of the same fossil (based on distinct hypotheses and following different paths) must result in similar outcomes.

## 6.3 PHYSICAL AND VIRTUAL RECONSTRUCTION

Before we consider how these principles are implemented in practical applications, it is necessary to compare manual reconstruction techniques with virtual reconstruction. Classic physical fossil reconstruction is a process during which fragments are isolated from rock matrix and then assembled and completed to yield a best approximation of the in vivo state of skeletal morphology. This procedure seems relatively straightforward, but it is fraught with difficulties and limitations that have consequences for the interpretation of the inferred morphology. This is of special significance in paleoanthropology, where inferences drawn from fossil morphology are traditionally far-reaching with respect to the scarcity of the available material (Tattersall, 1999).

Virtual fossil reconstruction, on the other hand, combines computational and anatomic considerations throughout the whole process and attempts to establish criteria of reliability for the reconstructed morphology (Zollikofer et al., 1998). Box 6-1 shows that the concepts of virtual reconstruction are similar to those of experimental science; every single reconstructive step is planned according to predefined criteria that assume the function of null hypotheses (for example, symplesiomorphy is a central null hypothesis). Because every consecutive step is hypothesis-based, the reconstruction as a whole is reproducible and can be modified specifically, if null hypotheses at any stage must be rejected.

## 6.4 PREPARING AND RESTORING FOSSILS ON THE COMPUTER SCREEN

Physical preparation is a central but tedious sine qua non preceding the scientific analysis of a fossil specimen. We argued above that the study of a real fossil specimen can never be fully substituted by examination of its virtual counterpart, but that virtual methods augment classic analysis. Likewise, virtual fossil preparation cannot fully replace physical preparation, but—as we will show here—virtual procedures complement and enhance physical procedures.

During physical preparation, it is almost impossible to avoid breakage of minor parts, which must then be glued back to the original, and it is often impractical to free matrix-filled cavities, whose morphology thus remains unexplored. The obvious advantage of CT-based electronic preparation lies in its noninvasive nature and the possibility of reaching physically inaccessible regions. This potential was already exploited in the earliest CT-based analyses of the structure of the endocranial and internal otic cavities of fossils (Conroy and Vannier, 1984; Wind, 1984; Conroy and Vannier, 1985; Wind and Zonneveld, 1985).

How can computer tools be utilized to free a fossil from matrix? X-ray densities of fossil bone and surrounding matrix may differ markedly, such that automated data segmentation procedures—for example thresholding (see Chapter 4)—can be applied to extract the fossil. In many instances, however, density-based extraction yields unsatisfactory results, such that user interaction is required. At

## PHYSICAL AND VIRTUAL FOSSIL RECONSTRUCTION: A COMPARISON OF POTENTIALS AND LIMITATIONS

| Physical Reconstruction | Virtual Reconstruction |
| --- | --- |

### *Preparation and Replication*

When fossil fragments are freed from matrix, they are exposed to a considerable risk of damage, notably when the surrounding material exhibits a higher density than the fossilized bone. Moreover, convoluted structures, cavities, and internal structures are not accessible to physical preparation.

With CT image data, the matrix adhering to specimens can be removed by semiautomated segmentation procedures (see Chapter 4). Likewise, earlier reconstructions can be disintegrated into original fragments without physical interference. Visualization tools give access to both external and internal anatomic structures, thus expanding the array of morphologic features yielding potentially relevant anatomic clues for reconstruction.

Similar limitations are imposed on the production of casts from fossil fragments.

Rapid prototyping technology (see Chapter 7) can be used to produce hard copies of virtual fossils of arbitrary topological complexity.

### *Reconstruction*

Many fossil specimens are recovered as large numbers of isolated pieces. During recomposition, the positioning and fixation of these fragments relative to one another poses a number of practical difficulties. First, because of gravitation, fragments must be stabilized and fixed to each other with struts, plaster, and glue. Second, any changes effected in the position of one single fragment entail positional changes in neighboring fragments, potentially leading to "reconstructive error propagation" over the specimen as a whole. Error

Electronically isolated fragments can be positioned and oriented in anatomic space without the need for stabilization. To define the position of isolated fragments in anatomic space, the following criteria and procedures are applied: Fragments with preserved anatomic correspondences are assembled on the computer screen. Isolated fragments are placed according to the hypothesized ancestral anatomic condition, using both external and internal anatomic clues and reestablishing basic anatomic properties such as bilateral symmetry. Anatomic

BOX *6-1*

**PHYSICAL AND VIRTUAL FOSSIL RECONSTRUCTION: A COMPARISON OF POTENTIALS AND LIMITATIONS (*Continued*)**

| Physical Reconstruction | Virtual Reconstruction |
|---|---|
| propagation can only be controlled and minimized by repeated accommodation of the relative positions of isolated fragments. However, once a fossil has been assembled, subsequent attempts at correction and improvement of the present state are likely to degrade the original material. As a consequence, many fossils remain in an unsatisfactory state of reconstruction, and others have been seriously damaged during repeated reconstructive efforts (Ponce de León and Zollikofer, 1999). | inconsistencies can be eliminated by iterating these steps until a consistent solution is found. Because each reconstructive step is executed according to explicit hypotheses and predefined quantitative criteria, virtual reconstruction is a reproducible procedure that approaches the standards of experimental science. |

*Completion of Missing Parts*

| | |
|---|---|
| Although the production of mirror-imaged counterparts from existing fragments is a laborious process, the reconstructive inference of parts lacking on both sides of the symmetry plane depends entirely on the manual and artistic skills of the preparator. | Missing regions are generally completed by using mirror-symmetric autologous parts. Regions that lack anatomic evidence from both sides can be completed by application of 3D morphing procedures (e.g., thin plate spline morphing, see Chapter 8), using data from similar fossil or modern skeletal material. The morphing proceeds by matching homologous landmarks on preserved adjacent structures. This procedure also evaluates morphometric data (e.g., endocranial volumes) from incomplete structures. |

## BOX 6-1

### PHYSICAL AND VIRTUAL FOSSIL RECONSTRUCTION: A COMPARISON OF POTENTIALS AND LIMITATIONS (*Continued*)

| Physical Reconstruction | Virtual Reconstruction |
|---|---|

#### Reliability and Reproducibility

Fossil reconstruction is based on a number of implicit or explicit hypotheses about the range of likely original morphologies. These hypotheses should direct the execution of each reconstructive step without presupposing the outcome. Because of the essentially manual character of the procedure of reconstruction, it is difficult to plan, quantify, and protocol a sequence of reconstructive steps.

Reconstructive steps can be iterated until a consistent state of reconstruction is attained. To account for alternative anatomic assumptions as well as for fundamental reconstructive uncertainties, it is necessary to perform multiple independent reconstructions: Virtual fossil reconstruction results in a range of possible morphologies rather than one single solution.

The reliability of a particular reconstruction can be cross-checked by performing parallel reconstructions on electronically fragmented complete specimens, whose "virtual remains" are recomposed on the computer screen following exactly the same procedures as for the fossil counterpart. Comparison of the pseudo-reconstruction with the complete original yields an estimate of the reconstructive error.

#### Correction of Deformation

Deformation is a major hindrance during reconstruction, as its effects cannot be corrected on the original material. Generally, deformation results in misorientation of fragments relative to each other and skewed reconstructions.

In certain cases, the diagenetic causes of deformation can be inferred and specific computational procedures can be devised to reverse deformation (see Box 6-5).

### BOX 6-1

**PHYSICAL AND VIRTUAL FOSSIL RECONSTRUCTION: A COMPARISON OF POTENTIALS AND LIMITATIONS (*Continued*)**

| Physical Reconstruction | Virtual Reconstruction |
|---|---|
| *Interpolation and Extrapolation* | |
| Missing regions are typically completed manually with various filling materials. The criteria used to define these regions are often. not specified explicitly. | Information not present in the original fossil material can be completed by interpolation, using reference data sets of similar objects or applying purely geometric criteria. In any case, it should always be kept in mind that a technically feasible method does not necessarily yield biologically sensible results. |

this point, segmentation tools such as deformable models and morphologic operators come into play (see Chapter 4). In view of the enormous diversity of fossil material regarding object complexity, preservation, state of mineralization, and other properties, it is not possible to establish general rules on how data segmentation is optimally performed. Here, we consider two examples (Boxes 6-2 and 6-3); the Further Reading section at the end of this chapter contains references to a wider spectrum of possible approaches.

## 6.5 RECONSTRUCTING FOSSIL MORPHOLOGIES

### 6.5.1 Recovering Implicit Anatomic Information

How can we reconstruct a specimen's three-dimensional morphology from a collection of isolated virtual fossil fragments? We already are acquainted with the major challenges of this task: Each fragment has six degrees of freedom of movement, which results in a considerable number of possible ways in which parts can be assembled. The problem of having too many options of positioning and orienting fragments in space can be paralleled with the task of solving a system of equations with more unknowns than equations. Such systems may become solvable if we impose boundary conditions, for example, if we state how subsets of variables are related to each other, and try to find a stable solution through iter-

## SEE YOU LATER, ALLIGATOR

While prospecting Paleocene marine sediments in the Pyrenean mountains at an altitude of 1800 m, Spanish paleontologists Angel Galobart and Marcel Costa found a partially weathered fossil alligator skull embedded upside down in a carbonate sediment. After recovery of the 50-kg matrix block containing the fossil and backpack transportation to a base camp, the specimen was brought to the Crusafont Institute of Paleontology in Sabadell, Spain, for further analysis (Fig. 6-3). Initial attempts at physical and/or acid bath preparation proved difficult for two reasons: The matrix was more compact than the fossil itself, and the latter was partially eroded and preserved only in the form of cavities.

This is where virtual preparation came into play. The matrix block was analyzed with an industrial CT scanner (X-ray energy of medical scanners was insufficient to penetrate the block; see Fig. 3-15), resulting in a stack of 700 contiguous cross-sectional images, each one millimeter thick. Bone/matrix separation with automated thresholding did not yield satisfactory results because of the small density differences between fossil remains and surrounding sediment, and because of strong local density fluctuations within the fossilized bone. The data set was thus segmented manually to extract the skeletal structures and separate upper and lower jaws, which were preserved in anatomic articulation. The resulting three-dimensional virtual fossil is currently being described and classified.

ation. By analogy, during fossil reconstruction, it is often necessary to impose anatomic constraints that reduce the degrees of freedom of movement of fragments relative to each other. Subsequently, the process of positioning fragments can be iterated until a stable solution is found.

Adhering to the four rules of parsimony, this procedure works as follows. First, a frame of reference—an anatomic space—is needed within which we can place and orient fragments. Next, positional and orientational information contained in the fossil fragments (the abovementioned intrinsic information) is used to place them relative to each other and relative to the anatomic system of coordinates. Finally, still unresolved positional/orientational questions can be tackled

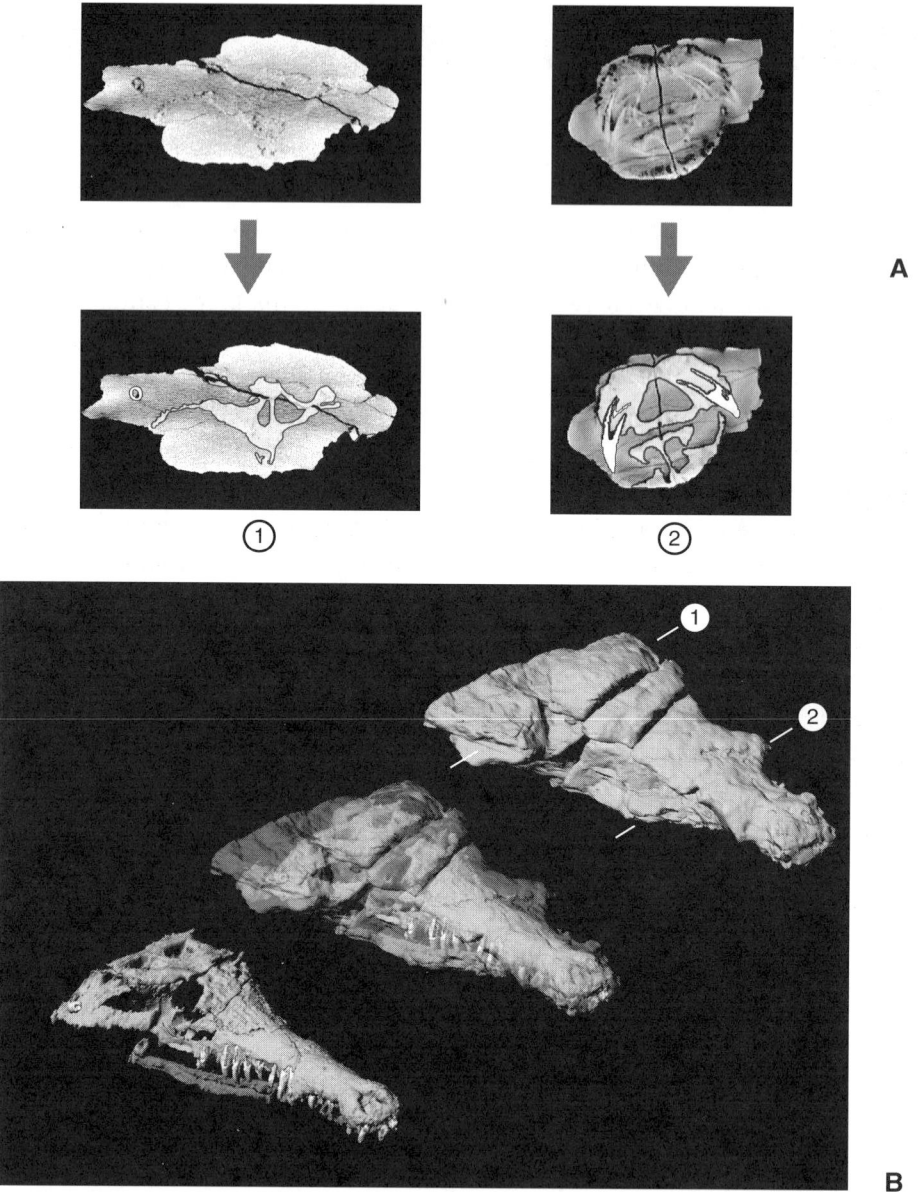

**FIGURE 6-3**   Recovery of a Paleocene alligator. **A**: Two CT slices before (top) and after (bottom) segmentation. **B**: The virtual specimen is freed from matrix.

tentatively by imposing extrinsic boundary conditions, that is, using symplesiomorphic anatomic constraints.

## 6.5.2 Combining Computer Graphics and Anatomy: The Globe Paradigm

Let us recall at this point the concepts of visualization and object manipulation used to create a virtual environment that facilitates the execution of these tasks

BOX 6-3

## FOSSIL NEO-DECONSTRUCTIVISM

The Le Moustier 1 Neanderthal fossil has a complex history (Hoffmann, 1997). In 1908, Swiss archeologist Otto Hauser excavated a virtually complete skeleton of an adolescent Neanderthal at the site of Le Moustier (Dordogne, France) (Hauser, 1909). To finance further excavations, he sold the fossil and associated stone artifacts to the Museum für Völkerkunde in Berlin, where the remains were on display until the 1940s. The skull was bunkered together with other museum items during World War II, later transferred to the Soviet Union, and reinstated to the German Democratic Republic in the 1960s. The postcranial skeleton, which remained in the Museum, had a less favorable fate, as the building went up into flames. Nevertheless, during a postwar rescue dig in the rubble of the Museum, Heberer was able to recover fragments belonging to the Le Moustier skeleton (Heberer, 1957).

The Le Moustier's history of reconstruction is similarly tortuous (Fig. 6-4). During recovery, the brittle cranium was fragmented into a puzzle of more than 50 pieces; taphonomic deformation pervading almost all fragments caused misfit of associated regions of the skull during recomposition. As an effect, the Le Moustier skull underwent at least five physical reconstructions, during which the original fossil material was steadily deteriorated, altered, and partially lost (Ponce de León and Zollikofer, 1999).

We decided to start noninvasive virtual reconstruction from scratch (Ponce de León and Zollikofer, 1999). As a first step, the skull's CT-based virtual duplicate was disassembled electronically. Removal of the filling material revealed that many bone fragments remained isolated in anatomic space, making it necessary to revise the current reconstruction as a whole (see Box 6-5 and Figs. 6-5, 6-7B).

(see Chapter 5). The basic spatial reference in computer graphics is the system of world coordinates. This system of reference is independent of ourselves and of the virtual objects that we manipulate. CAD users such as architects are accustomed to refer themselves and their constructions to world coordinates, as architectural objects have a unique location and orientation in geographic coordinates. In contrast, as "bioscientific operators", we do not locate biological objects and

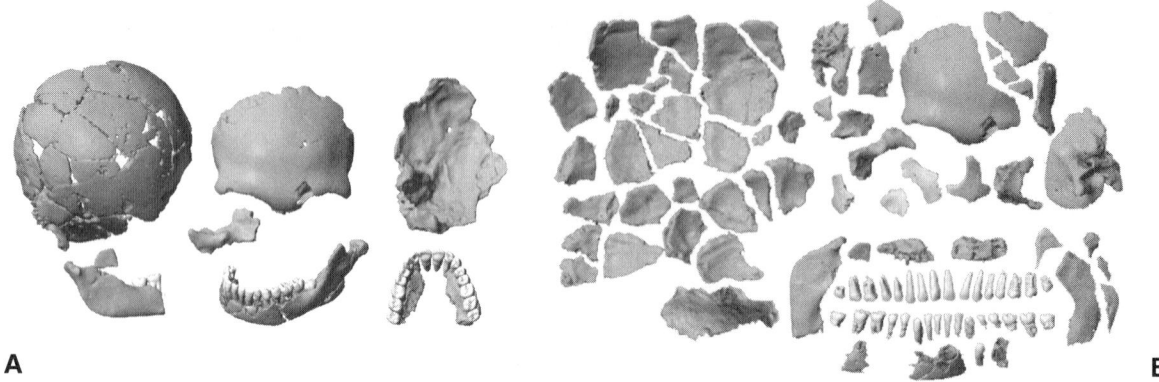

**FIGURE 6-4**  Le Moustier 1 cranial deconstruction. **A**: Present physical state of the cranial remains (each piece is composed of several smaller fragments). **B**: Decomposition into original fragments.

ourselves with respect to world coordinates. For example, we do not assemble fossil fragments with reference to the edges of the table (the world coordinates), but with reference to an imaginary system of anatomic coordinates, into which we fit the fragments. Moreover, to examine and handle virtual objects, we tend to switch continuously between the anatomic *object-centered* reference system and a *subject-centered* visual space, whose coordinate axes are defined by our actual line of sight, left/right, and top/bottom directions. During object manipulation, visually guided hand movements are carried out with reference to the subject-centered system, whereas the object-centered system is used as a reference when several objects are positioned in relation to each other. Obviously, neither reference system used by bioscientific operators relates to world coordinates.

The premises of object- and subject-centered spaces can be combined in the globe paradigm (Fig. 6-5). We sit in front of a virtual table (the world coordinates), onto which we place a globe with an inscribed system of axes (the anatomic coordinates). The globe serves as a reference structure within which we reconstruct the fossil. Because it is a virtual globe on a transparent desktop in a gravitation-free space, we can rotate and translate it arbitrarily, inspect it from all sides, and position it relative to ourselves without losing the anatomic frame of reference, and without losing our own orientation relative to the desk.

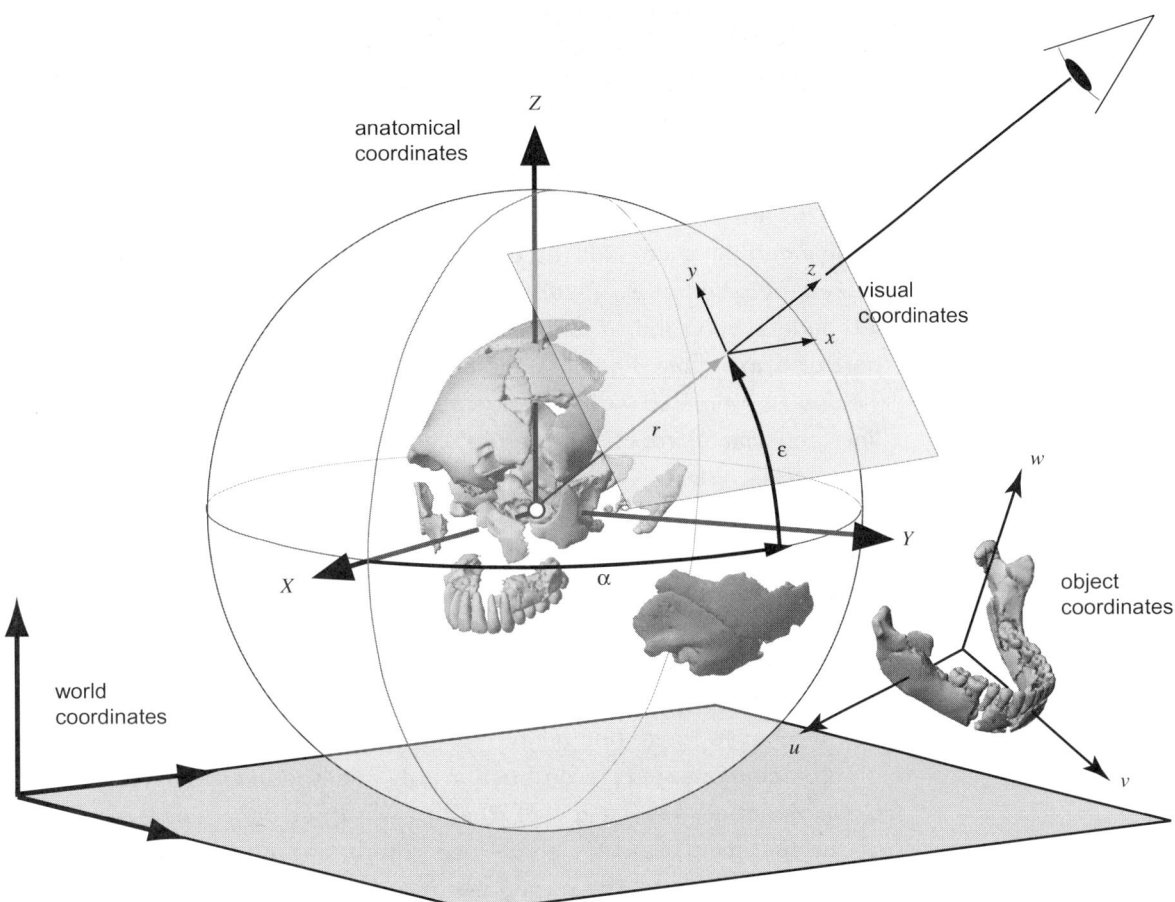

**FIGURE 6-5**  The globe paradigm of spatial interaction. Object manipulations may be effected either in anatomic coordinates (rotations and translations on axes *X,Y,Z*) or in subject-centered visual coordinates [rotations and translations on axes *x,y,z; x* and *y* correspond to the axes of the computer screen (shaded rectangle), *z* is the line of sight]. To inspect the object from different sides, the user typically uses a *gaze sphere* (azimuth α, elevation *r*, radius *r*), which establishes an intuitive spatial link between anatomic and visual coordinates. Any object (e.g., the isolated mandible) has its own system of coordinates (*u,v,w*) defining its physical dimensions. The figure shows parts of the Le Moustier 1 Neanderthal skull during assembly.

With the globe paradigm, it is possible to assemble fragments in virtual anatomic space by reestablishing anatomic contacts between neighboring parts and using morphologic clues hinting at the position and orientation of isolated fragments within the anatomic frame of reference (Zollikofer and Ponce de León, 1995). During this process, seemingly unrelated or disparate spatial anatomic information preserved in the scattered fragments is recovered and reconstituted as a three-dimensional object. A practical example is given in Box 6-4.

**BOX 6-4**

## VIRTUAL RECONSTRUCTION OF THE DEVIL'S TOWER NEANDERTHAL CHILD FROM GIBRALTAR

In 1926, archeologist Dorothy Garrod found the scattered fossil remains of a Neanderthal child skull at the limestone quarry site of "Devil's Tower," Gibraltar (Garrod et al., 1928). The excavations yielded a frontal and associated left parietal bone and, at some distance, right maxillary and temporal bones and part of a mandible. Radiographic assessment of dental developmental status yielded an estimated individual age at death of five years. With dental growth line (perikymata) counts, this figure was later corrected to three to four years (Dean et al., 1986).

Because no direct contacts are preserved between the temporal bone and the cranial vault fragments, nor between lower and upper teeth, Skinner and Sperber (1982) produced a composite lateral X-ray image of the fragments at their approximate anatomic position, thus providing the first "virtual" reconstruction of the specimen. The lack of anatomic contacts between the fragments, and apparent differences in their developmental stage, led to the hypothesis that the Devil's Tower find might represent two individuals (Tillier, 1982).

Computer-assisted reconstruction of this specimen served three aims (Zollikofer et al., 1995) (Fig. 6-6). The first was three-dimensional reconstitution of the specimen based on predefined criteria and quantitative, reproducible object manipulations. The second was investigation of internal structures, such as the unerupted dentition (whose developmental state could not be fully resolved with conventional X-ray techniques; Garrod et al., 1928) and the cavities of the inner ear. The third aim was to test the hypothesis that the five fragments represent one single rather than two individuals (Tillier, 1982).

In the following table and in Figure 6-6, each reconstructive step is described and illustrated together with the geometric and biological criteria on which it is based.

**BOX *6-4***

## VIRTUAL RECONSTRUCTION OF THE DEVIL'S TOWER NEANDERTHAL CHILD FROM GIBRALTAR (*Continued*)

| Step | Action | Geometric Criteria (constraints) | Biological Criteria and Null Hypotheses |
|------|--------|----------------------------------|------------------------------------------|
| 1 | Place mandibular fragment in anatomic space such that infradentale (the point between the left and right first incisors) is located on midsagittal plane. | Reduce degrees of freedom (DF) of mandibular translation to 0. | H0: bilateral organization of the mammalian cranium (symplesiomorphy). |
| 2 | Create mirror image of mandibular fragment; rotate original mandible, and counterrotate mirror image correspondingly until symphyseal regions match. | Reduce mandibular rotational DF from 3 to 1 (rotation about mandibular joint); this step also permits assessment of asymmetry in the preserved symphyseal region. | H0: general mirror symmetry of cranial bones (symplesiomorphy). |
| 3 | Extract preserved left deciduous molars and insert mirror-imaged counterparts into right alveoli. Reconstruct missing left ramus with a mirror image from the preserved right side. | Reconstruct missing anatomic landmarks that will be used during subsequent reconstructive steps. | Idem |
| 4 | Position and orient the right maxillary fragment: (1) Establish dental occlusion between upper and lower (reconstructed) deciduous molars; (2) rotate until maxillary midplane coincides with midsagittal plane. | Intercalation of upper and lower molar cusp patterns reduces translational DF to 0; likewise, two rotational DF [around transversal (pitch) and longitudinal (head) axes] can be reduced, whereas the inclination of the maxillary tooth row [rotation around an anteroposterior (roll) axis] must be fixed by making reference to the maxilla's anatomic midplane (definition of rotation axes see Appendix F.1). | Intrinsic anatomic correspondence; bilateral symmetry. |
| 5 | Position the right temporal bone on the mandibular condyle. | Establishment of the temporomandibular joint reduces translational DF to 0. | Anatomic correspondence. |

**BOX 6-4**

**VIRTUAL RECONSTRUCTION OF THE DEVIL'S TOWER NEANDERTHAL
CHILD FROM GIBRALTAR (Continued)**

| Step | Action | Geometric Criteria (constraints) | Biological Criteria and Null Hypotheses |
|------|--------|----------------------------------|------------------------------------------|
| 6 | Extract the internal otic structures; use semicircular canals as an compass to orient temporal bone in anatomic space. | Establishing the anatomic orientation of the semicircular canals permits reduction of the rotational DF of the temporal bone to 0. | H0: the superior and posterior semicircular canals assume 45° angles relative to the midsagittal plane (mammalian symplesiomorphy). H0: the lateral canal is inclined between ±12° relative to a horizontal plane (hominoid symplesiomorphy). H0: the plane defined by the lateral canal (i.e., the vestibular plane) is at an angle of approx. 30° relative to the alveolar plane of the maxilla (mammalian symplesiomorphy: Delattre and Fenart, 1960). |
| 7 | Mirror image positioned and oriented temporal and maxillary bones to left side. | Construction of placeholders for subsequent reconstructive steps. | H0: bilateral symmetry. |
| 8 | Position and orient the frontal bone in anatomic space. | Reduce translational DF to 0; reduce roll (anteroposterior) and head (longitudinal) rotational DF by matching the frontal bone with its mirror-imaged counterpart. One rotational degree of freedom (pitch) remains open. Mirror-matching is also used to assess intrinsic asymmetry of this bone. | H0: bilateral symmetry. |

**BOX 6-4**

**VIRTUAL RECONSTRUCTION OF THE DEVIL'S TOWER NEANDERTHAL CHILD FROM GIBRALTAR (Continued)**

| Step | Action | Geometric Criteria (constraints) | Biological Criteria and Null Hypotheses |
|---|---|---|---|
| 9 + 10 | Join left parietal bone to frontal bone; complete with mirror-imaged counterpart. | Reduce rotational and translational DF of parietal relative to frontal bone. | Original anatomic contact; bilateral symmetry. |
| 11 | Dock reconstructed cranial vault (steps 8–10) to reconstructed cranial base and face (steps 1-7). | Reduce rotational and translational DF of cranial vault relative to base. | Reconstructed anatomic contact between right temporal (original) and parietal (mirror-imaged) bones. |
| 12 | Check match between upper and lower parts of the reconstruction. | Independent reconstruction of subregions and subsequent assembly guarantees minimal error propagation over the recon struction as a whole. | Close match between the two subregions suggests that all fragments belong to the same individual. |
| 13 | Verification: perform steps 1–12 with modern skulls decomposed into fragments corresponding to those recovered from the Devil's Tower site. | Reconstruction of a simulated fossil whose original morphology is known permits assessment of repeatability of the virtual reconstruction. | Test effects of natural deviations from symmetry versus geometric symmetry of recon-struction. (H0: diff-erences between repeated reconstruc-tions are in the same range of variation as left/right differences in natural skulls). |
| 14 | Extrapolate the shape of the occipital bone. | Use thin plate spline morphing approach and reference sample of modern human/Neanderthal skulls. | Because no direct information on this cranial region is avai-lable, it is necessary to find a range of possible shapes, using modern human varia-tion as a reference. |

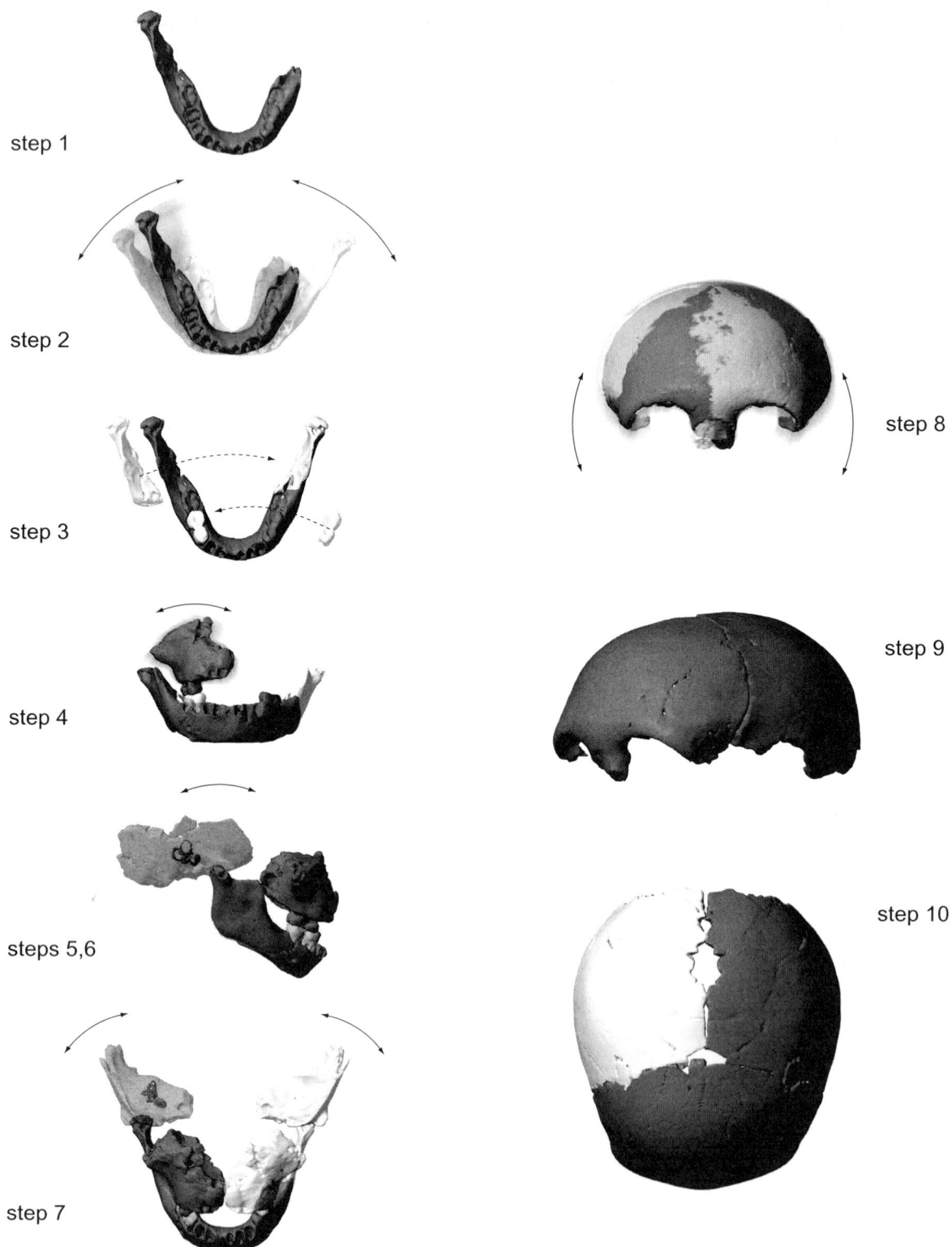

step 1

step 2

step 3

step 4

steps 5,6

step 7

step 8

step 9

step 10

**FIGURE 6-6**   Computer reconstruction of the Gibraltar 2 Neanderthal child.

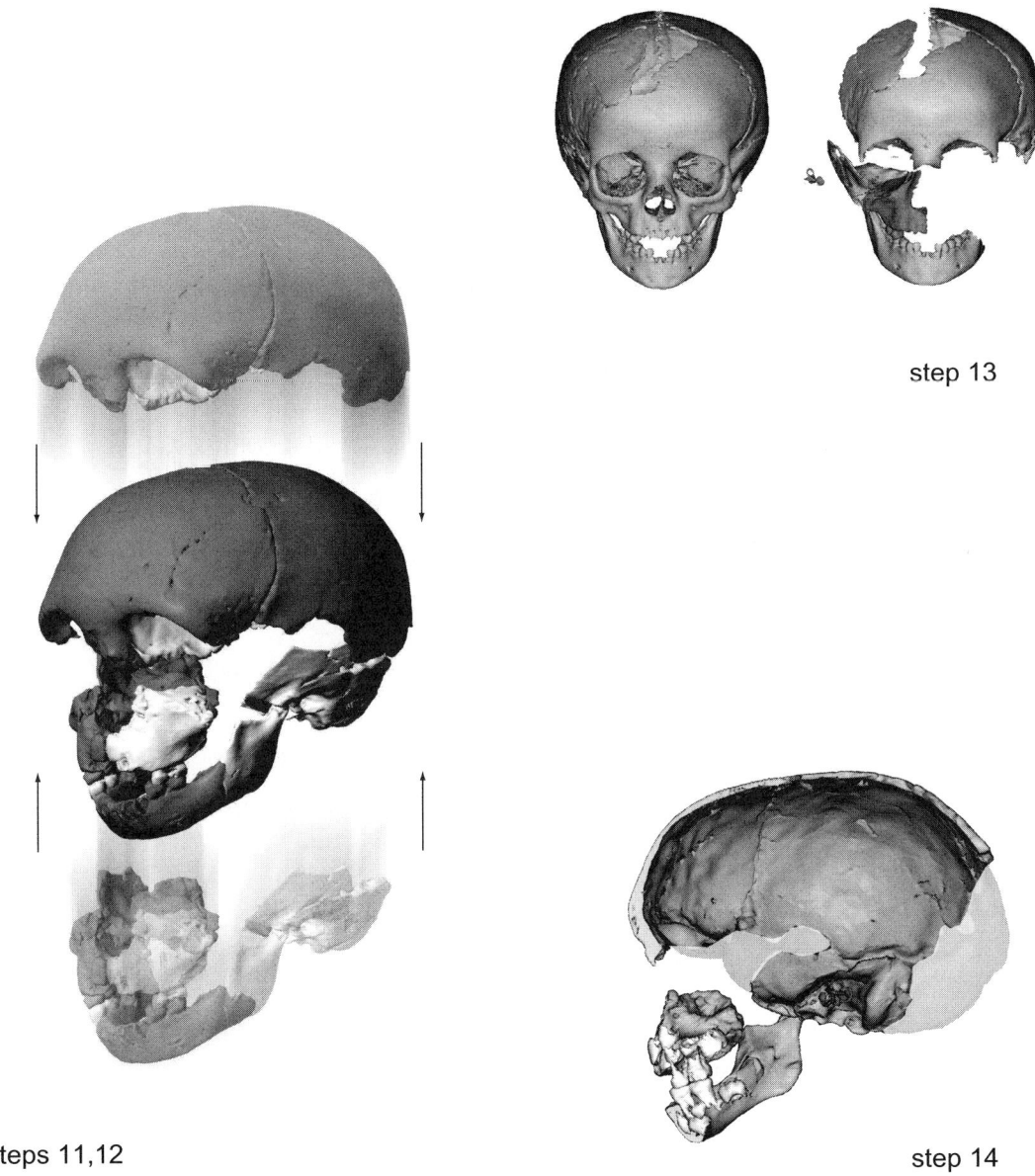

step 13

steps 11,12                                    step 14

**FIGURE 6-6**  (*Continued*).

### 6.5.3 Inferring Missing Information

#### *Aims, Data Types, and Methods*

In many fossils, significant portions of the anatomy are not preserved. It is one of the major challenges of virtual reconstruction to infer missing information in order to achieve a more complete picture of what a fossil specimen looked like at the time of its death. Here, we consider various strategies of inference and illustrate them with practical examples. During this endeavor, we must take into account three aspects of recovering missing information: the aims and purpose of the reconstruction, the type of data that we want to recover, and the methods of inference. We may discern three principal aims.

- *Inference of anatomic clues for virtual reconstruction.* During virtual reconstruction, isolated fragments "floating" in virtual anatomic space must be positioned and oriented with reference to various anatomic clues. The aim here is to reduce the degrees of freedom of motion of isolated fragments by imposing additional anatomic constraints.
- *Inference of morphometric data for comparative analyses.* To include incomplete fossil specimens in comparative morphometric analyses, it is often necessary to provide estimates for missing data of relevant parameters. We may distinguish between interpolation, that is, filling gaps between neighboring fragments, and extrapolation, that is, reconstruction of missing regions outside the range of preserved morphology. The important point here is to state explicitly the geometric and biological hypotheses used to infer missing data.
- *Inference of morphologies for visual rendering.* Computer-based virtual reconstruction makes it possible to reconstruct missing morphologic features comprehensively, notably for museum display and forensic applications. For the visual reification of hypotheses about what a fossil's morphology or a individual's face looked like, it is often helpful to complete missing skeletal regions or soft tissue with three-dimensional morphing procedures. Here, we tend to move from reconstruction to construction.

With respect to data types and methods of inference, let us distinguish between two basic situations. In the first, missing information can be reconstructed with fair reliability from intrinsic information contained in the fragmentary specimen; in the second, the information necessary to complete missing regions cannot be recovered from the available pieces of evidence, so that extrinsic information must be utilized. The corresponding reconstructive strategies are summarized in Table 6-1 and described in depth in the following two subsections.

### Scenario I: Reconstructing Missing Data with Intrinsic Information

As an example, we consider completion of missing regions in a bilaterally symmetrical organism. Parts preserved on one side can be mirror-imaged to the other side (see Box 6-4 for applications of this method during the virtual reconstruction

TABLE 6.1 Reconstructive Strategies to Infer Missing Data

|  |  | Intrinsic Information | Extrinsic Information |
| --- | --- | --- | --- |
| Extensional data |  | Duplication of left/right data in bilaterally organized structures | Inference based on classic multivariate statistics (e.g., with regression analysis) of a comparative symplesiomorphic reference sample |
| Relational data | Morphologic approach | Completion of bilateral structures with autologous mirror images | Inference based on geometric-morphometric analysis of comparative symplesiomorphic reference sample |
|  | Geometric approach | Interpolation and/or extrapolation via morphing and/or deformable templates |  |

of the Gibraltar Neanderthal child skull). Technically, mirror-imaging a virtual object is equivalent to multiplication with a scaling factor of −1 (see Appendix D).[1]

Completion of missing regions with autologous mirror images complies with the reconstructive parsimony rules. Specifically, the procedure is based on the null hypothesis that the left and right sides of the same individual are more similar to each other than left and right sides of any two different individuals (Zollikofer et al., 1998).

Although mirror image completion yields the best estimates of missing parts in a bilaterally organized organism, it is important to be aware of the fact that natural deviations from mathematical symmetry cannot be recovered with this method. This has two consequences. First, we must check in which way mirror image-based reconstructions deviate from biological reality. This can be verified by electronic decomposition of complete specimens into several virtual fragments and subsequent reconstruction of artificially "missing" regions with autologous mirror images. Comparisons between artificial reconstructions and original specimens show that the range of morphometric variation resulting from repeated reconstructions of a given specimen is similar to the range of variation reflecting natural left-right asymmetry within that specimen (Zollikofer et al., 1998). As a result, we may state that virtual reconstruction using mirror-imaged complement parts yields unbiased results.

---

[1] Negative scaling inverts the inside/outside orientation of free-form object surfaces, such that this effect must be corrected accordingly.

Bilateral symmetry must be considered from yet another perspective. In living organisms, deviations from symmetry are typically described with a two-component model postulating random deviations from mathematical symmetry (fluctuating asymmetry; the variance about the symmetric mean form of a sample) and biased deviations (directional asymmetry; deviations of left and right sample mean forms from the symmetric average; see Palmer, 1994, and Mardia et al., 2000).[2]

In fossils, it is difficult to recover natural patterns of asymmetry, because taphonomy-induced postmortem deviations from symmetry are superimposed onto the biologically relevant in vivo deviations. As we show below, however, geometric analysis of asymmetry can be used to infer specific taphonomic processes and correct distortion accordingly.

## Scenario II: Reconstructing Missing Data with Extrinsic Information

Reconstruction of missing information with comparative samples has a long tradition in paleontology, paleoanthropology, and forensics. For example, it is common practice to estimate key parameters such as body size from linear dimensions of long bones (for examples, see Ruff et al., 1997 and references therein) and endocranial volume from linear dimensions of the skull (e.g., Stringer et al., 1990). The principal design of such approaches is as follows:

Provide an array of measurements $X_1, X_2, X_3, \ldots$ on the incomplete specimen that may serve as predictors for the missing measurement, $Y$.

Set up a comparative sample of well-preserved specimens in which both "known" $(X_1, X_2, X_3, \ldots)$ and "unknown" $(Y)$ variables can be measured.

Estimate missing values for the fossil specimen with methods relating $Y$ to $X_1, X_2, X_3, \ldots$, for example with multiple regression (Sokal and Rohlf, 1995). Provide multiple estimates based on various independent predictors.

The fundamental premise of this approach is that the pattern of morphologic covariation observed in the comparative sample determines morphologic covariation in the fossil specimen. In other words, it is presupposed that (a) the fossil belongs to the comparative sample and (b) the measurements taken from the sample and from the incomplete specimen are homologous. Because we cannot verify these assumptions, we must apply reconstructive parsimony to avoid a bias

---

[2]Note that the two-component model of asymmetry is descriptive rather than explanatory, as more than two processes might be responsible for the observed pattern of asymmetry, and one and the same process may contribute to both fluctuating and directional components.

toward preconceived morphologies. This implies that the comparative sample must represent the shared ancestral pattern of variation rather than patterns of variation characteristic for derived taxa.—In sum, it is sensible to use "neutral," taxon-independent strategies of inference of missing data.

Here, we focus on how virtual three-dimensional reconstruction can build upon these principles and enhance the power of morphometric inference. As an illustration, let us return to the Gibraltar child example (see Box 6-4). The occipital region and most of the basicranium is not preserved in this specimen; nevertheless, we are interested in an estimate of the endocranial volume and in three-dimensional reconstruction of the missing region as a whole. In both cases, we must make reference to a comparative sample to infer the missing information. How must this sample be structured to comply with the rules of reconstructive parsimony? And how much extrinsic information is necessary/acceptable to obtain a reasonably unbiased reconstruction?

Recalling the data typology proposed in Chapter 2 (Fig. 2-14), we note that endocranial volume is an extensional measure, that is, a scalar devoid of specific shape information (except the fact that it is measured in volumetric units). Being a simple number, the endocranial volume can therefore be inferred from preserved linear dimensions of the cranial vault bones, without the necessity to specify cranial shape in detail. Applying modern human standards (probably not an optimum neutral comparative sample), estimates of cranial capacity of the Devil's Tower Neanderthal yield values above 1400 ccm (Stringer et al., 1990).

Virtual reconstruction complements this estimate in several independent ways. In addition to the known linear dimensions of the isolated cranial vault bones, the virtual three-dimensional specimen provides data for neurocranial height and breadth. These measurements permit an alternative estimate of cranial capacity, yielding a figure of 1360 ccm.

As a further approach, we reconstruct the entire braincase and determine its volume directly (Fig. 6-7A), which yields endocranial volumes in the range of 1370–1420 ccm (Zollikofer et al., 1998). To this end, we must infer the three-dimensional shape of the missing occipital and basal regions. We may discern morphology-driven and geometry-driven methods to reconstruct the missing shape information.

Morphology-driven methods are similar to those used during inference of extensional data in that they make reference to a comparative sample. For example, the position of missing anatomic landmarks in the basioccipital region of the Gibraltar child skull can be inferred from a reference sample by techniques of geometric morphometric analysis and multiple regression (see Chapter 8).

However, there is an important difference between the inference of endocranial *volume* and endocranial *shape*. Data characterizing the shape of the braincase denote spatial relationships between three-dimensional anatomic points of refer-

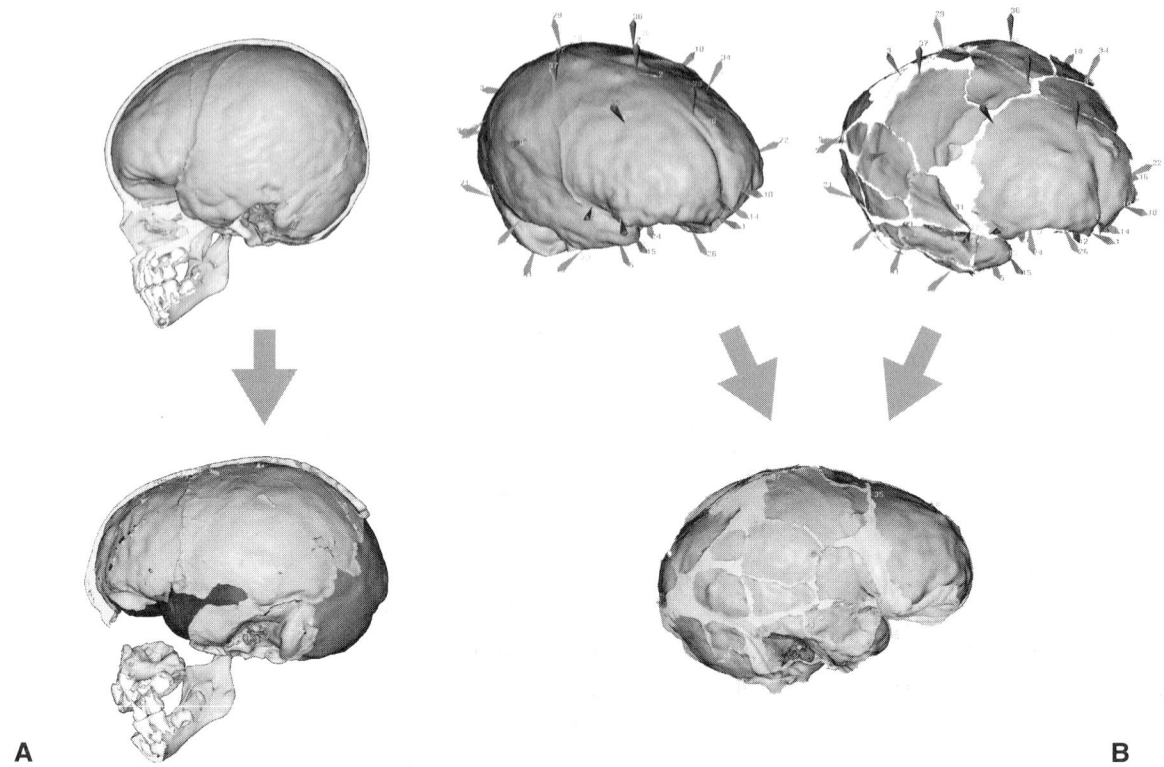

**FIGURE 6-7** Estimating endocranial volumes. **A**: Extrapolation of the occipital and basal endocranial shape of the Devil's Tower Neanderthal child skull using data from a modern human skull of comparable dental age. **B**: Reconstruction of the incomplete endocast of the Le Moustier 1 Neanderthal adolescent. Corresponding landmark locations on both specimens define a thin plate spline interpolation function (see Chapter 8) that morphs the modern human into the Neanderthal endocranial volume.

ence (relational data; see Chapter 2). Accordingly, although inference of extensional data is based on relatively loosely defined homology relations, inference of relational data must be based on more specific homology relations. This leads to a potential conflict between inference of missing shape information and reconstructive parsimony rules. The more information we infer, the more the reconstruction becomes hypothesis-driven and tends to be biased toward morphologic types represented in the sample.

To circumvent these problems, an alternative approach consists in geometry-based inference. Rather than using a comparative sample, inference of missing information is made by reference to the geometric properties of preserved regions. For example, the overall form of the Le Moustier 1 braincase is relatively well represented by original fossil material but "leaky" (Fig. 6-7B). A direct estimate of the endocranial volume can be obtained by closing the open areas between adjacent fragments. In terms of computer graphics, a deformable template (see Chapter 4) is used that adapts itself to the preserved endocranial surface and bridges the gaps according to a minimum-energy criterion.

## 6.5.4 Interpolation and Extrapolation

To conclude our discussion of methods of inference, we summarize the proposed strategies of reconstruction from a more formal perspective. Whether we follow morphology- or geometry-driven approaches, the inference of missing data is carried out within a framework of predefined constraints. These constraints come as a pattern of covariation within a reference sample or as an energy-minimum criterion imposed on the shape of a deformable template. In both cases, the inferred data depend on predefined constraints and functions that are applied to the measurements available from the incomplete specimen.

This brings us to the notion of interpolation and extrapolation. Generally speaking, interpolation means inference or prediction of a missing value within a given range of known values, whereas extrapolation means inference outside such boundaries. Reconstructive parsimony requires that inferences be "on the safe side" in a *biological* sense; hence we must interpolate rather than extrapolate missing data. However, geometric/statistical interpolation does not automatically entail biological interpolation.[3] For example, statistical prediction (interpolation) of missing values corresponds to biological interpolation if the reference sample is symplesiomorphic, otherwise, it results in biological extrapolation. Similarly, using a geometric interpolation function (e.g., the thin plate spline function; see Chapter 8) to morph one shape into another results in morphologic interpolation within the convex hull defined by the spline nodes and in morphologic extrapolation if we are outside this region (see Fig. 6-7). As a consequence, methods of morphologic inference must be specified not only statistically and geometrically, but also in terms of their biological foundations and implications.

## 6.6 CORRECTING FOSSIL DEFORMATION

In the previous sections, we concentrated on how anatomic information is used to recover and/or infer the three-dimensional morphology of fragmentary fossil specimens. In addition to anatomic clues, fossils also carry taphonomic information that can be utilized to reconstruct their postmortem history. This is especially important during the correction of taphonomic deformation, because our general aim is to separate postmortem morphologic "noise" from the ontogenetic and phylogenetic morphologic "signal" conveyed by the fossil at the time of its death (see Fig. 6-2).

---

[3] Here again, we encounter parallel terminologies with slightly different geometric and biological meanings.

**TABLE 6.2  Possible Alterations Detectable on Hominid Fossil Skeletal Remains**

*In vivo Alterations*

Deformation resulting from developmental processes (e.g., deviation from bilateral symmetry) or behavioral/functional modification (e.g., asymmetric mastication).

Pathological deformations resulting from congenital malformation, disease, and/or trauma.

Artificial deformation resulting from culturally motivated actions (e.g., artificial cranial deformation).

*Postmortem Alterations*

Biological deformation resulting from culturally motivated actions (such as application of fire, most often leading to destruction and loss of skeletal parts).

Taphonomic deformation resulting from geophysical diagenetic processes such as de- and remineralization, compaction of strata, and tectonic events.

Modifications resulting from isolation of the fossil from surrounding material (e.g., changes in pressure and humidity) and preservation (e.g., chemical treatment).

To reach this goal, it is first necessary to make an inventory of potential causes and effects of deformation and to associate them with characteristic patterns of morphologic alteration. Table 6-2 provides such a list for the main types of deformation occurring in fossil hominids.[4]

### 6.6.1 Taphonomic Scenarios

Among postmortem effects, we will focus here on identification and correction of taphonomic deformation. From a physical point of view, the way in which a fossil is deformed depends on the mechanical properties of fossilizing bone and the pattern of external forces applied to the fossil. We may distinguish between three basic scenarios of deformation induced by forces exerted by the surrounding strata (Fig. 6-8) (Ponce de León and Zollikofer, 1999).

In the first case—*fracturing deformation*—the fossil breaks apart early during diagenesis (Fig. 6-8A). The formation of fractures induced by load indicates resistance against plastic deformation through localization of peak forces within the cracks and subsequent dislocation of parts. As a consequence, fossils that are heavily fragmented in situ tend to exhibit little overall deformation after virtual reconstruction.

---

[4] We use the term *deformation* here rather than *distortion*, because deformation has a precise technical meaning—change in shape resulting from the application of stress—while distortion is relatively loosely defined mechanically.

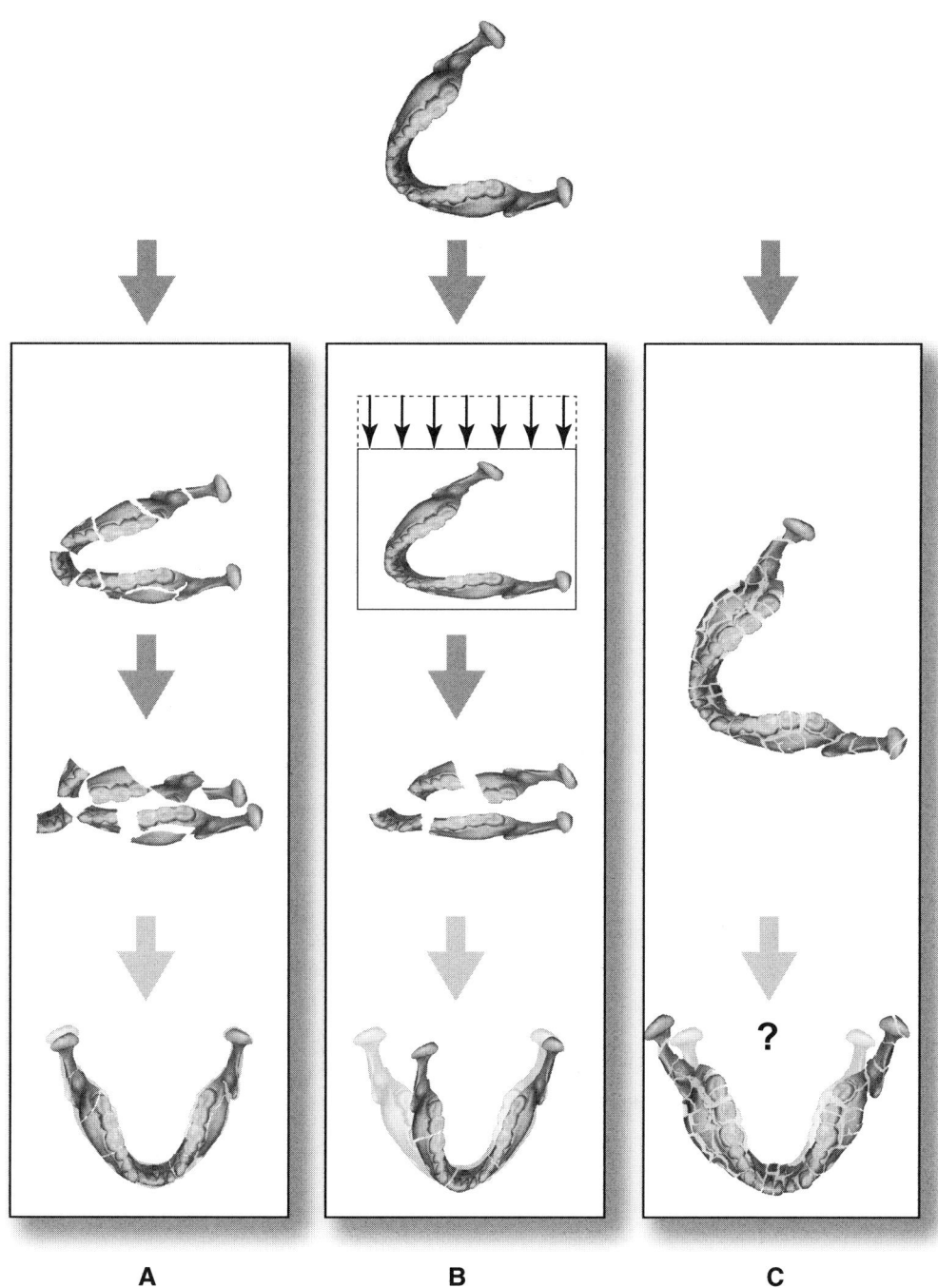

**FIGURE 6-8** Scenarios of taphonomic deformation and possible correction. **A**: During fracturing deformation, the fossil disintegrates, and fragments are displaced relative to each other without being deformed; recomposition of isolated fragments results in an undistorted morphology. **B**: The fossil undergoes plastic deformation early during diagenesis and disintegrates in a later phase. Part recomposition results in a distorted morphology, which can be corrected with virtual procedures (see Fig. 6-9). **C**: Expanding matrix distortion (White, 2003) leads to pervasive disintegration, such that the original morphology cannot be recovered reliably.

In the second case, the fossil undergoes *plastic deformation*. Specimens deformed in this way tend to be recovered in relatively few pieces and retain distortions after reconstruction (Fig. 6-8B). Plastic deformation results from a complex interplay between compression by overlaying strata, and de- and remineralization of fossilizing bone. As a long-term consequence, bone behaves like a ductile material and may change its micro- and macrostructure considerably without breaking.

Whereas taphonomic forces tend to compress fossils, a third scenario of deformation results in expansion of the original anatomic structures (Fig. 6-8C). This scenario was described as *expanding matrix distortion* (EMD) (White, 2003); fossils that have undergone EMD are pervaded by a network of small cracks, and fragments appear to be driven apart through expansion. How can this effect be explained? During fossilization, bone-matrix interactions such as de- and remineralization involve alterations at a microstructural level. When these processes are in equilibrium, the fossil's overall structure may remain unchanged or undergo plastic deformation. However, perturbation of the equilibrium results in macroscopic change. Whereas net material loss leads to eventual dissolution of the fossil, net material uptake results in expanding matrix distortion.

To correct deformation and infer the original anatomic status of a fossil at the time of its death, it is necessary to reconstruct its diagenetic history. This involves hypotheses about the potential causes of deformation, the temporal sequence of diagenetic events, and the spatial pattern of forces acting on the fossil. Although correction of fracturing taphonomic deformation is relatively straightforward (see Fig. 6-8B), fossils affected by plastic deformation need further consideration. Attempts at inferring the original shape of a fossil from its deformed state remain tentative in this case, because the effects of successive diagenetic events are superimposed in the fossil's morphology and cannot be resolved in full temporal detail.

## 6.6.2 Correcting Plastic Deformation

Nevertheless, the problem is tractable for one fairly common diagenetic scenario, vertical compression of fossil-bearing sediments induced by the load of overlaying strata. Modeling compression for bilaterally symmetrical objects shows that the outcome of deformation depends on the orientation of the embedded fossil relative to the direction of compression (Fig. 6-9).

Although compression along directions parallel or orthogonal to the midsagittal plane of the model skull does not affect its overall symmetry, any deviation from this condition will result in apparent skewing (shearing) of the entire fossil geometry. From a computational point of view, compression is an affine transformation. This class of linear transformations turns squares into parallelo-

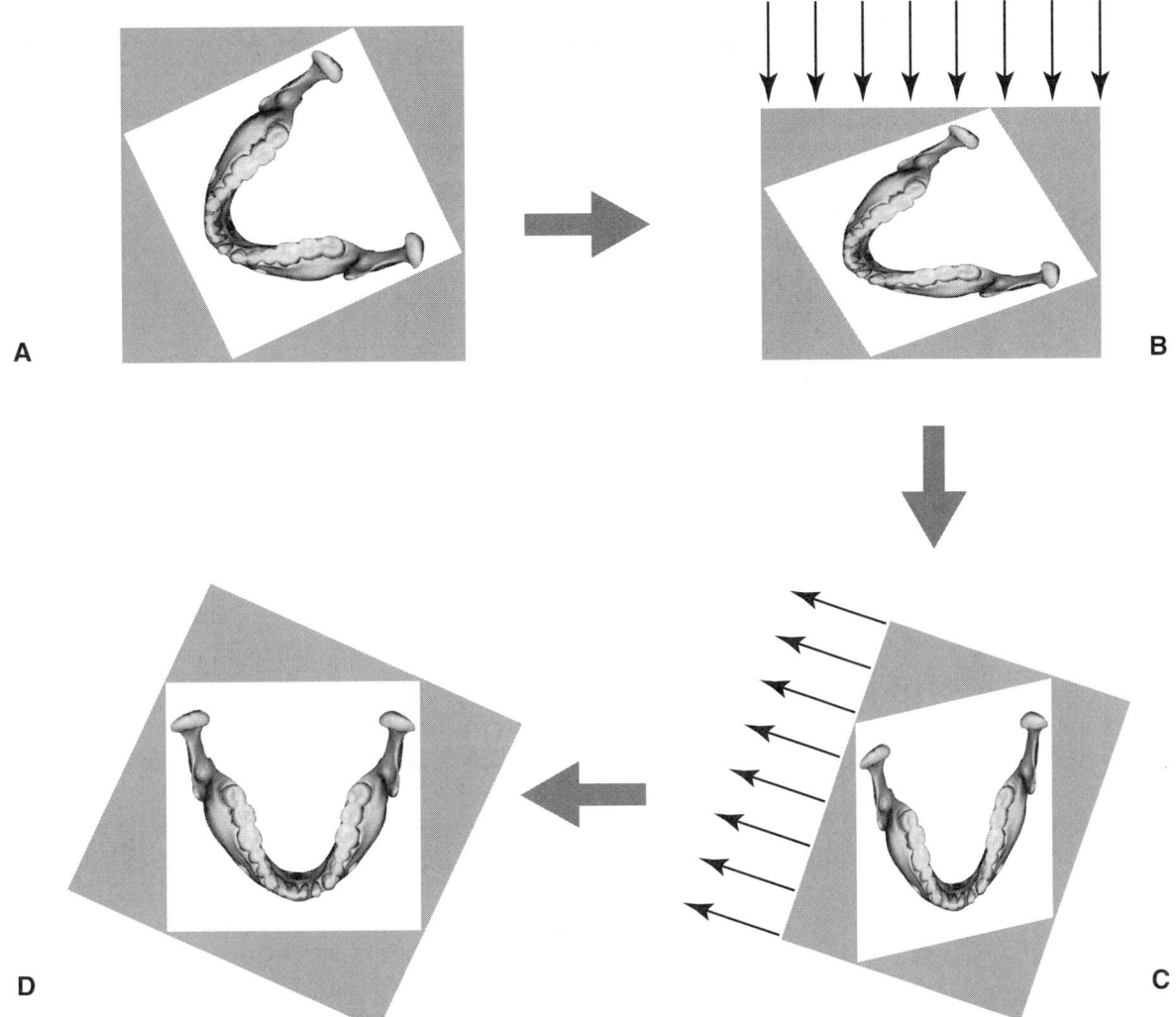

**FIGURE 6-9** A model for the correction of plastic compressive deformation. **A, B**: Compression of the entire morphology is equivalent to scaling the fossil by a factor <100% along a given direction. **C**: After recovery and reassembly of the fossil, compressive deformation reveals itself by overall skewing of the morphology relative to the object's midplane. **C, D**: Deformation can be corrected by rescaling the fossil along the inferred direction of compression.

grams, but otherwise leaves parallel lines or planes parallel. A further important property of affine transformations is that they affect each point in space equally, that is, they act globally rather than locally. Under these particular conditions, identification and tentative correction of deformation is possible. The general approach is to find an optimum affine function that retransforms parallelograms into squares. Examples are given in Box 6-5, and a more rigorous treatment of the formalism is presented in Appendix E.

**BOX 6-5**

## CORRECTING TAPHONOMIC COMPRESSION: EXAMPLES

### Example I: Le Moustier Neanderthal skull

The virtual reconstruction of the Le Moustier 1 Neanderthal skull from its decomposed fragments (see Box 6-3) resulted in a slightly slanted morphology (Fig. 6-10). Connections between corresponding left/right anatomic landmarks show similar skewing over the entire morphology, hinting at global deviation from symmetry resulting from overall compression of the skull by overlaying strata. To correct deformation, two parameters were determined, the average slanting direction of the transversal lines and the direction of compressive forces (Ponce de León and Zollikofer, 1999). The latter direction could be derived from historical photographs documenting various stages of the excavation and recovery of the skull. Realignment of deformation proceeded as follows: The virtual skull was positioned in situ on the computer screen and decompressed in the vertical direction until slanting of transversal connections relative to the skull's midplane vanished. Compressive deformation was in the range of no more than 4%. When uncorrected, this small amount of plastic deformation causes substantial misfit between neighboring fragments during any attempt at restoring cranial bilateral symmetry.

### Example II: Notharctus skull

The skull and mandible of this Eocene North American primate are well-preserved but deformed. Overall compression is evinced by the slanted lines connecting contralateral landmarks (Fig. 6-11A, B). The principal direction of compression is unknown; however, it is possible to infer a direction that requires minimum compression to yield the observed effects. This direction is a line bisecting the mean slanting direction and the midsagittal plane (arrow in Fig. 6-11C; see Appendix E). Decompression is realized by scaling along the principal direction of compression (Fig. 6-11D, E). The corrected morphology is no longer slanted but still exhibits some nonlinear deformation (Fig. 6-11F).

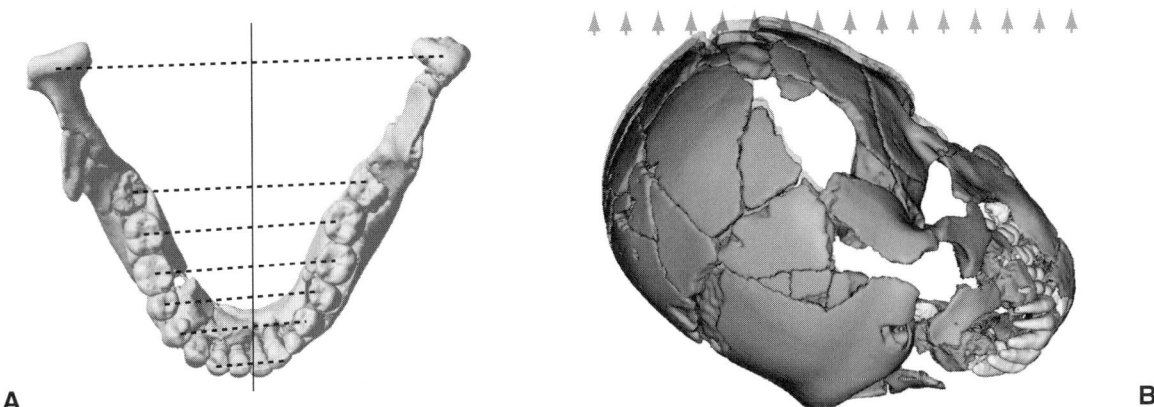

**FIGURE 6-10**  Correction of plastic deformation in the Le Moustier 1 Neanderthal skull. **A**: Global deformation of the mandible revealed by similar slanting directions of transversal lines (dashed) relative to the midplane (solid). **B**: Decompression of the skull positioned in situ according to site photographs.

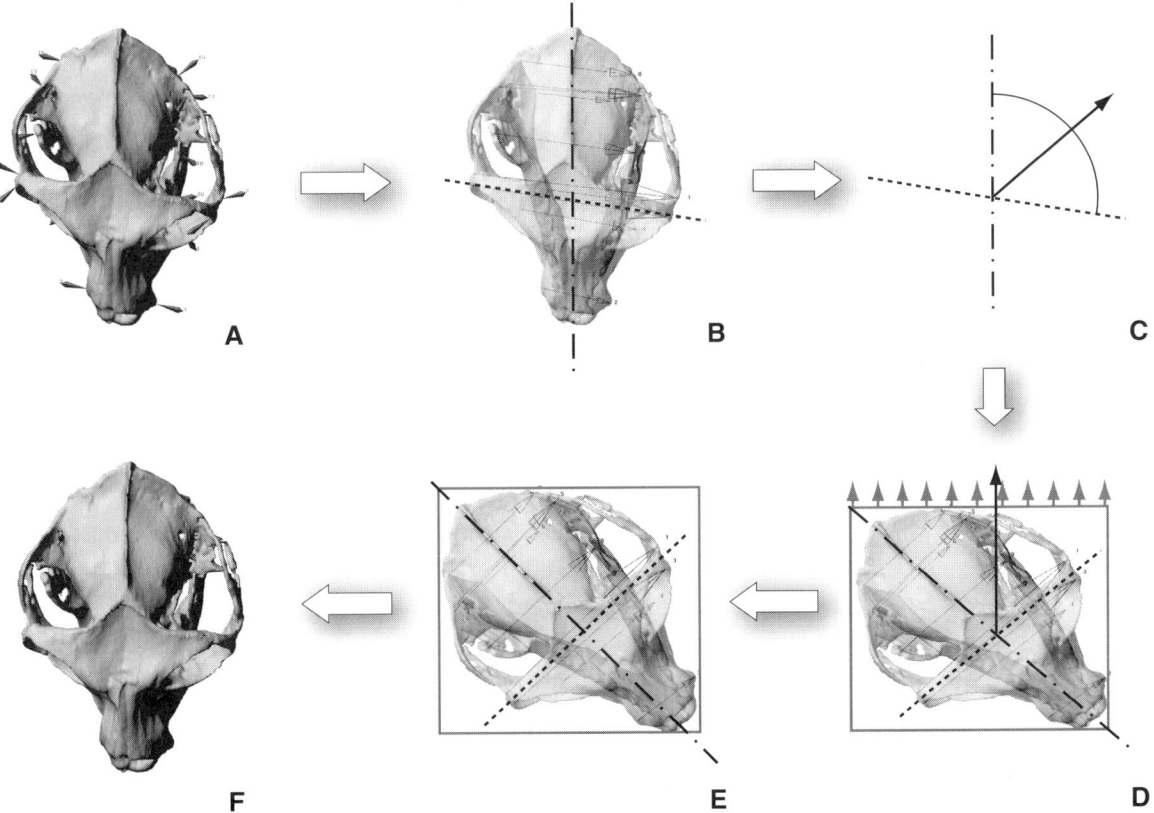

**FIGURE 6-11**  Correction of plastic deformation in a *Notharctus* skull. **A, B**: Slanted lines connecting opposite anatomic landmarks indicate overall compressive deformation. **C**: Evaluation of the direction corresponding to minimum compression. **D–F**: Correction of deformation by linear decompression.

Nonlinear deformation affects fossil morphologies locally rather than globally, such that the task of inferring the spatiotemporal pattern of taphonomic forces becomes more hypothetical and no general rules can be stated regarding correction. In many specimens, however, intrinsic anatomic information can be used to tentatively recover the original anatomy. An example is given in Box 6-6.

## BOX 6-6

### CORRECTING NONLINEAR DEFORMATION: THE LAGAR VELHO CASE

In 1998, the Lagar Velho rock shelter, near Lapedo (Portugal) yielded a well-preserved child burial site from the early Upper Paleolithic (Duarte et al., 1999). By applying physical and virtual reconstruction techniques, it was possible to reestablish anatomic continuity between nearly 100 isolated cranial fragments (Zollikofer et al., 2002b).

The preliminary physical reconstruction exhibits extensive nonlinear taphonomic deformation, especially on its left side (Fig. 6-12A). To correct deformation, a two-step approach was followed: First, the vault was decomposed along in situ cracks into several independent units. These units were then recomposed and readjusted, imposing various anatomic constraints and following reconstructive parsimony rules. For example, the breadth of the cranial vault had to fit the reconstructed breadth of the undeformed mandible, and the interparietal suture had to be relocated in the midplane of the vault. By following these procedures, bilateral symmetry could be restored under preservation of most anatomic contacts between fragments.

In a second step, geometric continuity between the deformed fragments of the cranial vault was restored. To this end, a set of landmarks was placed on the cranial vault, and their three-dimensional coordinates were determined before and after the reconstruction of Step 1. Positional changes of the landmarks were used to define a thin plate spline function (see Chapter 8), which performs a smooth interpolation of the original into the reconstructed morphology, thus correcting the effects of nonlinear deformation. The results of this procedure are shown in Figure 6-12. Given the hypothetical nature of any correction of nonlinear deformation, it should be kept in mind that the recovered shape of the cranial vault represents a tentative reconstruction.

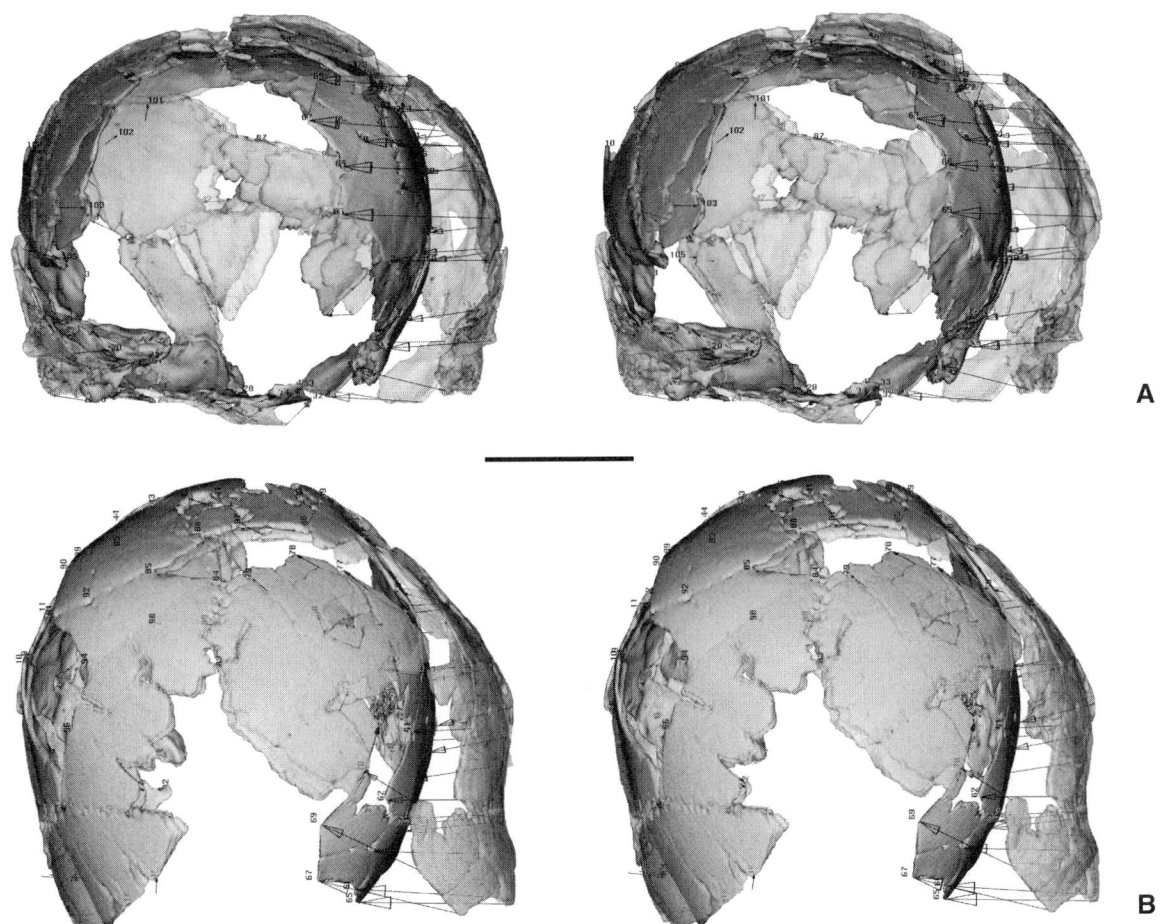

**FIGURE 6-12** Correction of plastic deformation in the cranial vault of the Lagar Velho child. **A, B**: Frontal (top) and superior (bottom) views (crossed stereo pairs). The deformed cranial vault (transparent) was restored (solid) with a TPS interpolation function (see Chapter 8) defined by positional changes (arrows) of 105 anatomic landmarks before and after virtual recomposition of the skull from isolated fragments. Scale bar is 5 cm.

## 6.7 VALIDATING VIRTUAL RECONSTRUCTIONS

In previous sections, we recognized that virtual reconstruction is a process of inference of hidden and/or lost information, involving methods of interpolation and extrapolation to reconstruct an individual's three-dimensional morphology at the time of death. We stated that a principal difference exists between the reconstruction of the fragmentary Roman ground plan and the reconstruction of a fossil morphology. Rome is unique, such that it is possible to make reference to various independent sources of information during map reconstruction, and to dig up additional direct evidence by on-site archeological studies. During fossil reconstruction, the relationship between object and plan are inverse; our goal is to infer the *bauplan* of a fossil taxon from various reconstructed specimens. However, there is no reference according to which the reconstruction of this plan can be verified. Likewise, during reconstruction of a given specimen, it is not possible to verify the individual's morphology with reference to an existing bauplan.—How can we then validate the results of virtual fossil reconstruction?

Let us consider that reconstruction of fossil morphologies from fragments has many things in common with reconstruction tasks discussed earlier, for example, reconstruction of three-dimensional objects from two-dimensional images and reconstruction of two-dimensional images from one-dimensional projections. In each case, we are confronted with an ill-posed problem (a problem with more unknowns than equations; see Chapters 3 and 4) that requires specific assumptions in order to be solved and whose outcome can be validated in terms of accuracy and precision.

Accordingly, to determine the accuracy of a virtual fossil reconstruction, we must verify that the null hypotheses on which the reconstruction is based do not introduce a reconstructive bias. And to assess precision, we must estimate limits of confidence for the reconstruction of a given specimen.

Because the "true" morphology will never be known, the only practicable way to assess accuracy and precision is to produce a range of possible reconstructions rather than one single solution. This can be achieved with a combination of three different procedures:

- *Repeatability*: Perform the entire process of reconstruction several times, involving various trials of the same operator (intraperson bias and scatter) and of different operators (interperson bias and scatter).
- *Robusticity*: Carry out alternative reconstructions based on different null hypotheses. This procedure reveals how stable the reconstruction of a given specimen is against differences in the basic anatomic assumptions used as guidelines.
- *Verification*: Because a given fossil specimen is represented by a unique set of fragments representing only part of the complete anatomy, we also need to verify the reconstruction with extrinsic information. This can be achieved by complementing the fossil reconstruction with simulated reconstructions of complete specimens. To this end, complete specimens are decomposed into a set of fragments equivalent to those preserved in the fossil specimen and recomposed again, observing the criteria established for virtual fossil reconstruction. Deviations between the original (complete) and reconstructed morphologies yield an estimate of the potential biases inherent in the applied procedures.

Applying these procedures provides a range of reconstructed morphologies. Accuracy and precision of the virtual reconstruction are most relevant when seen in relation to morphologic variability of a comparative sample. In comparative geometric-morphometric analyses (see Chapter 8), for example, reconstructive variants appear as a "cloud" of points in morphospace. The precision of the virtual reconstruction is specified by the dimensions of the cloud relative to morphologic variability displayed by other specimens in the sample, and reconstructive accuracy can be assessed by positional changes of the cloud's center under different reconstructive assumptions.

The relationship between the reconstructive cloud and the comparative sample ultimately determines the outcome of comparative analyses. For example, reconstructive variants of the Lagar Velho child skull (Box 6-6) show that, despite reconstructive uncertainty in the position of the facial region relative to the brain-case, and despite the tentative character of the corrected nonlinear deformation, the individual is clearly associated with *Homo sapiens* cranial morphologies (Fig. 6-13).

A final important issue concerns verification of virtual reconstructions with anatomic criteria not used during the process of reconstruction. In addition to the evolutionary conservative orientation of the internal otic cavity (Delattre and Fenart, 1960), various anatomic constraints characterize cranial morphology over a wide range of primate taxa (Enlow, 1990; McCarthy and Lieberman, 2001). Although the functional and/or developmental significance of many of these

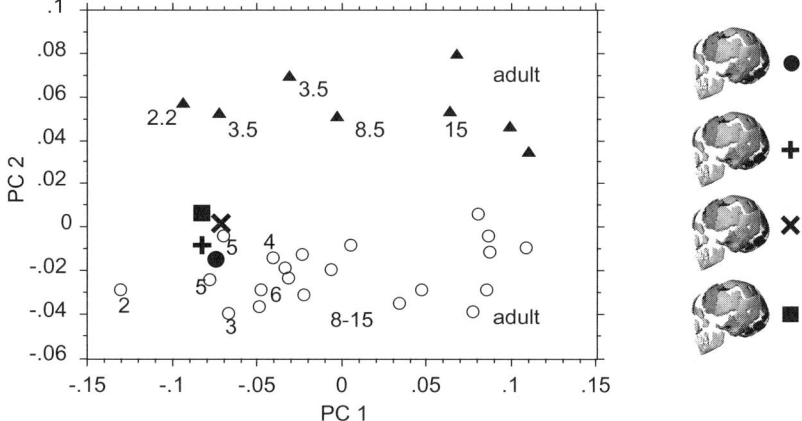

**FIGURE 6-13** Reconstructive variants of the Lagar Velho Gravettian child skull. In this specimen, the orientation of the preserved facial remains relative to the neurocranium remains inconclusive, because no direct anatomic contacts are preserved. Reconstructive variants (right) modify the specimen's position in shape space (shape components PC1 and PC2; see Chapter 8). However, the Lagar Velho reconstructive "cloud" is clearly associated with the *Homo sapiens* (open circles) rather than with the *Homo neanderthalensis* (black triangles) subsample. Moreover, its position along the first shape axis (PC1), which correlates with individual age (figures in the graph) and size, is relatively stable (after Zollikofer et al., 2002b).

biological constants is still unknown, they serve as important independent criteria to assess the anatomic consistency of reconstructions.

## 6.8 PALEODIAGNOSTICS AND PALEOFORENSICS

In addition to identification and correction of postmortem skeletal alterations, methods of virtual reconstruction can also be applied to investigate in vivo modifications of the skeleton (Fig. 6-14). At the outset of in vivo modification

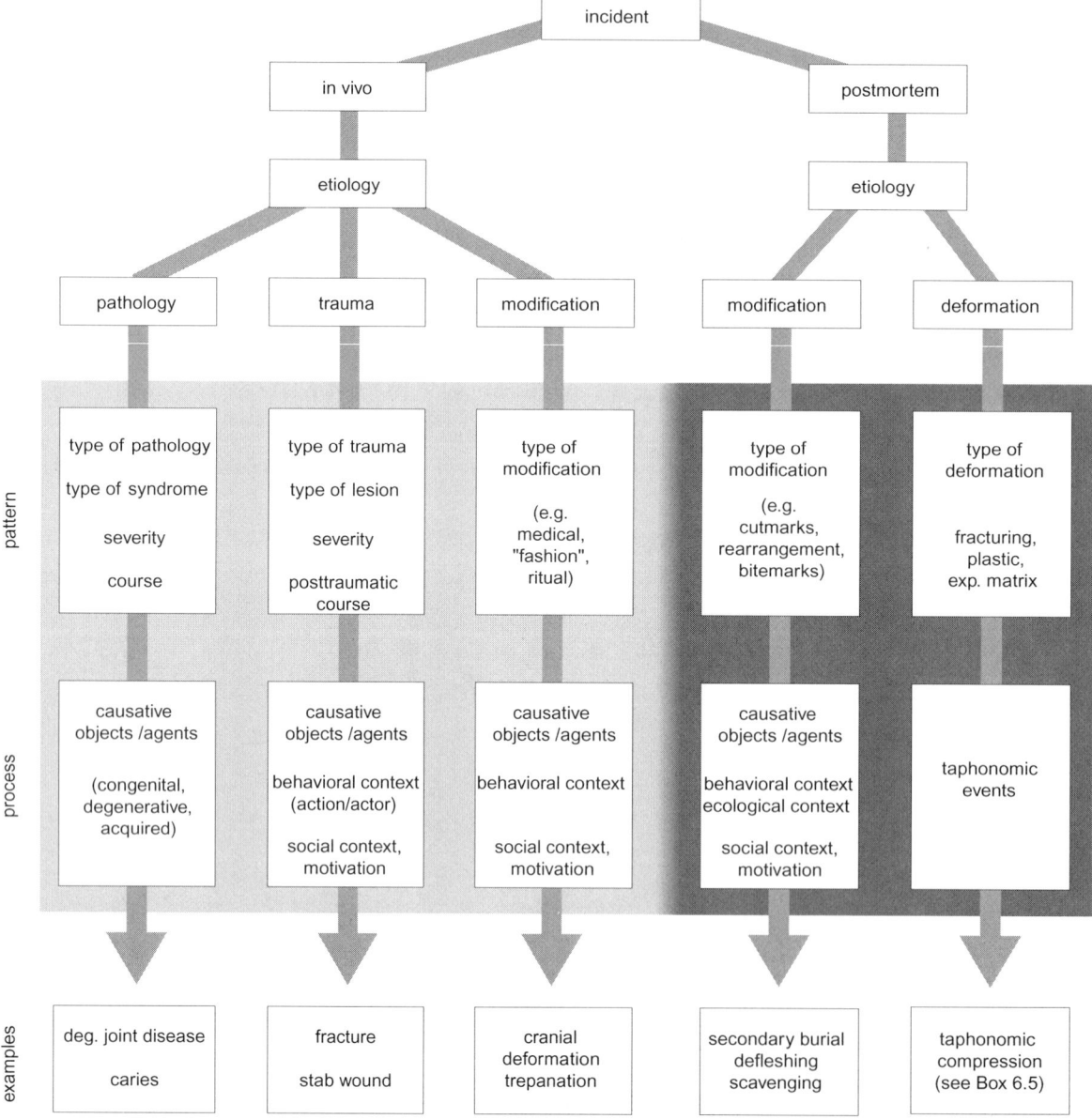

**FIGURE 6-14**    A classificatory proposition of in vivo and postmortem modifications observable in hominid fossil remains.

analysis, it is crucial to discern between pathology and taxonomic relevance. In taxa represented by only few specimens it is often difficult to discern between taxon-specific traits and skeletal alterations that occurred during an individual's lifetime. Neanderthals are the paramount example of recurrent attempts at explaining taxonomy in terms of pathology (Dobson and Geelhoed, 2001), whereas recent CT-based analyses of one of the few well-preserved T-Rex skeletons revealed that features thought to be taxonomically relevant most probably represent pathologies (Brochu and Ketcham, 2003).

During reconstruction of in vivo modifications, we adhere to the criteria established for virtual fossil reconstruction and discriminate between intrinsic versus extrinsic information and inference by interpolation versus extrapolation, respectively. It is often possible to infer patterns of trauma or pathology when evidence from the reconstructed fossil morphology is combined with microstructural bone analysis and comparative archeological and clinical data. Box 6-7 gives an example of trauma analysis in a Neanderthal skull.

In any case, inferences regarding the processes and causes leading to the observed in vivo modifications are of an extrapolatory nature. Whereas proximate causes such as mechanical actions causing trauma or biotic and abiotic agents causing pathology can often be identified with comparative clinical and forensic evidence, the inference of the motivational and cultural background of in vivo modifications must remain tentative. For example, the cultural context of one of the most intriguing hominid behavioral patterns—defleshing conspecifics—can only tentatively be reconstructed from the available hard evidence (Defleur et al., 1999).

Paleodiagnostics and paleoforensics has still another aspect. As shown by Rowe and colleagues in the "*Archaeoraptor* case," the combination of CT-based analysis and three-dimensional reconstruction is highly effective in debunking fake fossils (Dalton, 2000a; Rowe et al., 2001b; Zhou et al., 2002).

## 6.9 INFERRING SOFT TISSUE STRUCTURES

### 6.9.1 Motivation

Fleshing out skeletal reconstructions is visually appealing and thus often thought of as a primarily artistic task. However, the quantitative reconstruction of soft tissue structures also has a scientific motivation. In paleontology and paleoanthropology, the reconstruction of soft tissue elements forces scientists to think of the skeletal remains of an individual as forming part of a developmentally and functionally integrated ensemble. From a paleontological perspective, the skeleton appears less ephemeral than soft tissue. However, from a developmental perspective, it is the soft tissue that shapes skeletal structures during growth and through function (Moss and Young, 1960). Consequently, the reconstruction of soft

## BOX 6-7

### THE ST. CÉSAIRE CASE

The St. Césaire 1 skeleton represents a "late" Neanderthal (dated to 36,000 years ago), associated with Châtelperronian rather than Mousterian stone implements, and probably contemporary with early modern humans diffusing from Africa into Europe. This setting makes the specimen's analysis in terms of morphologic and cultural change particularly interesting.

We focus here on the cranial remains. Because only the right side of the skull is preserved (the upward-facing left side was weathered away), virtual reconstruction (Fig. 6-15A) relied on mirror image completion of missing parts, following the parsimony rules established above. An apical vault fragment, whose medial border was originally thought to represent the midline of the cranial vault, turned out to be incompatible with standard anatomic conditions and, likewise, with natural anatomic variants in this region of the skull. Using comparative material and applying the classification scheme for in vivo modifications (Fig. 6-14), we identified the structure as a healed slash, most probably resulting from sharp trauma (Zollikofer et al., 2002a).

The morphologic pattern displayed by the slash permits inferences regarding the traumatic scenario. The linear shape, apical position, and anteroposterior orientation of the scar indicate an intentional action effected with a sharp implement rather than an accidental injury. Further inferences regarding the nature of the implement, the agents, and the behavioral setting must remain tentative. The most likely scenario is relatively unspectacular but common: interpersonal violence within a Neanderthal group, followed by some assistance during healing of the wound.

tissue structures is equivalent to stating and reifying hypotheses about developmental and functional integration of reconstructed skeletal parts.

In clinical medicine, forensics, and museology, a central motivation for inferring soft tissue structures from skeletal data is rendering reconstructive results in a form that is easy to communicate to the patient or the public (Vanezis and Vanezis, 2000). In forensics, for example, there is a strong interest in generating facial reconstructions from skeletal parts that help identify an unknown individual (Taylor, 2001). In clinical applications, such as maxillofacial surgery, it is impor-

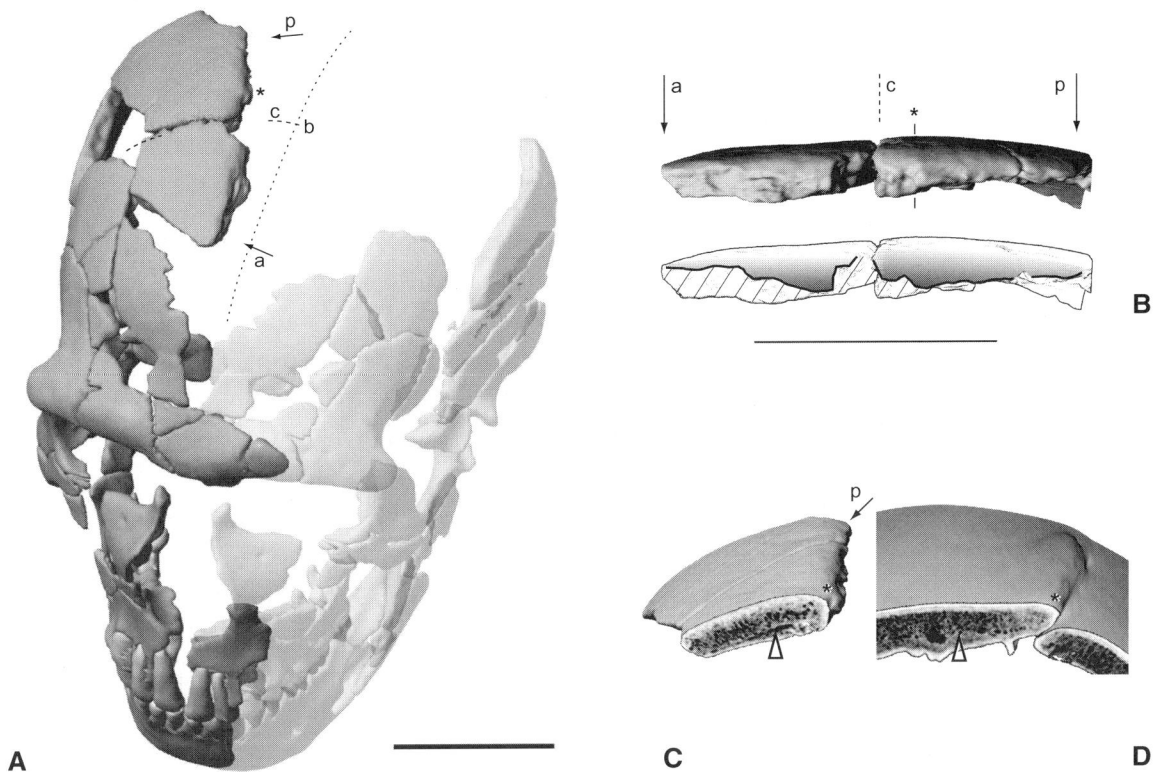

**FIGURE 6-15** St. Césaire virtual reconstruction and paleodiagnostics. **A**: Virtual reconstruction of the skull (transparent parts are completed) showing the lateral border of a healed scar (a → p: anteroposterior extent; c: coronal suture; dotted line: midsagittal plane; scale bar is 5 cm). **B**: Mediolateral view of the scar; the drawing indicates areas of bone remodeling (gray) and postmortem fracture (hatched). **C**: cross-sectional CT image (at position * in **B**). **D**: Comparative cross-sectional morphology of a healed scar resulting from a sharp blow in an archeological specimen. Note bone remodeling at the impact site (*) and dislocation of internal bone lamina resulting in a gap (Δ).

tant to anticipate the effects of a complex intervention on the visual appearance of the patient's face (see Section 6.10). And finally, because a typical museum visitor is more familiar with the external aspect of animals and humans than with skeletal morphology, characteristic features of fossil taxa are more readily recognized when skeletal reconstructions are complemented with soft tissue reconstructions (see Box 6-8).

The same principles apply during soft tissue reconstruction as during hard tissue reconstruction: First, as much information as possible is inferred from anatomic clues preserved in the fossil. Second, this information is combined with general anatomic data about the relationship between hard and soft tissue structures. Third, missing information is completed by various interpolation and extrapolation techniques. In the following examples, we consider how these principles are combined in practical paleontological and clinical applications.

### 6.9.2 Fossil Soft Tissue Reconstruction: Classic and Virtual Approaches

Soft tissue reconstruction in paleoanthropology has a long tradition that is closely related to the tradition of forensic reconstruction (Kollmann and Buchly, 1898; Gerasimov, 1945, 1971; Tyrrell et al., 1997; Nelson and Michael, 1998). Both intrinsic and extrinsic anatomic information are used during reconstruction (Lebedinskaya et al., 1993); in fossil specimens, direct evidence of soft tissue structures is preserved mostly in the form of skeletal superstructures and tuberosities indicating regions of muscular attachment. Exploiting this information, anatomic structures, notably muscles, are modeled layer by layer. However, because muscular insertion marks are relatively loosely correlated with actual muscular volumes (Weiss, 2003), and because superficial tissue layers do not leave traces on the skeletal system, extrinsic anatomic information is needed to infer the dimensions of these structures. Notably for facial reconstruction, therefore, population- and sex-specific data on overall soft tissue thickness measured at various anatomic points of reference are used to complement direct muscular modeling (Rhine and Campell, 1980; Rhine and Moore, 1984).

In using virtual tools and applying reconstructive parsimony rules, the following new approaches to soft tissue reconstruction emerge (Fig. 6-16):

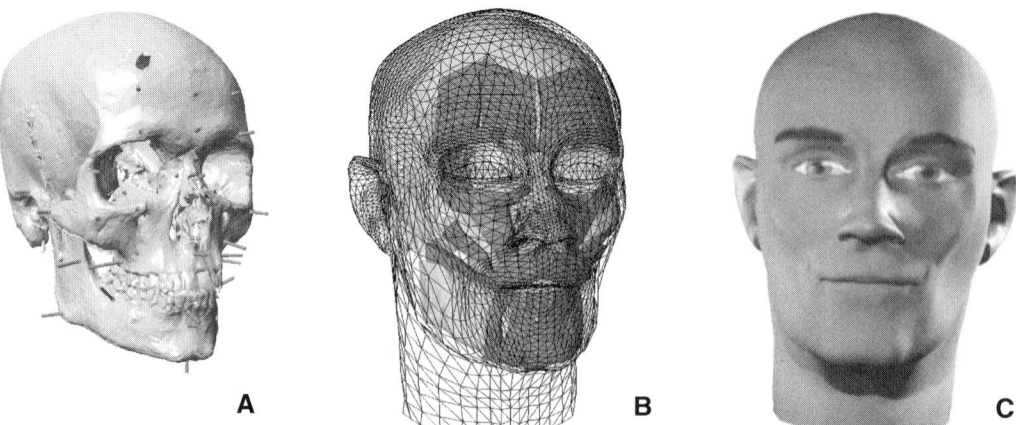

A         B         C

**FIGURE 6-16** A virtual approach to soft tissue reconstruction. Reference data of tissue thickness (**A**) are combined with CAD-based modeling of muscles (**B**) and morphing procedures to model facial expression (**C**). From Kähler et al. (2003), © 2003 ACM, Inc. Reprinted by permission.

## RECONSTRUCTING THE FACE OF THE DEVIL'S TOWER NEANDERTHAL CHILD

How can the peculiarities of fossil morphology be conveyed in an intuitive and comprehensive manner to a nonexpert, for example, a standard museum visitor? One possibility is to bring into play the difference between the familiar (our own external morphology) and the nonfamiliar (the fossil's inferred external morphology). Here, we use a statistical approach to infer facial appearance of the finalized virtual reconstruction of the Gibraltar Neanderthal child skull (see Box 6-4) (Fig. 6-17). As an external reference sample, we used composite clinical CT data sets of two 4-year-old modern human children to determine a set of anatomic landmarks on the skeletal surface. The same set of landmarks was determined for the Neanderthal skull. To transfer the modern soft tissue thickness data to the Neanderthal child skull, we used a three-dimensional thin-plate spline (TPS) interpolation function (see Chapter 8). Based on the modern-to-Neanderthal skeletal landmark homology, the TPS function morphs the entire modern average cranial skeleton into the Neanderthal cranial skeleton. During this process, the modern average soft tissue is allowed to "flow" together with the skeletal data and is transformed into the inferred Neanderthal facial morphology. A similar approach, based on an alternative volume morphing method (Chen et al., 1996), has been proposed for forensic reconstruction (Nelson and Michael, 1998).

The resulting soft tissue reconstruction is "parsimonious" in the sense that no specific assumptions were made about Neanderthal soft tissue peculiarities; it is based on the hypothesis that the relation between skeletal and soft tissue structures was essentially the same in Neanderthals as in modern humans. Parsimony results in a trend toward "modern humanness," but leaving open intentionally any questions regarding the inference of "archaic" features distinguishes this reconstruction from classic approaches.

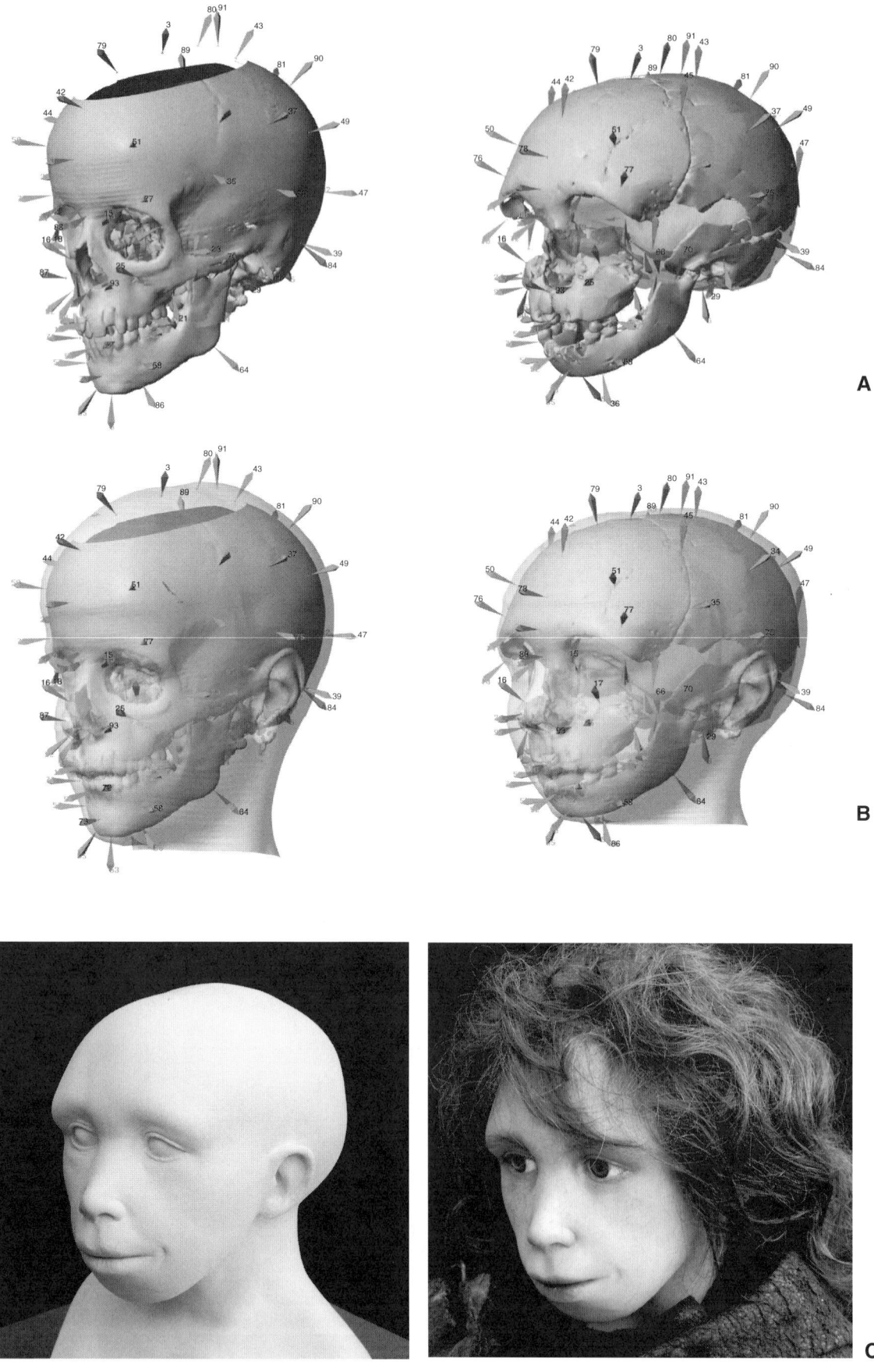

- *CAD approach*: In analogy to the shaping and sculpting of soft tissue structures with plaster and clay, a CAD approach can be used as a tool to create spline-based muscular models and to attach them to virtual skeletal elements (Quatrehomme et al., 1997; Kähler et al., 2003). The major advantages of this method are the same as those of virtual skeletal reconstruction: the possibility of planning, documenting, and quantifying each step of the reconstruction in terms of underlying hypotheses regarding muscular dimensions.

- *Statistical approach*: The classic idea of using soft tissue data from known reference samples can be transposed into the digital world. Whereas traditional measurements were obtained by point-by-point sampling from cadavers, today's medical imaging technology (CT and MRI) yields complete volume data sets from living subjects, from which three-dimensional relationships between hard and soft tissues can be derived at any level of complexity (Tyrrell et al., 1997). For example, in a first approach, correlative data on the relationship between the facial skeleton and the external appearance of the face are gathered, without specifying details of the intermediate anatomic structures.

- *Engineering approach*: The above methods can be expanded to incorporate data on the mechanical properties of various tissue types. For example, finite element modeling (FEM)—a method used to model stress and strain in engineering parts—can be used to simulate how skin spans over muscular tissue and how it moves during muscular contraction (Koch et al., 1998). This approach is especially appealing to add dynamics (e.g., for animations) to the otherwise static reconstruction of the musculoskeletal system (Kähler et al., 2003)(Fig. 6-16). FEM methods are also used to simulate changes in soft tissue morphology following a surgical intervention (Remmler et al., 1998).

## 6.9.3 What Shall Be Reconstructed?

One central question of fossil hominid soft tissue reconstruction, particularly of the face, concerns perceptual rather than anatomic/morphologic issues. The human visual system is tuned to "classify" faces with respect to various features, such as sex, age, ethnicity, and personal acquaintance. Although forensic reconstruction aims at inference of individual facial parameters and draws on our

◄

**FIGURE 6-17** Reconstruction of the Gibraltar Neanderthal child's face. **A:** Corresponding anatomic landmarks on the Neanderthal child skull (right side; see Box 6-4) and a modern human child skull (left side; composite clinical CT data from two subjects). **B:** Morphing of modern soft tissue structures based on skeletal landmarks. **C:** Plaster reconstruction (left) and finished model (right) of the Neanderthal child.

recognition and classifying skills, the same classifying mechanisms often interfere with the aims of fossil soft tissue reconstruction. For example, we tend to perceive the reconstructed Gibraltar Neanderthal child (Box 6-8) as an individual rather than as a specimen representing the species characteristics of *Homo neanderthalensis*. Likewise, the question of whether this individual's dark skin color represents suntan or genetic disposition typically occupies the minds of museum visitors more than the morphologic differences it exhibits in comparison with modern humans. On the other hand, the soft tissue reconstruction immediately raises the question why this 3- to 4-year-old child looks "older" than a modern human of the same age. Drawing on our ability to estimate age from facial appearance leads us to think about differences in developmental timing between species that are hard to conceive on a purely skeletal basis.

How can we find a balance between the specimen representing an individual and the specimen representing a fossil taxon? There is no unique solution to this dilemma, but virtual reconstruction provides several ways to tackle it. One possible solution is shown in Box 6-8. Following the reconstructive range concept developed for skeletal reconstruction, we may show various stages of the reconstruction and propose several (virtual and real) final variants.

### 6.9.4 Soft Tissue Reconstruction and Measurement

Most of our skeleton can be thought of as being contained within soft tissues that determine the external appearance of our body. In contrast, the braincase is a skeletal structure that envelops significant portions of soft tissue. In fact, one of the most prominent applications of soft tissue reconstruction—although rarely declared as such—is the evaluation of endocranial volumes. We have already discussed various practical approaches to virtual endocranial reconstruction (Fig. 6-7), but how do the above considerations on soft tissue reconstruction impinge on this?

Above in this section, we stated that the correlation between skeletal and muscular elements is relatively loose, such that reconstruction by reference to extrinsic information represents extrapolation rather than interpolation. It is often overlooked that the same is true for the relationship between the braincase and the brain (Zollikofer and Ponce de León, 2001a). Endocranial surface structures are only loosely correlated with brain volume and cortical surface structures, because various layers of tissue separate the surface of the brain from the endocranial surface of the bone (see Fig. 8-30). To obtain reliable estimates of brain rather than endocranial volumes, more quantitative data on the relationship between the brain and its case are required.

Overall, although the topographic anatomy of humans and apes is well known, we are only beginning to understand the *quantitative* three-dimensional relationships between hard and soft tissues (Nelson and Michael, 1998). Further research in soft tissue reconstruction requires establishment of extended comparative volumetric databases with medical imaging technologies such as CT and MRI. The Visible Human project and related projects in virtual animal anatomy represent an important step toward this goal (Spitzer and Whitlock, 1998).

## 6.10 VIRTUAL SURGERY: A PALEOANTHROPOLOGIST'S EYE VIEW

### 6.10.1 Motivation

Computer-assisted surgery (CAS) is a burgeoning field of research, whose topics and current developments are amply covered in specialized journals and conference proceedings volumes. CAS permits precise planning and rehearsal of complex interventions, facilitates the design of custom-tailored prostheses, anticipates the esthetic result of interventions, and supports delicate manipulatory tasks during the real intervention. By improving surgical efficiency, these procedures help to improve patient safety on short-term and long-term scales.

It is beyond the scope of this book to delve into details (general references are given at the end of this chapter). Nevertheless, it might be instructive to realize that the same algorithms, computer tools, and parsimony rules that we used to de- and reconstruct virtual fossil morphologies can be applied to patient data to plan and simulate surgical interventions and their outcome. In the following sections, we envisage potential clinical applications in a series of case studies in virtual surgery.

### 6.10.2 Virtual Planning and Simulation of Surgical Interventions

Computer-based planning of cranio-maxillofacial surgical interventions has become an important complement to classic planning methods, particularly because helical CT offers the potential of acquiring entire cranial volume data sets within one breathhold. Commercial medical imaging software now offers the possibility of planning and carrying out virtual osteotomy, of relocating skeletal elements according to the surgeon's indications, and of visualizing the outcome of these virtual operations. Notably in reconstructive surgery after trauma, conceptual parallelisms exist with fossil reconstruction. A central question is, How can preserved skeletal and soft tissue parts be relocated and complemented with prosthetic parts so as to achieve a most parsimonious reconstructive solution?

Let us consider a scenario in which reconstructive principles established for fossils come to bear (Fig. 6-18). This patient miraculously survived an attack by a bear but suffered almost complete destruction of the midface. After emergency surgery, CT revealed that most of the osseous midfacial structures had been lost

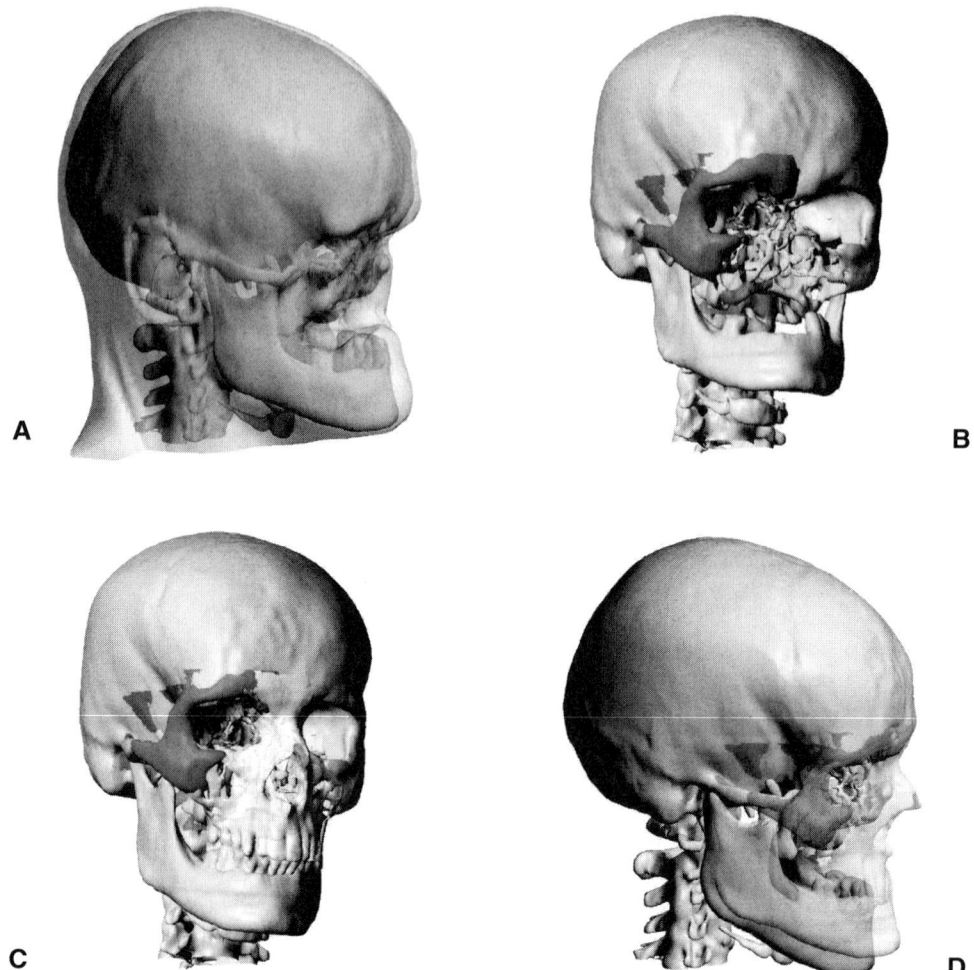

**FIGURE 6-18** Computer-assisted planning for reconstructive surgery. The original dimensions of the severed midface of this patient (**A**) were recovered by mirror-image completion of lesioned parts (**B**, dark), matching comparative virtual skeletal elements from a healthy individual (**C**, light) and corresponding downward rotation of the preserved mandible (**D**).

and had to be reconstructed with soft tissue/bone autografts and a maxillary prosthesis. Before reconstructive surgery, it was necessary to infer the original dimensions of the patient's face to reestablish the physiological position of the mandible and to design the prosthetic parts. Facial height had been shortened considerably as a consequence of upward rotation of the mandible because of the absence of maxillary counterparts; the right zygomatic arch had been crushed and bent inwards (Fig. 6-18A).

We used two different approaches to evaluate the original height and width of the face. By mirror-imaging the preserved left orbital cavity and zygomatic bone to the right side and adjusting for asymmetry, it was possible to reconstruct the transverse dimensions of the face (Fig. 6-18B). Linear measurements taken from these parts were used as predictors of facial height, using Howells's craniometric data (Howells, 1989, 1996). In a complementary approach, maxillofacial structures from a skeletal sample were adapted to the patient's skull, using landmark-based morphing procedures as described in Box 6-8 (Fig. 6-18C). Combination of the results of the craniometric and morphing approaches indicated that the mandible had to be rotated downward by 10° to reconstitute normal facial proportions (Fig. 6-18D). These planning steps provided the quantitative basis for a subsequent sequence of five surgical interventions.

## 6.10.3 Custom Implant Design

Surgical reconstruction of large traumatic defects often involves implantation of prosthetic plates. Whereas interventional demands are comparatively moderate, reconstruction is delicate from an esthetic point of view. Notably in the frontal region of the cranial vault, even subtle form changes affect the individual appearance of the patient.

Custom implant design, that is, modeling patient-specific prostheses, is similar to complementing missing parts in a fragmentary fossil. Figure 6-19 shows the case of a child with a large traumatic defect in the right frontoparietal region and extended lesions of the midface (Buitrago-Téllez et al., 1997a,b). Whereas the midface was reconstructed with autografts, the gap in the cranial vault was filled with a ceramic custom implant. Implant design involved several stages. CT showed that the anterior margin of the right parietal bone, the entire right frontal bone, as well as parts of the left frontal were lacking (Fig. 6-19A,B). Three-dimensional mirror copies of the preserved bones on the left side of the cranial vault were used to cover approximately two-thirds of the lesion (Fig. 6-19C). The remaining openings in the frontal bone, where no original counterparts were available, were filled with CAD procedures. We designed different variants of the midfrontal region and assessed their potential effect on the patient's general appearance until a satisfactory solution was found (Fig. 6-19D). The finalized custom implant consisted of two parts divided along the coronal suture. This design permits accommodation of changes in frontal size and shape during growth. The actual ceramic implant was produced on the basis of stereolithographic replicas of the custom implant (see Fig. 7-9).

## 6.10.4 Soft Tissue Reconstruction

Soft tissue reconstruction in patients starts from different preconditions than soft tissue reconstruction in fossils, because clinical CT and MRI data sets provide detailed information about a patient's preoperative soft-to-hard tissue

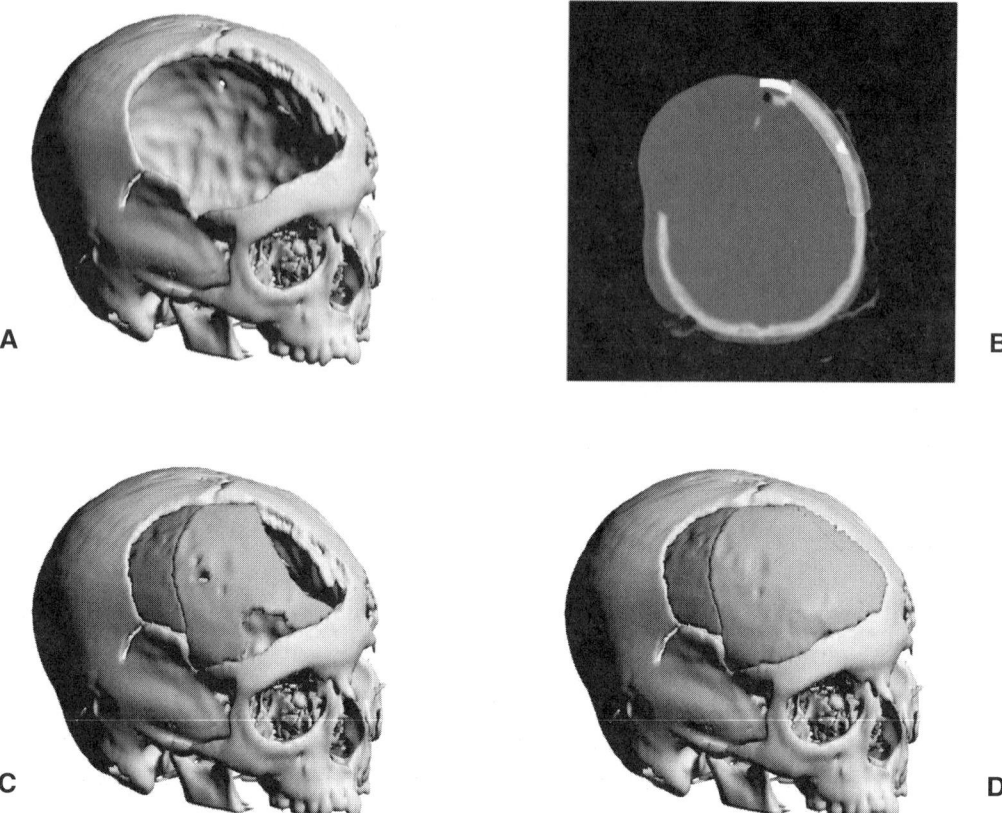

**FIGURE 6-19** Custom implant design in a patient with a large frontoparietal traumatic lesion (**A**). Three-dimensional mirror images of preserved regions (**B**, highlighted area) of the right parietal and frontal bones were used to cover the defect (**C**). CAD modeling techniques were used to fill regions where no original bone structures were available (**B**, white areas). The two pieces of the final prosthetic structure (**D**) are separated along the coronal suture to allow adaptation to form changes during skull growth. Stereolithographic replicas of these pieces were used as templates for the production of a ceramic implant (after Buitrago-Téllez et al., 1997a).

relationships. On the basis of these data, it is possible to infer postoperative morphologic change. Typically, the virtual surgical intervention is carried out on the skeletal elements and its impact on soft tissue structures is inferred as a function of spatial skeletal alterations and of the mechanical properties of the soft tissue (Remmler et al., 1998).

Surface data of the face are of special interest, because cranio-maxillofacial interventions aim not only at correction of a pathological state but also at esthetic improvement of the facial appearance. Planning esthetic corrections is a process during which the ideas and wishes of the patient are combined with the medical needs to achieve an optimum solution. Virtual reconstruction offers the possibility of simulating the transition from preoperative to postoperative appearance interactively, ideally under direct supervision of the surgeon and the patient.

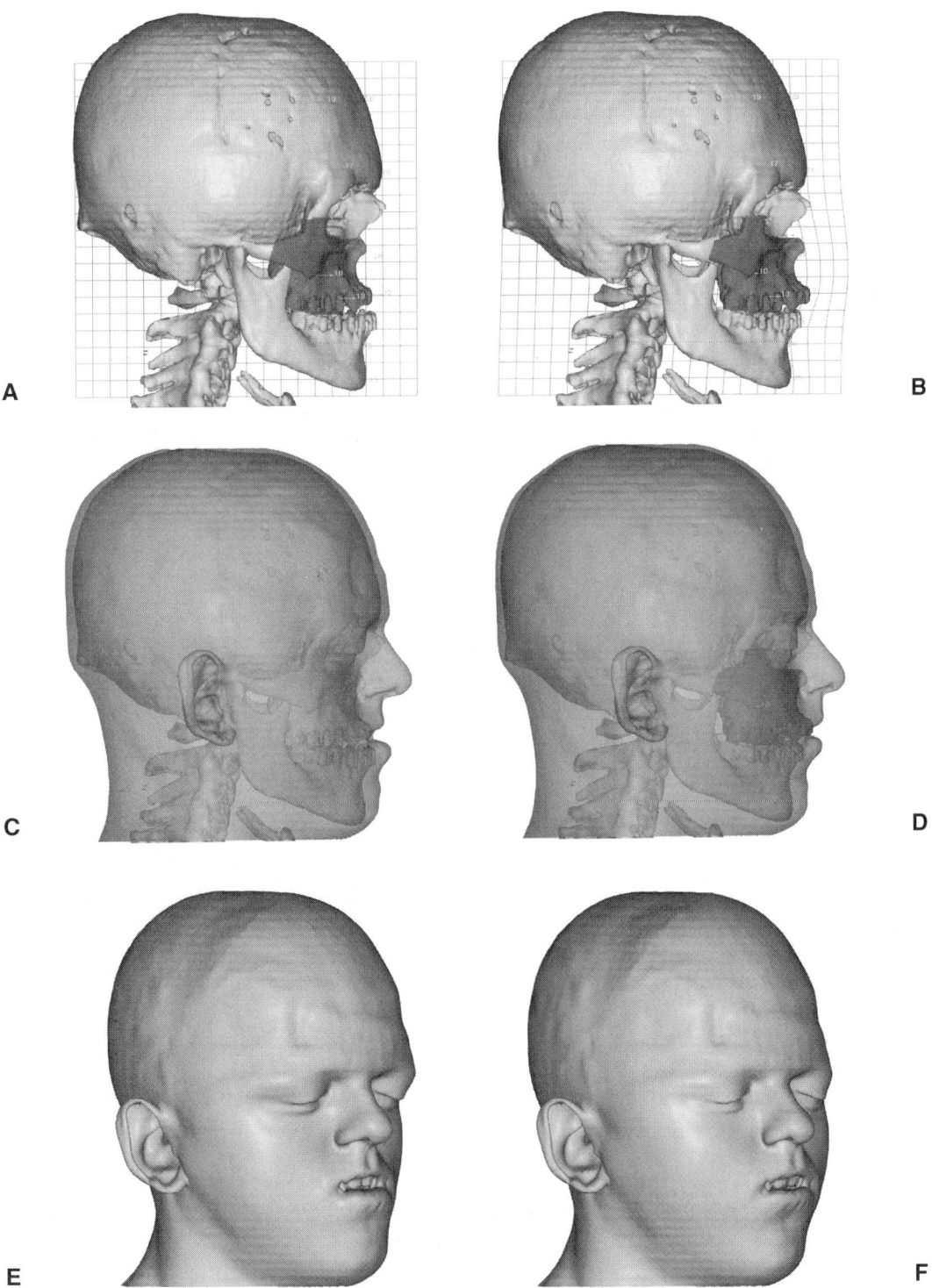

**FIGURE 6-20** Computer-assisted planning of midfacial advancement and concomitant soft tissue changes in a patient with Crouzon syndrome. To achieve normal dental occlusion, the midface is advanced by 10 mm (**A, B**). The positional shift of skeletal anatomic landmarks (numbered arrows) defines a three-dimensional TPS morphing function, whose effects are visualized with a midsagittal grid. The interpolant is applied to the soft tissue parts (**C, D**) in order to predict changes in facial appearance effected by skeletal advancement (**E, F**).

Consider the case of a young patient (Fig. 6-20) suffering from Crouzon disease, a congenital developmental disorder involving lagged growth of the midface and resulting in impairment of respiration, speech, and dentognathic articulation (Jones, 1988). Surgical treatment is complex, involving mobilization and advancement of the maxilla to adjust dental occlusion. The surgical intervention was simulated on the virtual skeletal parts (Fig. 6-20B). To infer postoperational changes in soft tissue morphology, we proceeded analogously to the soft tissue reconstruction of the Neanderthal child (Box 6-8). The dislocation of skeletal landmarks from pre- to postoperative positions was used to define a three-dimensional morphing function, which yielded an estimate of the postoperative face according to the displacement of the skeletal parts. The predicted facial data can ultimately be compared with actual postoperative data to evaluate the goodness of fit between predicted and actual facial topography and to improve methods of inference.

## 6.11 FURTHER READING

Early attempts at reconstructing fragmentary fossils in a virtual environment are described by Kalvin et al. (1995) and Zollikofer et al. (1995). Concepts of virtual reconstruction are discussed and exemplified in Zollikofer et al. (1998) and Ponce de León and Zollikofer (1999). The importance of virtual fossil reconstruction as a prerequisite for three-dimensional morphometric analyses is shown in Ponce de León and Zollikofer (2001).

Studies dedicated to the reconstruction and/or inference of missing data abound in the paleontological and paleoanthropological literature; rather than giving an overview, we leave the reader with the task of verifying reconstructive parsimony (notably reference to symplesiomorphic comparative samples) in each single case. One of the most hotly debated topics of reconstruction in paleoanthropology is brain size, which is typically estimated by using endocranial capacity as a proxy. Again, the literature abounds. As a starting point, see Ralph L. Holloway's pioneering early studies (1975) (see also http://www.columbia.edu/~rlh2). Regarding virtual reconstruction, CT-based inference of endocranial volumes was initiated by Glenn C. Conroy (Conroy and Vannier, 1984; 1985; Conroy et al., 1990), and new data are produced on a continuous basis (Zollikofer et al., 1995; Conroy et al., 1998, 2000, 2002). Interestingly, many debates about endocrania arise from uncertainties about the criteria used during reconstruction (Lockwood and Kimbel, 1998). Virtual fossil reconstruction, as proposed here, will help clarify many of these issues.

Recalling the references discussed at the end of Chapter 3, it is important to note that computer tomography has become a standard tool of *paleodiagnostics*. In this area, the literature is growing exponentially. We just provide a selection of references related to the following subjects:

- *Primates*: Vannier et al. (1985); Spoor et al. (1994); Flynn et al. (1995); Spoor and Zonneveld (1995); Rae and Koppe (2000); MacLatchy and Muller (2002); Ryan and Ketcham (2002) .
- *Archeological specimens, notably mummies*: One of the pioneering projects in this field is the "virtual mummy" of K.-H. Hoehne's research group (`http://www.uke.uni-hamburg.de/institute/imdm/idv/forschung/mumie/index.en.html`), and the notorious iceman, "Oetzi" (Pickering et al., 1990; Yasuda et al., 1992; Melcher et al., 1997).
- *Dinosaurs*: Jones et al. (1998), Stokstad (2000), Golder and Christian (2002). A suite of papers dealing with a dinosaur "heart" show the limitations of the method (Dalton, 2000b; Fisher et al., 2000; Rayfield et al., 2001; Rowe et al., 2001a: Morell, 2000).
- *Fossil vertebrates in general*: Rogers 1999; Maisey 2001; Spoor et al. 2002; Clack et al., 2003.
- *Microstructural and macrostructural biomechanics*: Müller and Rüegsegger 1995; Merz et al., 1996; Rüegsegger et al., 1996; Borah et al., 2000; Borah et al., 2001; Rayfield et al., 2001; van Rietbergen 2001; Cooper et al., 2003.

Links to related websites are provided on the Web Companion.

One study on fossil mollusks is worth being noted here, because it demonstrates that classic and virtual methods can be combined during virtual reconstruction. Sutton et al. (2001) use an invasive data acquisition method (creating cross-sectional images by grinding away a fossil layer by layer) and three-dimensional reconstruction to reveal the morphology of fossil mollusks at unprecedented levels of detail. Paleoanthropologists typically do not mill their specimens into dust; however, even prominent specimens have to pay their tribute once in a while to fossil DNA analysis (see Krings et al., 1997; Lindahl, 1997)—after having been CT-scanned, of course.

An aspect of computer-assisted paleoanthropology that deserves special consideration is the fact that virtual fossil data can easily be exchanged via the Internet, whereas access to originals is typically limited by various regulations. Virtual fossils on the Internet represent an vast potential as they enable free access to important scientific resource material. However, web-based scientific freedom is relative; most countries from which important fossils are recovered do not possess the technological, infrastructural, and personal resources to make scientific use of this information, at least not at the pace dictated by modern science. The resulting unidirectional knowledge transfer—though unintentional—creates problems that are similar to those associated with "gene trading." A snapshot of the current state of affairs is given by Gibbons (2002).

In the biomedical and clinical area, the Visible Human Project may serve as a prototype of virtual reconstruction/diagnostics (Höhne et al., 1992; Spitzer and Whitlock, 1998). Introductory textbooks on computer-assisted surgery are Robb (1998), and Mudry et al. (2003). Consult the Web Companion for further links.

# FROM VIRTUAL REALITY TO REAL VIRTUALITY

Reality is merely an illusion, albeit a very persistent one.
—ALBERT EINSTEIN (1879–1955)

## 7.1 REIFYING VIRTUAL OBJECTS

One of oldest known stories about the wish to reify a virtual image comes from Greek mythology: Narcissus saw his own image reflected in water. Not realizing that it was himself, he found it so beautiful that he immediately fell in love and tried to grasp this fascinating face with his hands. What a disappointment when he had to realize that, as soon as he touched the water surface, the creature vanished! Similarly unhappy was Pygmalion, a skillful sculptor who fell in love with his most beautiful sculpture and desperately tried to animate it.

The wish to materialize virtual objects that exist in our imagination, and to re-create natural objects including ourselves, is as old as humanity, as evinced by early Aurignacian artwork (see Fig. 8-15).[1] As we recognized in previous chapters, virtual reconstruction does part of the job, and "animation" has become a standard term of virtual reality applications. In this chapter we focus on methods that allow us to materialize the results of virtual reconstruction. In other words, we are on the way back from virtual reality to real virtuality, which represents the final step in a reverse engineering process (see Chapter 1).

Why is it appropriate or even necessary to reify the results of virtual reconstruction and return to real physical models? Recall that a major goal of virtual reality consists of establishing a perceptually equivalent world, that is, a world that conveys multimodal sensory feedback in real time. We recognized that, with increasing object complexity, perceptual equivalence is increasingly difficult to achieve. Rather than attempting replication of real-world visual, acoustic, and

---

[1] It is a short way from Narcissus' mirror image to human clones.

*Virtual Reconstruction: A Primer in Computer-Assisted Paleontology and Biomedicine.*
By Christoph P. E. Zollikofer and Marcia S. Ponce de León.

touch-and-feel feedback systems, therefore, it seemed sensible to follow the path of enhanced reality and provide the user with a peek at "invisible" properties of the objects under investigation.

As expressed by the concept of "real virtuality" (RV) (Bresenham et al., 1993), hard copies of virtual objects combine the advantages of virtual and enhanced reality with the advantages of physical reality. From this perspective, RV objects appear as compact analog three-dimensional data storage and representation media of great robusticity that permit device-independent, multimodal, and delay-free manipulation of 3D objects, giving the user high perceptual equivalence (Zollikofer and Ponce de León, 1995).

Rapid prototyping (RP) models are therefore useful complements to virtual reconstruction, and they have become important constituents of practical clinical applications such as surgical planning and rehearsal. However, it should be kept in mind that RP is a technology rather than a research method. Accordingly, its significance for bioscientific research primarily consists of providing new means of communication of scientific results, rather than providing new results.

## 7.2 PRINCIPLES OF RAPID PROTOTYPING

Rapid prototyping comprises various technologies for automated production of physical hard copies of virtual objects. A major distinction can be made between subtractive and additive fabrication techniques. In subtractive fabrication, the model is carved out of a material block by a computer-guided milling machine, whereas during additive fabrication, the model is built up layer by layer from various materials.[2] Common to both techniques is the principal task of transforming geometric data of a virtual object into machine tool instructions that specify how to remove or add material.

Additive and subtractive techniques have their strengths and weaknesses that result from trade-offs between material properties and geometric properties (i.e., form). Whereas subtractive fabrication offers freedom of choice regarding materials, additive fabrication provides freedom of complexity regarding form. With milling technology, virtually every material can be worked. For example, it is possible to produce functional custom implants consisting of biocompatible, corrosion-proof, and stress-resistant materials. However, the degrees of freedom of movement of the machine tool and the dimensions of the milling head typically restrict part complexity. With additive technologies, on the other hand,

---

[2] In architecture, a similar distinction can be made. Construction typically follows the additive approach; the subtractive approach is most beautifully implemented in the famous early Ethiopian rock churches, and in many rock dwellings all over the world.

objects of arbitrary topological complexity can be built in one single production process. For example, intricate structures of the human skull such as foramina, bony bridges, canals, and cavities, all of which are beyond for the reach of a milling machine's mechanical arm, can be reproduced with layer-by-layer additive technology. However, the mechanical properties of materials suitable for additive fabrication are typically less favorable, and biocompatibility remains a challenge. In the following sections, we will concentrate on additive technology, because emphasis is put on model complexity rather than functioning prototypes.

As mentioned in Chapter 1, the basic principle of additive fabrication is similar to a procedure that readers may know from manual model building. Architects, for example, cut out cardboard layers and stack them on top of each other to obtain three-dimensional terrain models. It seems straightforward to automate this process and materialize a stack of patient CT images by layered fabrication. In fact, this technology has been transposed into the digital world. Known as *laminated object manufacturing* (LOM), it is especially well-suited to producing large-scale models. Typically, sheets of adhesive-coated paper are stacked on top of each other. A high-power infrared ($CO_2$) laser beam cuts each cross section's boundary structures into the paper sheet, such that excess material outside the cross-sectional object area can be removed (note that this latter step is reminiscent of subtractive fabrication).

Over the past two decades, an wide variety of further automated layer-by-layer building technologies have been implemented, most of which use prime materials that are solidified locally. We will concentrate here on those technologies that are relevant for bioscientific prototyping.

- *Fused deposition modeling* (FDM) is based on the principle of extrusion of thermoplastic materials. Typically, a filament of thermoplastic material is extruded through a computer-guided nozzle that builds layers by depositing material along lines (similar to extrusion of toothpaste onto a toothbrush). The resulting models are comparatively coarse, but the technology is easy to handle and available in desktop format.
- *3D-printing* technology expands the principle of inkjet printing. Instead of printing with ink on consecutive sheets of paper, thin layers of paraffin-based materials are printed on top of each other. As in inkjet technology, the material is temporarily heated and blown through an array, or matrix, of fine nozzles. Layer thickness is in the range of 0.02 mm, such that the resulting models reproduce fine structural detail. However, because of constraints imposed on building materials (short melt-and-solidify cycles), 3D-printing models are comparatively soft and fragile.
- *Selective laser sintering* (SLS) is at the other end of the scale of material stability. Here, the building material comes in form of a powder. A high-energy

laser beam "draws" like a pen on the surface of the powder. Through heating, particles are baked together, resulting in local *sintering* (the technical term for this type of particle solidification). Consecutive layers are built by covering the sintered cross-sectional surface with a layer of powder of defined thickness, which is subsequently solidified.

- *Laser stereolithography* uses similar procedures. The idea here is to polymerize a liquid photosensitive resin by using UV light as a trigger (sensitivity to visible light would be impractical). In a typical technical implementation (Fig. 7-1), a computer-guided laser beam draws on the liquid's surface, triggering local hardening through polymerization. In analogy to the process of drawing an object on a sheet of paper, the laser first outlines object contours and then cross-hatches solid areas. Compared to high-energy lasers needed for SLS, lasers used for photopolymerization are weak, because their only function consists of providing a modest amount of activation energy that initiates the polymerization reaction. Current stereolithographic resins are based on acrylate, epoxy, or vinylether components.

- *Solid ground curing* (SGC) is based on factual 2D polymerization. Rather than "drawing" an object layer with a point-shaped laser beam, each layer is

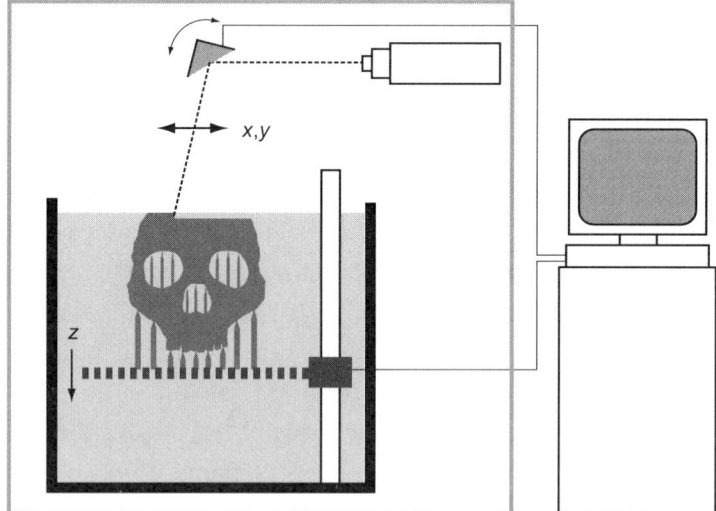

**FIGURE 7-1** Principle of laser stereolithography. A computer-guided UV laser beam polymerizes the surface of a liquid photosensitive resin contained in a vat. The $x,y$ movements of the laser are mediated by a set of deflecting mirrors. After completion of each layer, the object is flooded with resin, and the liquid surface is leveled with a sweeper; downward movement of the elevator in the $z$-direction determines layer thickness, $\Delta z$. Model production includes generation of supporting structures that prevent sagging of isolated parts. At the end of the production process, the part is completely immersed in the vat. Postprocessing involves skipping off supporting structures, removing excess resin, and polymerization of remnant resin with a UV flood light.

photo-polymerized with a strong UV flood light in one single building step. This technique requires complex preparatory steps; regions outside the object are shielded from incident UV light with a mask, which is applied to a glass plate with xerographic techniques.

From a practical point of view, it is helpful to compare these techniques with respect to space requirements and ease of handling. FDM devices come as desktop printers and are thus easy to handle. 3D-printing devices are the size of a large deep-freezer. Because of the chemical inertness of the building materials, they can be installed and run like office-type standard 2D printers. Laser stereolithography is more demanding; liquid photopolymers contain reactive components that require specific safety measures during handling and special-purpose technical installations for postprocessing. SLS is even more demanding, not least because of the energy requirements of the laser sintering process. And finally, the room-filling SGC machines can only be run by specialized technicians.

How are three-dimensional object data transferred from a virtual environment to a rapid prototyping machine? The de facto standard format for data exchange and transmission is the *stl* (*stereolithography*) format, which describes three-dimensional free-form objects as triangulated surfaces (see Chapter 2). Before part building, stl data sets are converted into device-specific data formats. To this end, the object is cut into virtual slices that correspond to the successive layers of the building process. Ideally, RP devices are integrated into computer networks such that they can be operated like standard printers, permitting object selection and specification of the number of copies and of optional scaling factors.

In contrast to virtual-reality objects, real-virtuality objects must comply with the laws of physics, both during part building and during subsequent manipulation. A specific issue resulting from real-world constraints is the construction of suspended and overhanging object regions. As in real construction, supporting structures must be integrated during the building process to prevent sagging of parts. These supports must be stable but relatively slender such that they can be easily removed after construction (Figs. 7-2, 7-3D). Furthermore, individual parts "floating" in virtual space must be connected to each other to obtain a functional ensemble (Fig. 7-3A).

User control over building parameters such as layer thickness, material density and support design depends on the type of RP system used. With respect to user friendliness, 3D printers come closest to the comfort of a modern 2D printer; data conversion from the stl file format to device-specific formats, evaluation of optimum building parameters, and support construction are fully automated. Laser stereolithography, on the other hand, permits full control over every detail of part and support construction but requires special-purpose software to perform these prebuilding operations.

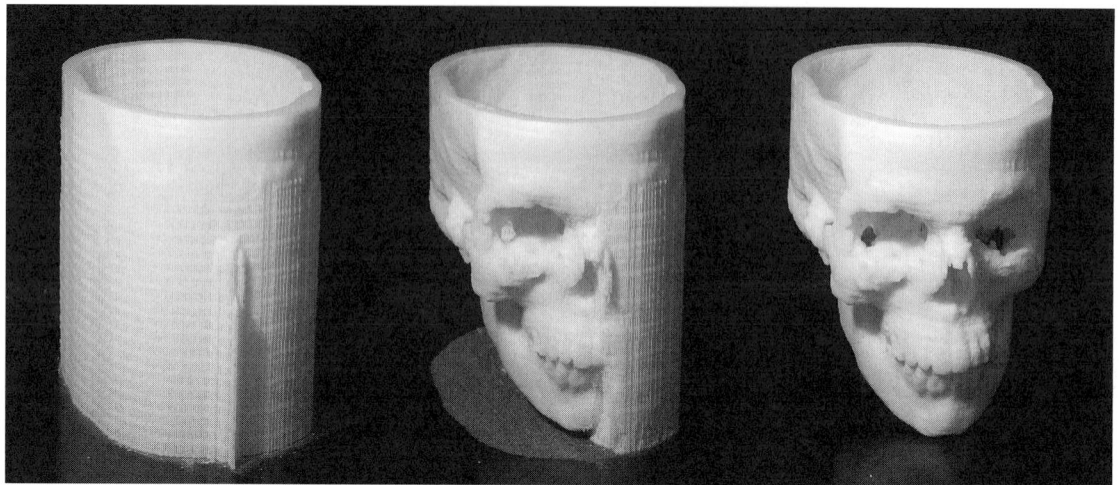

**FIGURE 7-2** 3D printing of a patient skull model. The raw model is contained in a framework of supporting structures, which can be removed with a soft brush.

In most RP systems, the actual part building process is fully automated. The total build time depends on various parameters. It can be reduced by using chemical components that permit faster polymerization, by building fewer but thicker layers, and by reducing the amount of material used per layer, for example, by creating hollow rather than solid (filled) objects. With laser stereolithography or 3D-printing technology, prototyping a skull model in natural size (see Figs. 7-2, 7-3, and 7-5) takes between 12 and 24 hours.

Parts produced with RP technologies typically require postprocessing to obtain a finalized model. First of all, supporting structures and excess material must be removed from the part. 3D printers produce supporting webs that can be removed from the finalized object with a toothbrush (Fig. 7-2). In stereolithography, postprocessing is more labor-intensive (Fig. 7-3). After construction, the model is removed from the resin vat, and excess resin is allowed to drip off. Supports are then skipped off, and nonpolymerized resin is washed out with organic solvents. Finally, the model is postcured in a UV floodlight chamber to polymerize remnants of resin and harden its surface.

Concepts of resolution, accuracy, and precision developed in Chapter 3 can be applied to assess the performance of RP technologies in part replication and to compare these technologies with classic cast-and-mold replication techniques. Such comparisons must involve the entire reverse engineering process, from CT data acquisition through three-dimensional reconstruction to model production. As a matter of fact, the layer thickness of a stereolithographic replica imposes a lower limit on spatial detail resolution. Before production, it is therefore sensible to adjust the layer thickness to values that are smaller than the actual pixel or

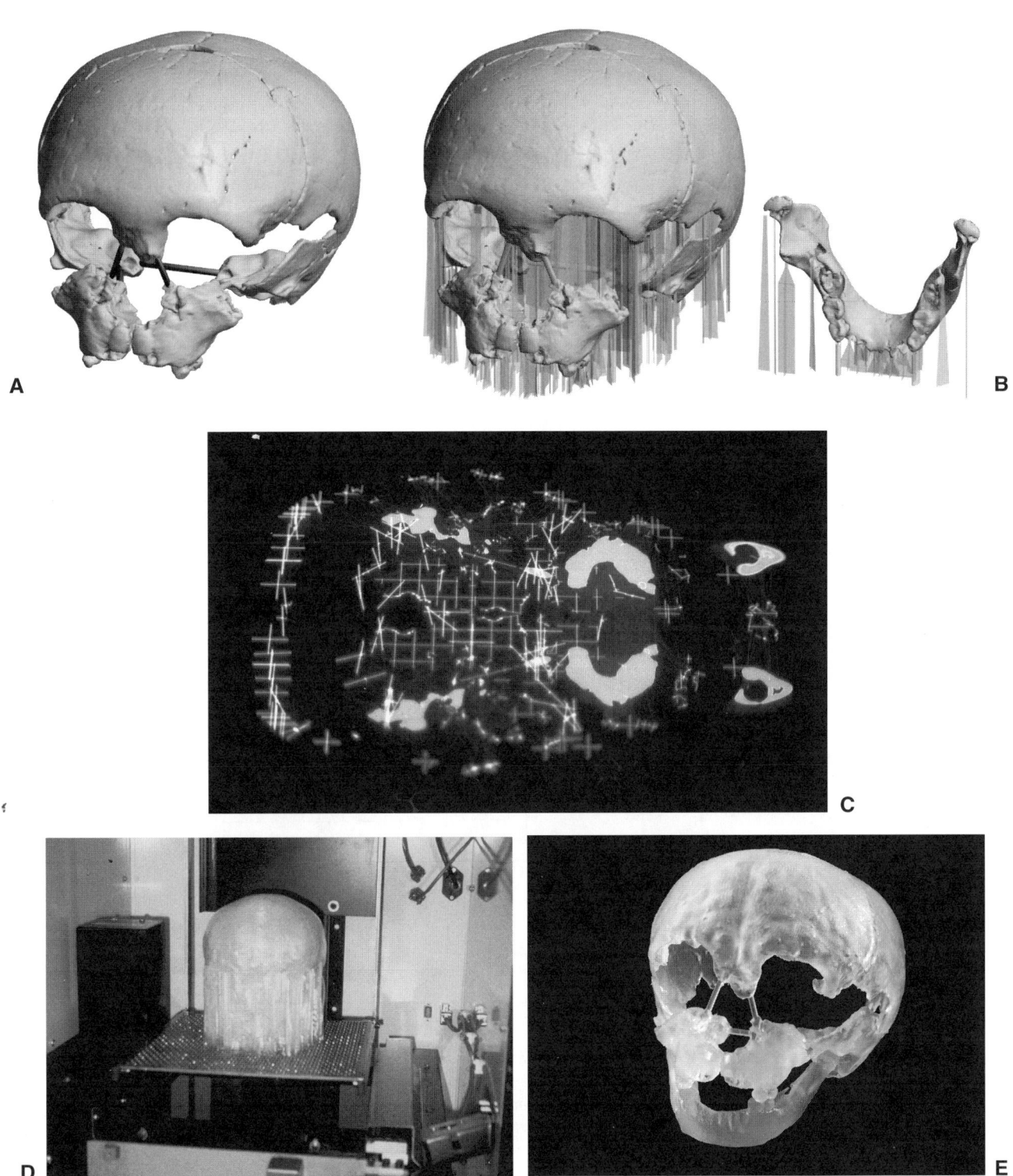

**FIGURE 7-3** Stereolithographic replication of the virtual reconstruction of the Devil's Tower Neanderthal child skull. **A**: Construction of connecting struts between isolated parts. **B**: Generation of supporting structures. **C**: Stereolithographic part building through layer-by-layer polymerization. This photograph with an exposure time of 2 minutes shows how the laser beam draws one cross-sectional layer onto the liquid resin surface (the crosshair shapes represent supporting elements). **D**: The raw model with support structures. **E**: The final model.

voxel size of the original data volume. Obviously, this does not enhance spatial resolution beyond the limits of the data acquisition device, but it minimizes stairstep effects in the hard copy of the virtual object (Fig. 7-4A). Overall, under optimum data acquisition, reconstruction, and reproduction conditions, it is possible to achieve spatial detail resolution in the range of 0.1–0.2 mm.

Mold-and-cast replication is superior with respect to detail resolution in the submillimeter range; for example, the surface texture of a fossil tooth with its highly significant scratches, polished wear facets, and perikymata (dental growth lines) is clearly below the resolution of RP model production. Regarding the replication of linear measurements in the millimeter and centimeter ranges, however,

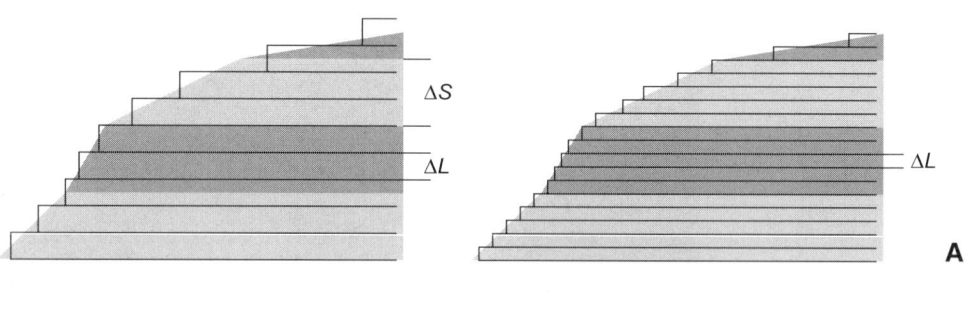

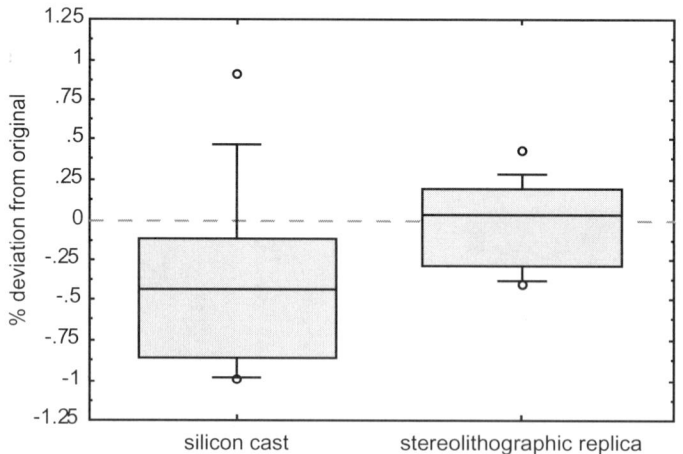

**FIGURE 7-4** Resolution, accuracy, and precision of stereolithographic replicas. **A**: Model resolution is limited by the resolution of the imaging device, for example, the between-slice distance of a CT ($\Delta S$; CT layers are indicated as light and dark gray bands) and by the layer thickness of the stereolithographic device ($\Delta L$). To avoid stairstep effects, stereolithographic layer thickness must be considerably smaller than CT interslice distance. **B**: Accuracy of conventional silicon casts and of stereolithographic replicas of the Gibraltar 2 Neanderthal child skull (after Zollikofer et al., 1998). The graph indicates deviation in percentage of measurements taken on the original ($N = 30$; measurements range from 10 mm to 150 mm); boxes and whiskers indicate 50th±25/±40 percentiles, respectively; dots indicate the range. Note overall shrinkage of silicon cast.

RP models are superior to classic casts (Zollikofer et al., 1998). Because RP production is based on digital data, and because part production is not constrained by model topology, stereolithographic replicas can be produced in one processing step, in large numbers, and at reproducible levels of quality. For example, material shrinkage, which represents a major problem in classic casting, and which also occurs in photopolymers, can be compensated for by adjusting the respective construction parameters.

## 7.3 COMBINING VIRTUAL REALITY AND REAL VIRTUALITY

Real bioscientific objects, their virtual-reality complements, and their real-virtuality replicas can be combined in various ways during virtual reconstruction. Table 7-1 gives an overview of potential scenarios. Here, we explore applications in paleoanthropology, medical sciences, and related areas.

**TABLE 7-1   Scenarios of RP Use in the Biosciences**

| Area | Type of Application | Remarks |
| --- | --- | --- |
| Paleoanthropology | 1:1 Replication | Noninvasive alternative to classic cast-and-mold techniques |
| | Replication of invisible structures | Complement to computer-based enhanced reality (endocast production) |
| | Monitoring virtual reconstruction | Complex docking tasks between anatomically corresponding parts are difficult to verify with graphical methods only |
| | Replication of virtual reconstructions | Final step of reverse engineering |
| Clinical | Surgical planning and rehearsal | Complement to computer-based surgical planning |
| | | Physical rehearsal with surgical instruments; preoperative adaptation of prosthetic parts |
| | Custom implant design | Production of patient-specific prostheses (biocompatibility of RP prime materials still a challenge) |
| | Case collection | Teaching collection of syndromes, trauma, standard surgical interventions (*real* cases in anatomic collections are very rare) |
| Teaching/ education/ museology | Replication of cultural goods | Noninvasive alternative to classic cast-and-mold techniques; production of scaled-up replicas of small or invisible objects |

In paleoanthropology, casts of fossil or archeological specimens are an important means of scientific communication. A straightforward application of RP technology therefore consists in one-to-one replication of fossil specimens. A noninvasive reverse engineering approach that combines CT data acquisition, 3D object reconstruction, and RP hard-copying techniques is in fact an ideal alternative to conventional plaster or silicon casting (Hjalgrim et al., 1995). Many fossils are too brittle to be subjected to physical casting, and even relatively well-preserved specimens are exposed to a certain risk of damage and surface contamination during casting procedures. The reverse engineering approach can be extended to noninvasive replication of archeological objects and cultural goods, for example, for museum exhibits (Fig. 7-5) (Powell, 1994).

As a complement to computer-based visualization, internal object structures can be physically replicated with RP technology in various ways. Stereolithographic models made from acrylate or epoxy resins are transparent and thus permit direct inspection of, for example, the paranasal sinuses in a skull, whereas cavities such as the internal surface of the brain case, or the labyrinth structures

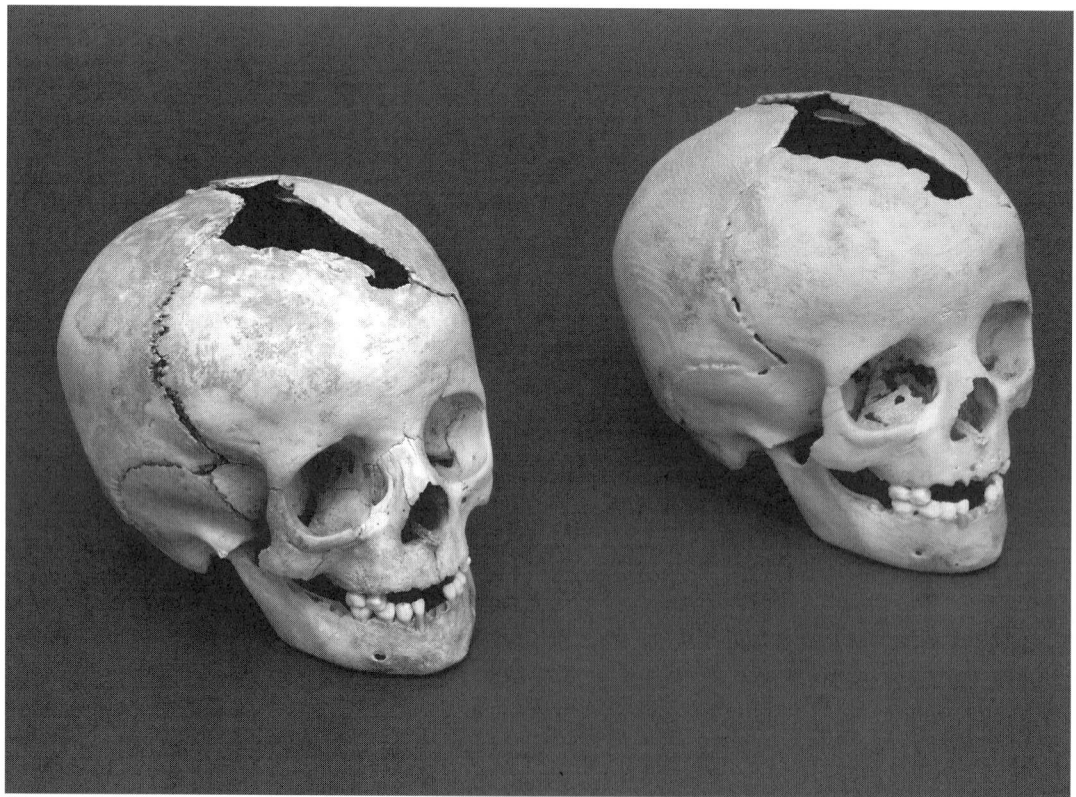

**FIGURE 7-5** Stereolithographic one-to-one-replication. This child skull (left) was recovered from a medieval graveyard during a rescue dig. Because of its brittle state of conservation, it was replicated with CT-based stereolithography (right).

of the inner ear, can be reproduced as hollow fillings, so-called *endocasts* (Fig. 7-6).

Building physical hard copies from virtual objects is not only a means of documenting the final result of virtual reconstruction but also offers the possibility to monitor spatially complex part assembly tasks that require multimodal rather than purely visual information. Let us return to the Devil's Tower Neanderthal skull as an example (see Box 6-4). Reestablishing dental occlusion between upper and lower teeth was a central step of the virtual reconstruction because the resulting anatomic fit imposes constraints on the relative positioning of the maxilla and mandible, which themselves influence the shape and positioning of the face relative to the cranial vault. A stereolithographic model of this stage of reconstruction was used to check the "goodness of fit" between upper and lower teeth and to explore associated degrees of freedom of motion of the jaws relative to each other (see Fig. 6-6). A similar but more intricate instance of dental reconstruction combining virtual and stereolithographic object manipulation is shown in Figure 7-7.

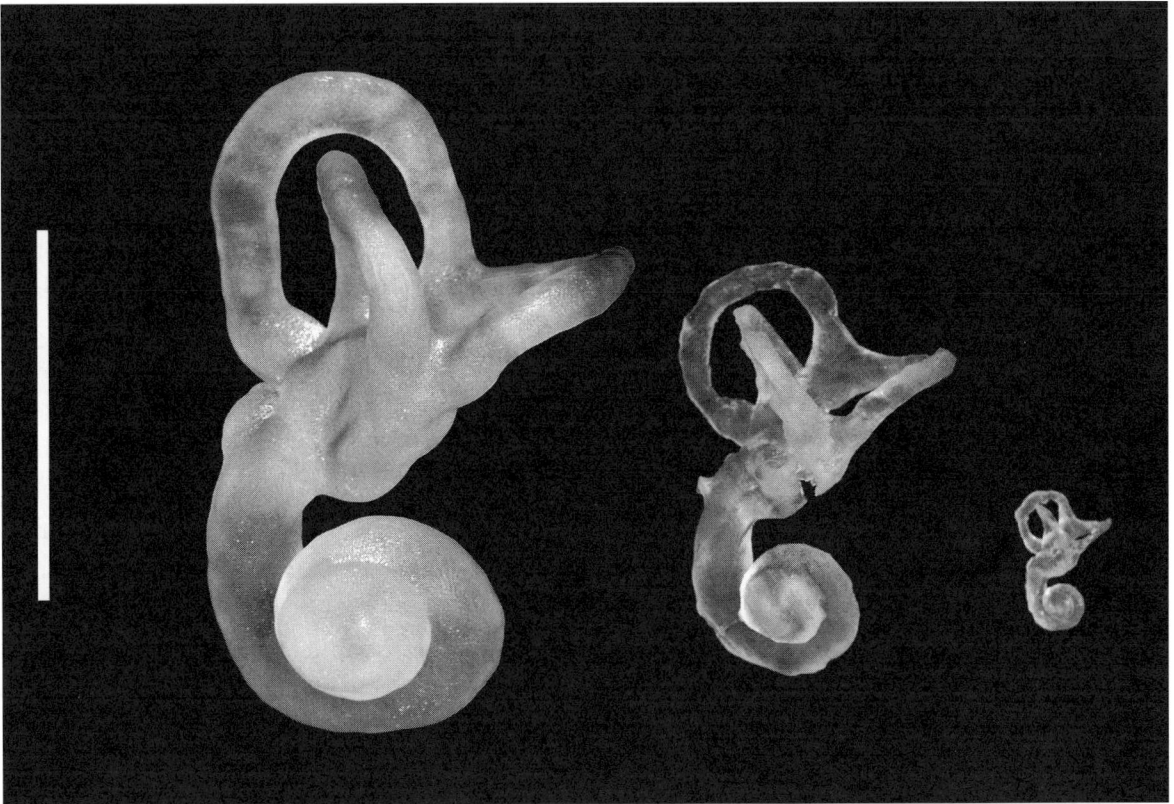

**FIGURE 7-6** Stereolithographic replication of an "invisible" structure. The senses of equilibrium, acceleration, and hearing are contained within the complex cavity system of the inner ear (length ~15 mm). It is replicated as a stereolithographic endocast (from left to right: modern human right inner ear, scale 8:1; Le Moustier 1 Neanderthal right inner ear at scale 3:1 and 1:1; scale bar is 50 mm).

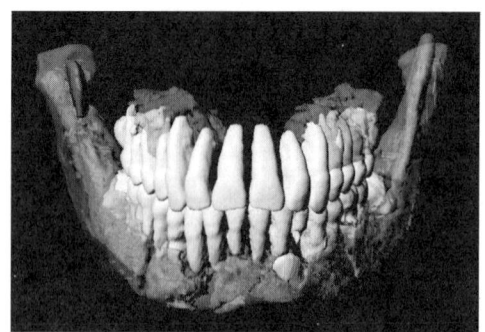
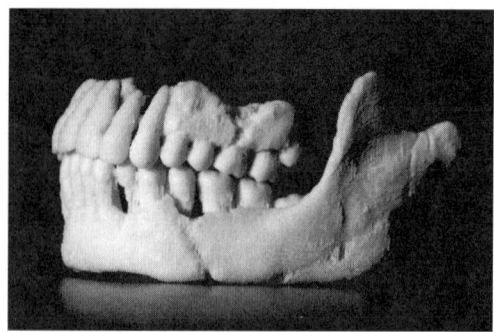

**A**                                                                                              **B**

**FIGURE 7-7**   Combining virtual reality and real virtuality during fossil reconstruction. **A**: Upper and lower jaws of the Le Moustier 1 adolescent Neanderthal are heavily fragmented, but the dentition is well preserved. Virtual reconstruction of the dental arcade thus proceeded via reenactment of dental occlusion. **B**: This spatially complex task was monitored with RP models produced with a 3D printer.

In many clinical applications, notably in the domain of maxillofacial surgery, the combination of computer-assisted surgical planning with custom RP modeling has become a standard procedure (Zonneveld and Noorman van der Dussen, 1992; Zonneveld, 1994; Petzold et al., 1999). Maxillofacial surgery typically deals with patients whose anatomy deviates significantly from standard conditions, be it through malformation or as a consequence of trauma. In both cases, detailed knowledge about the three-dimensional anatomic peculiarities of a patient's head is required to plan every step of a surgical intervention and anticipate potential difficulties. As in fossil reconstruction, a majority of surgical planning tasks can be performed in virtual reality; however, when it comes to surgical rehearsal and fitting of custom-made implants, virtual surgery is still far from physical reality. A number of specific surgical procedures thus can greatly profit from real-virtuality models. Many surgical interventions involve complex bone sectioning procedures followed by mobilization, rearrangement, repositioning, and fixation of separate elements. Furthermore, small prosthetic parts must typically be fitted to the patient's cranial anatomy during the surgical intervention. With a one-to-one patient model of the skeletal parts, the entire surgical procedure can be rehearsed manually, using the same instruments as during real surgery. Most notably, prosthetic parts can be tailored and adapted to the model before the actual intervention without intraoperational time constraints. Overall, manual rehearsal of operational steps and preadaptation of prosthetic parts increases surgical quality and helps to reduce operation time (Sailer et al., 1998).

Figure 7-8 shows the case of a patient affected by Crouzon disease, a growth disorder that leads to malformation of various parts of the skull and the limbs (Jones, 1988). Surgical correction involved complete mobilization and advancement of the midface. A stereolithographic model was used to rehearse each step

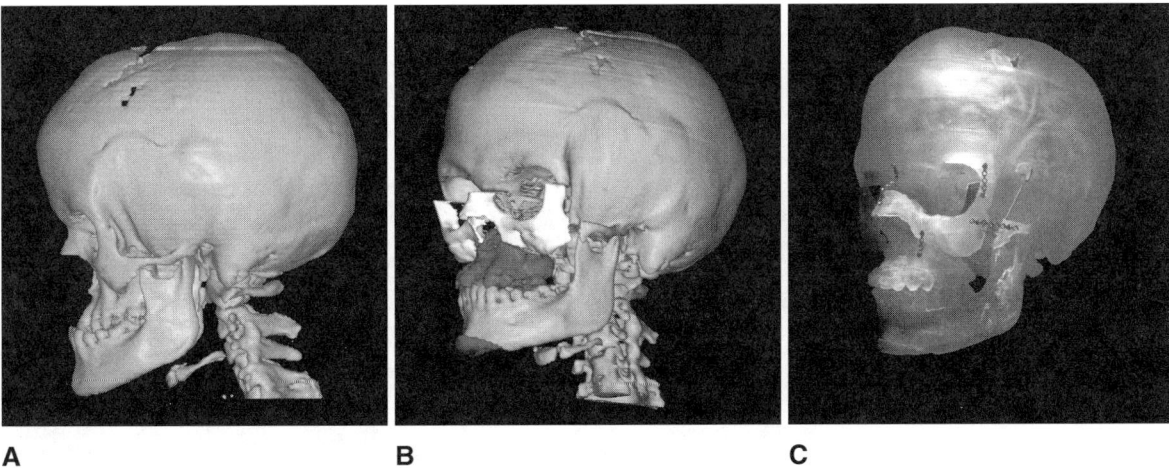

**A**              **B**              **C**

**FIGURE 7-8**   Surgical rehearsal. **A**: CT-based 3D reconstruction of a patient skull affected by Crouzon syndrome. **B**: Surgical planning on the computer screen. **C**: Stereolithographic model used for surgical rehearsal; note preadapted prosthetic bridges that fix isolated bony parts in their new positions.

of the intervention, to reposition the isolated parts, and to adapt small titanium fixation plates to the patient's individual anatomy.

Another major topic of RP application in medicine is the production of custom implants that meet the specific anatomic properties of an individual. With respect to mechanical properties and biocompatibility, current RP materials do not yet meet the standards for functional implants. However, RP models serve as suitable templates for mold-and-cast production of biocompatible custom parts.

Figure 7-9 illustrates a combination of virtual prosthesis design and custom implant prototyping. Surgical resection of a tumor in the left frontal bone was first planned and performed in virtual reality. Applying principles of virtual reconstruction, the resulting gap in the calotte was filled with a mirror-imaged copy of the unaffected right side of the frontal bone. A stereolithographic hard copy of this prosthetic part served as a matrix for mold-and-cast production of a biocompatible custom implant.

Further bioscientific applications of RP technology concern education and teaching. RP models are ideal objects to document and communicate specific clinical cases, for example, classic syndromes, pathologies, or various types of skeletal trauma. Furthermore, small anatomic structures such as the three-dimensional microstructure of cancellous bone, the cavity system of the inner ear, or isolated teeth, can be reproduced in magnified versions (Fig. 7-6) (Borah et al., 2001). As an extension, RP model building can also be utilized in artistic applications. For example, the soft tissue reconstruction of the Devil's Tower Neanderthal child (Fig. 6-17C) was modeled on top of a stereolithographic replica of the virtual skeletal reconstruction (Fig. 7-3E).

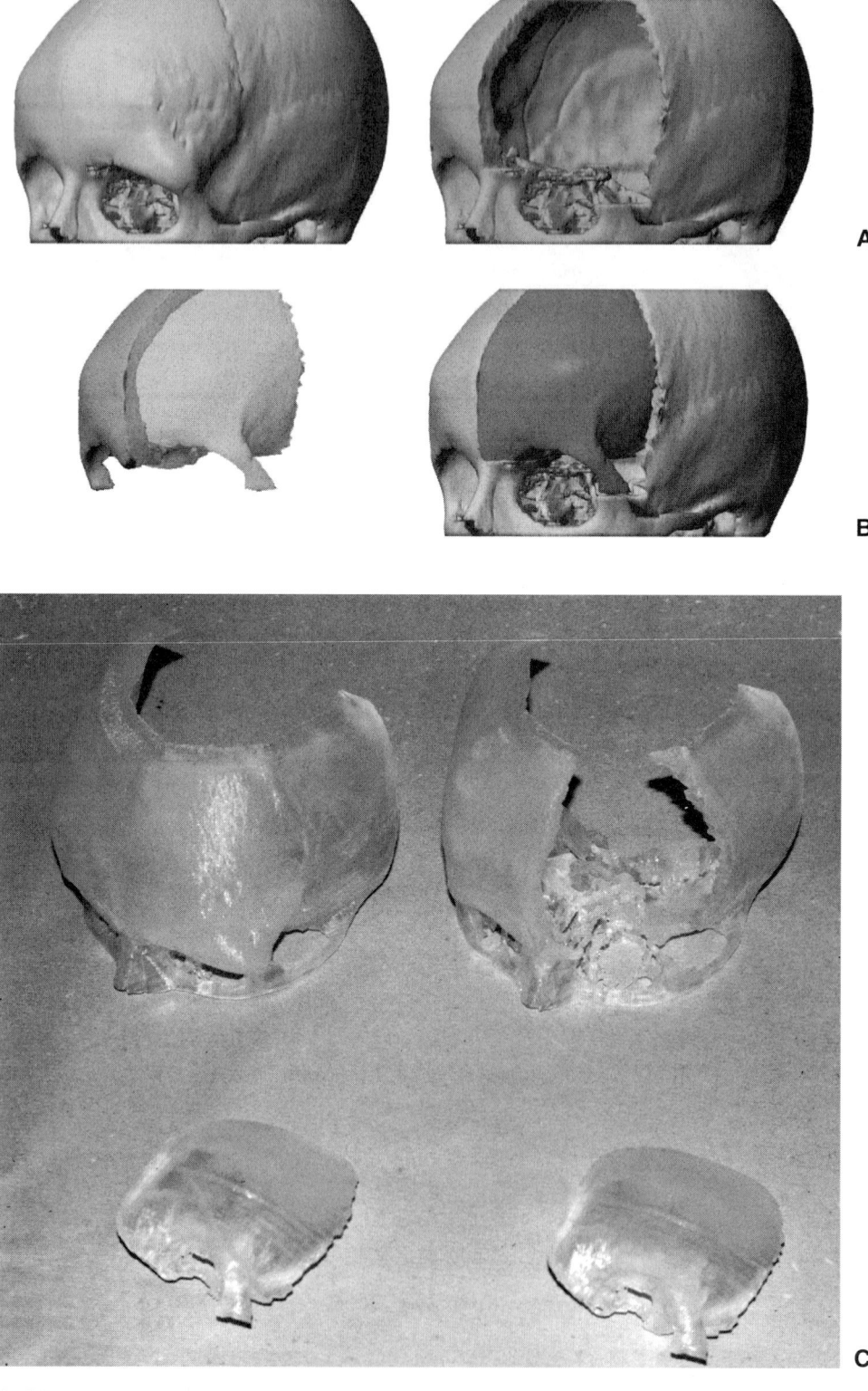

**FIGURE 7-9**  Custom implant design. **A**: CT-based 3D reconstruction of a patient skull affected by a bone tumor in the left frontal bone (see Fig. 8-13B) and virtual surgery. **B**: Prosthesis design by mirror image completion. **C**: Stereolithographic replicas used as templates for prosthesis production and fitting.

## 7.4 FURTHER READING

Jacobs (1996) and Gebhardt (2000) are textbooks that provide an overview of RP technologies. Reviews on early medical and forensic applications of RP technology are found in Zonneveld et al. (1992), Hemmy et al. (1994), Lambrecht et al. (1993), and Zonneveld (1994). The current state is reviewed by Petzold et al. (1999), Webb (2000), and Perez-Arjona et al. (2003). MicroCT-based reverse engineering approaches to bone microstructure are found in Borah et al. (2001) and Cooper et al. (2003). Research into additive generation of biocompatible custom implants from calcium phosphate glass-ceramic materials is documented in Klosterman et al., (1998), Steidle et al. (1997, 1999), and Griffin et al. (1997)

An interesting method for support-free generation of stereolithographic models is reported by Murakami et al. (2000).

# MORPHOMETRIC ANALYSIS

**8**

How odd it is that anyone should not see that all observation must be
for or against some view if it is to be of any service.

—CHARLES DARWIN IN A LETTER TO HENRY FAWCETT
(DARWIN AND SEWARD, 1902)

## 8.1 MORPHOMETRY AS RECONSTRUCTION

The relationship between geometry and real life is full of twists and turns.
Absorbed by a geometric problem, Archimedes (ca. 287–211/12 B.C.) was drawing
circles into the sandy ground of his quiet patio in Syracuse, ignoring the fact that
the city was being sacked by the Romans at that very moment. When a soldier
rushed into the patio, Archimedes uttered the famous sentence:

Don't disturb my circles!

for which he paid with his life.

Today, geometry—the science of mathematical form—is no longer a battle-
field, but morphometry—the science of organismic form—is still an arena in
which many different methodologies stand in opposition to each other. It is not
the intent of this chapter to support one school of thought over others. Rather, we
try to present the basic issues of morphometry from the biological perspective of
virtual reconstruction. Our aim is to give an overview of methods and topics, permitting the reader to form his or her own opinion about current developments
and potential applications.

But how is morphometry related to issues of virtual reconstruction? Let
us consider the aims and methods of morphometric analysis. The ultimate
goal of morphometry is to infer evolutionary, developmental, and functional/
adaptational processes from patterns of form variability. However, morphometric

*Virtual Reconstruction: A Primer in Computer-Assisted Paleontology and Biomedicine.*
By Christoph P. E. Zollikofer and Marcia S. Ponce de León.
Copyright © 2005 John Wiley & Sons, Inc.

analysis cannot track biological *processes* themselves; it can only observe *patterns* that result from such processes.[1]

Morphometric methods rely on definitions of organismic form that permit quantitative comparisons between specimens in a sample. At the outset of each morphometric analysis, therefore, we must ask, How can we define biologically relevant morphometric units that establish correspondence relations between the specimens of a sample? In other words, how can we quantify "the same" biological entity in every specimen? We have already noted that this is not an easy task. The two brains in Figure 2-16 represent corresponding biological structures, but how can we define corresponding points on their surfaces in terms of evolutionary, developmental, or functional equivalence?[2]

Intriguingly, the issue of defining biologically meaningful correspondence is inextricably intertwined with the issue of inferring process from pattern (Fig. 8-1). When we quantify organismic form, we need at least some "informed guess" about the processes that are supposed to generate the observed patterns of morphologic variability—applying Darwin's statement, morphometric analysis requires hypotheses about underlying processes. At the same time, the central aim of morphometric analysis is to shed light on processes shaping organismic form. These processes are difficult to track in most living organisms and can no longer be observed immediately in fossils; rather principles of organismic construction must be inferred by *reconstruction*.

It is important to recall that there are no "general" or "best" solutions to such problems. Nevertheless, viable approaches in the biosciences follow both empirical and theoretical tracks, combining data exploration and hypothesis testing with previous knowledge and heuristics (Feyerabend, 1975). The theoretical approach is used to establish hypotheses about potential biological processes bringing about morphologic variability and to determine how morphology should be quantified. The empirical approach is used to explore patterns of form variability before any specific hypotheses have been stated, or to test hypotheses stated explicitly at the outset of the analysis. Both approaches are connected through iteration, in the course of which methods of quantification are fine-tuned and hypotheses are stated more specifically by using preliminary results.

---

[1] Referring to Socrates' cave parable, we may state that patterns are appearance and processes are the underlying biological entities.

[2] Biological correspondence comes in two categories, *homology*—correspondence through evolutionary/developmental processes—and *homoplasy* (analogy)—correspondence through function (Owen, 1843). For example, bird wings and human arms represent evolutionarily modified vertebrate forelimbs; hence these structures are homologous to each other, whereas bird wings are analogous to butterfly wings. In practice, however, definitions of homology versus homoplasy are intricate because they depend on the biological context and level of organization at which they are applied (Rieppel, 1989).

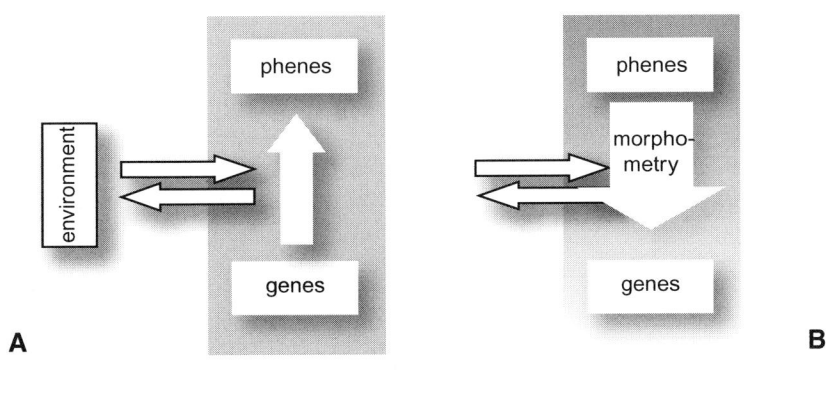

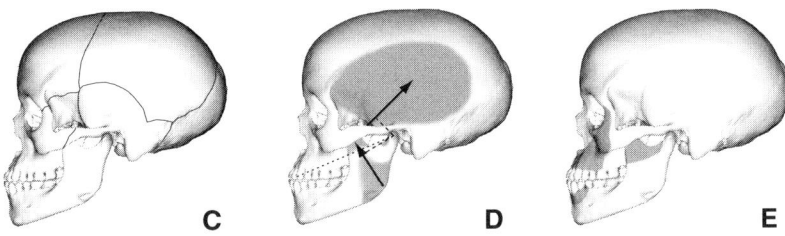

**FIGURE 8-1** Morphometric analysis as virtual reconstruction. **A**: Morphology (the phenotype of an organism) results from an interplay of developmental and environmental processes. **B**: Morphometry analyzes patterns of form variability and aims at inference of developmental, evolutionary, adaptational/functional, and environmental processes shaping organismal form. **C–E**: Hypotheses about the nature of these processes influence the definition of morphometric units and the interpretation of results. **C**: Comparative evolutionary and developmental analyses are based on homologous morphologic subunits, such as cranial bones. **D**: Functional analysis of mastication focuses on areas of muscle attachment (dark), lever arms and moments, and other biomechanically relevant parameters. **E**: In an evolutionary-developmental approach, patterns of form variability are understood in terms of differential activity of cranial growth fields (light/dark: depository/resorptive fields).

## 8.2 MORPHOMETRY AND GEOMETRY

### 8.2.1 The Role of Geometry

Let us propose a definition: Morphometry is the comparative analysis of biological form and change in form with the aid of mathematical tools that are designed to quantify form variability and reveal biologically relevant differences between forms. As we will realize in the course of this chapter, the notions of *change* and *difference* used in this definition have both mathematical and biological meanings. This is exactly the interface at which biological hypotheses about organismic form are transformed into concepts of geometry. To define a geometry, we need two things: a space and a metric, that is, we need a set of coordinate axes that represent spatial dimensions and a method to measure distances between points in such a space (Fig. 8-2).

One major operational goal of morphometry is to define natural geometries, that is, to define morphometric units in a geometry that makes explicit reference to biological hypotheses (Fig. 8-2). Accordingly, the interplay between biological and geometric/mathematical concepts determines which tools are used to measure and compare forms and how results are to be interpreted. Although in morphometry the connections between biology and mathematics are tight, a clear separation can be made between biological interpretation and computational procedure. Biological meaning comes into play at only two stages of morphometric analysis: during the definition and choice of morphometric units and during the conclusive interpretation of the results (Fig. 8-3). In the initial stage, morphometric units are always defined as comparative elements of a biological correspondence relation. Spatial correspondence between locations on organismic structures—in the sense of biological homology or homoplasy—pervades mor-

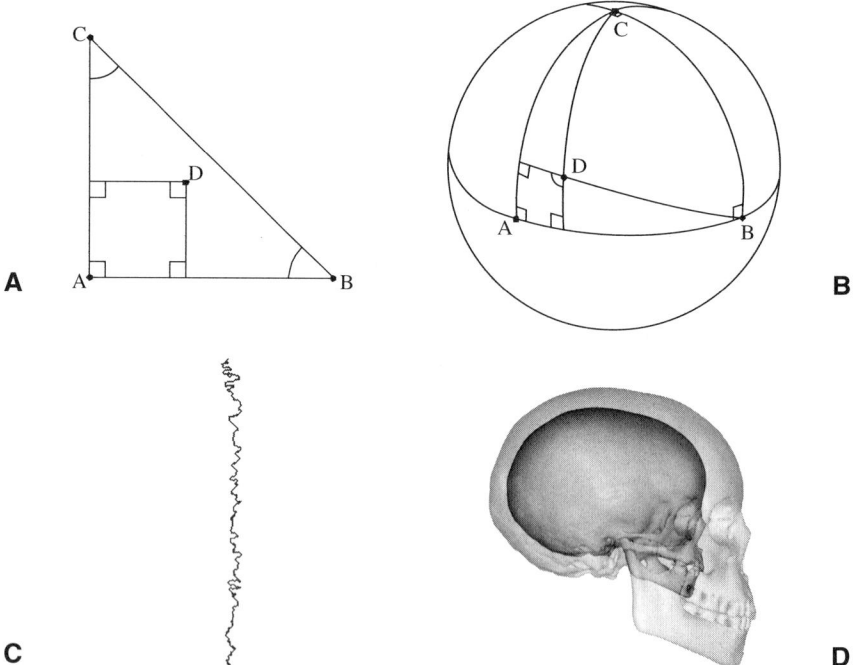

**FIGURE 8-2** A plethora of geometries. Euclidean geometry of plane (**A**) and spherical geometry (**B**) differ in axioms and metrics. In Euclidean geometry distances are evaluated along straight lines, whereas on a sphere distances are evaluated along circle segments. In the plane, triangles can subtend one right angle at maximum. On the sphere, triangles can subtend two or three right angles, whereas right-angled squares do not exist. The relation between spherical and euclidean geometry may serve as a metaphor for the potentially noneuclidean geometry of morphologic structures, such as a sutural interface between two growing bones (**C**) or a cranial geometry that defines a natural metric to measure the distance between an immature and adult human skull in morphospace (**D**).

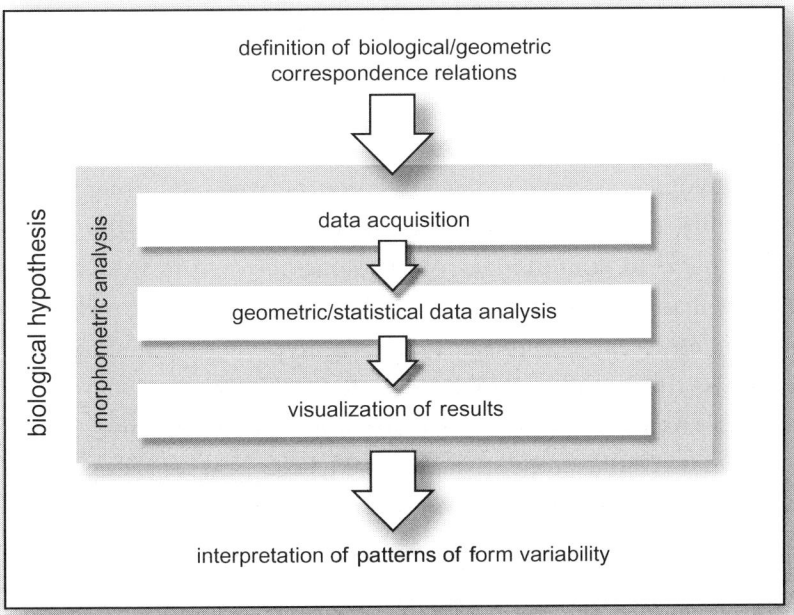

**FIGURE 8-3** Biological and computational aspects of morphometry. Morphometric analysis is a computational procedure that uses morphometric units expressing biologically defined correspondence relations between specimens and yields quantitative results that require interpretation in biological terms.

phometric analysis as the basic principle of comparison. Once correspondence relations have been established, morphometric tools can be utilized without further reference to biological meaning, such that the results of a morphometric analysis primarily represent the outcome of a mathematical procedure. The interpretation of these results depends again on the application of biological criteria.

It follows that biological hypotheses determine what is to be measured and how the results are to be interpreted. It is important to stress this point because there is an inherent tendency to overstate the biological explanatory power of morphometric and associated statistical procedures, notably because parallelisms that exist between mathematical/statistical and biological terminologies are a source of potential confusion. Explanation as a statistical term is purely metaphorical. From a biological perspective, morphometry is therefore an exploratory rather than an explanatory tool, although it establishes ties between the biological and mathematical significance of many of its concepts.

Let us recall Chapter 2 (Fig. 2-14), where we proposed a taxonomy of data types. We started with volume data, proceeded to surface data, and ended up with the extraction of landmark data. Furthermore, we made a discrimination between extensional and relational data types. This approach might appear cumbersome to the more classically oriented morphometrician who is accustomed to using a ruler to measure distances between certain well-defined morphologic points of reference. Nevertheless, it clarifies the relationship between geometric properties and associated biological properties of morphometric data. An anatomic landmark, for example, represents a geometric "singularity" and, at the same time, pinpoints a biological correspondence relation between specimens in a sample (Bookstein, 1991). In the same way, morphometric methods—the mathematical tools used to perform quantitative comparisons between forms—establish links between mathematical and biological models of correspondence relations. A direct connection therefore exists between biological hypotheses, the type of morphometric data, and the method of analysis. In the following sections, which are dedicated to a discussion of the current state of morphometric methodologies and of their biological implications, the proposed taxonomy of morphometric data may serve as a guide to a taxonomy of morphometric methods.

## 8.2.2 The Role of Size and Shape

As has been demonstrated in Chapter 2, extent-based variables are obtained by integration over the three-dimensional geometry of an entire organism, or of particular substructures, whereas relational data are obtained by differentiation. These opposing procedures are related to a fundamental problem of biology in general and of morphometry in particular, the dichotomy between *size* and *shape* (see Gould, 1966; McMahon and Tyler Bonner, 1983; Schmidt-Nielsen, 1984 for reviews). As with the concept of correspondence (homology vs. homoplasy), the notions of form, size, and shape have both a computational/geometric and a biological significance (Bookstein, 1989).

Let us first consider the computational side: *Size* and *shape* represent extensional and relational properties of one and the same *form*, respectively. Size represents the geometry-free component of form (a scaling factor), whereas shape represents the purely geometric component of form (Mosimann, 1988). Following the classical definition provided by Kendall (1981), shape can be defined specifically as a geometric homology between anatomic landmarks. In computational terms, two geometrically homologous forms have the same shape if and only if they can be superimposed landmark by landmark and appropriately fit each other after they have been shifted, rotated, and scaled. The reader may note that the computational definition of these notions largely corresponds to visual intuition, because our visual system tends to separate size and shape information and to assess size differences independently of shape differences (Fig. 8-4).

Just *because* these perceptual notions are so intuitive, it is wise not to attribute a biological meaning to them unless their origin has been clarified. But what is the biological relevance of the decomposition of form into size and shape? At least

**FIGURE 8-4** Form, size and shape. This Magdalenian cave painting (Panel 4 of the Niaux Cave, France; ca. 14,000 BP; redrawn after Clottes, 1995) shows 18 animal specimens belonging to three species (aurochs, horse, and Pyrenean ibex). Great care was taken by the painters to work out the taxon-specific, sex-specific, and age-specific shapes of the depicted animals. Nonetheless, animals exhibiting similar shape are drawn at different sizes.

since Galileo's consideration of the question of why giants cannot grow indefinitely (Galilei, 1638), it has been recognized that size and shape express different significant structural and functional aspects of biological form (Fig. 8-5).

Because this chapter has a major focus on the morphometric analysis of landmark data, let us consider a definition of size for a landmark configuration. *Centroid size* (Bookstein, 1991) is evaluated as follows (Fig. 8-6 and Appendix F.5): Calculate squared distances between the center of mass (the centroid) of a landmark configuration and each landmark, and take the square root of their sum. Centroid size represents a purely operational definition of size with no immediate biological meaning. Nevertheless, its evaluation is a convenient way to decompose form into size and shape; in a sample of forms exhibiting random distribution of landmark positions around a mean form, centroid size represents the only measure of size that is statistically independent from shape.

Although we are focused here on definitions of size and shape as the two components of a given form, it should be noted in passing that, in a wider biological context, definitions of size do not necessarily imply its being a component of form. For example, in Galileo's model of thighbone shape (Box 8-1), body mass could be used as a measure of "global" size instead of using bone length as a "local" measure of bone size.

Like size, shape can be measured in many different ways. As we will see in the following sections, defining and measuring shape is the tough part of morphometric analysis.

### BOX 8-1

### ON SIZE AND SHAPE

Both developmental and biomechanical hypotheses have been proposed to explain size-related changes in shape. Galileo (1638) considered the shape of femora in animals of different size (Fig. 8-5). Assuming that a long bone acts as a pillar, he concluded that, in order to support weight, its diameter must increase disproportionately relative to length. On biomechanical grounds, static equivalence requires that the cross-sectional area $A$ of a column be proportional to the weight $W$ it bears.

$$A \sim W \qquad (8\text{-}1)$$

Because area $A$ is proportional to the squared diameter, $d^2$, and weight $W$ can be assumed to be proportional to the cube of any linear dimension of the body, (e.g., to the cube of bone length, $l^3$), the Galilean disproportion becomes

$$d^2 \sim l^3, \text{ or } d \sim l^{3/2} \qquad (8\text{-}2)$$

In modern terms: The relationship between bone diameter and bone length has an allometric exponent of 3/2, expressing the fact that, with increasing length, long bones become more and more thick.

The term *allometry*, that is, "different metrics" goes back to Huxley, who postulated a morphogenetic background of size-related changes in shape (Huxley, 1924, 1932). Huxley's arguments can be expressed with Galileo's bones. We assume that bone length and bone diameter follow different *specific growth rates* (i.e., growth per unit time *and* per total amount already grown). Although the absolute values of specific growth rates are unknown, we assume that they are coupled to each other by a factor $k$ that is the exponent of the allometric equation (see Appendix F.3). For Galileo's model, $k$ equals 3/2. Let us now express the model in terms of size and shape. We may reformulate Equation 8-2, taking long bone length as a proxy of size and expressing long bone shape as the ratio between diameter and length

$$\frac{d}{l} \sim l^{1/2} \qquad (8\text{-}2a)$$

It emerges that shape ($d/l$) changes with the square root of size ($l$).

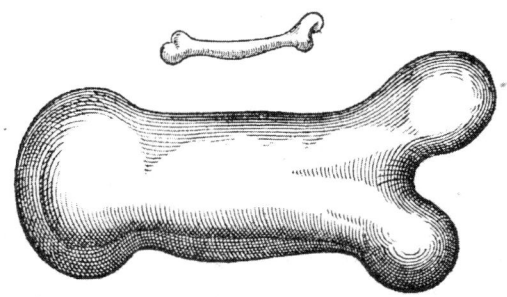

DEL  GALILEO.    129
*E per vn breue esempio di questo che dico disegnai già la figura
di vn' osso allungato solamente tre volte, & ingrossato con tal pro-
portione, che potesse nel suo animale grande far l' uffizio proporzio-*

*nato à quel dell' osso minore nell' animal più piccolo, e le figure son
queste: doue vedete sproporzionata figura, che diuiene quella dell'
osso ingrandito.  Dal che è manifesto, che chi volesse mantener in*

**FIGURE 8-5**  Galileo's original drawing of the femora of a dog and a giant (from *discorsi e dimostrazioni matematiche* [mathematical discourses and demonstrations], 1638; reproduced from the original by kind permission of the Library of the Eidgenössische Technische Hochschule, Zürich).

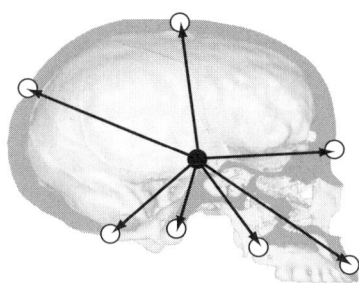

**FIGURE 8-6**  Centroid size of a landmark configuration is evaluated as the square root of the sum of squared distances between the configuration's center of mass (the centroid; black dot) and each landmark (white dots).

### 8.2.3 Multivariate Morphometry

Morphometric analysis relies to a great extent on anatomic landmarks as locations of correspondence between specimens and, accordingly, on measurements derived from landmark locations. However, it is often forgotten that *distances* between landmarks rather than landmark locations form the primary source of data in these studies. There are several reasons why interlandmark distances represent the preferred form of morphometric data. Measuring distances with a ruler or caliper is technically less demanding than sampling landmark coordinates in two or three dimensions with medical imaging and computer graphics procedures. Furthermore, techniques of linear multivariate analysis available in standard statistical software packages can be applied without further modification to sets of interlandmark distance data. As we will see in the following sections, a major challenge of geometric morphometrics is to apply multivariate analysis to real-space geometric properties of landmark data. But before we examine such approaches in more detail, some preliminary information about multivariate techniques is required.

*Multivariate morphometry* is morphometry based on more than two or three measurements per specimen. This is best illustrated with an example. Most readers will be familiar with *x-y* or *x-y-z* graphs depicting the relationship between two or three measurements taken on a sample of *N* specimens. Let us consider a study in which cranial length, height, and breadth of a human sample are plotted against each other (Fig. 8-7).

In Figure 8-7, each point represents a triplet of measurements taken from a specimen (more colloquially, we say that each point represents a specimen). Applying geometric terms, the graph's three axes define a euclidean space (see Fig. 8-2A), called *feature space*. As a matter of fact, Euclidean geometry provides the framework within which classic statistical entities such as mean values, standard deviations, correlations, regression coefficients, variances, covariances, Mahalanobis distances, etc. are evaluated. Although the reader is referred to Appendix F.4 and textbooks of statistics for a closer look at these features, (specifically, Reyment 1984; additional references are given at the end of this chapter),

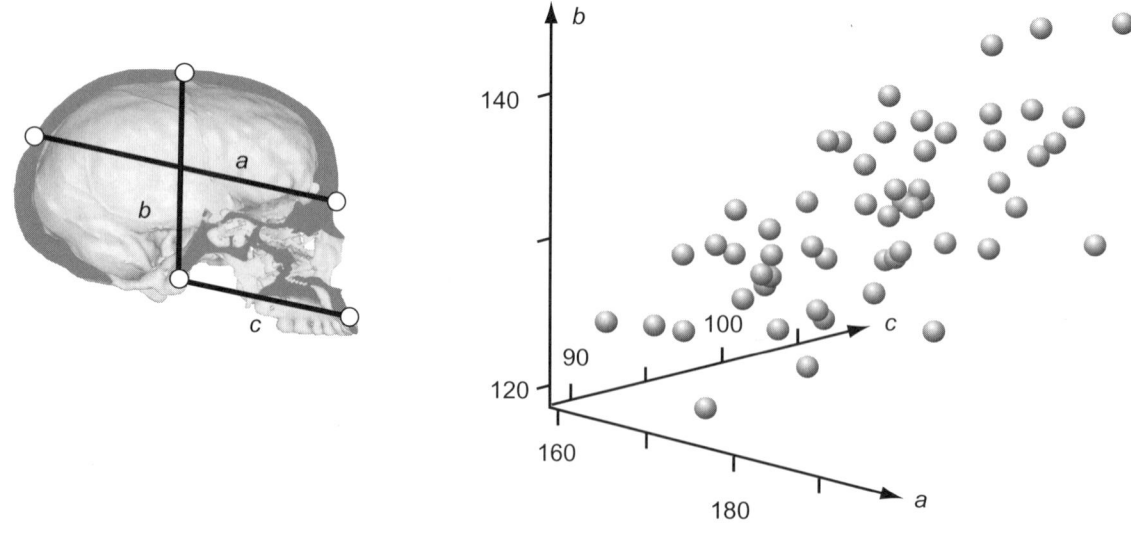

**FIGURE 8-7** A trivariate graph of neurocranial length (*a*), neurocranial height (*b*), and facial length (*c*) of a human sample.

we will focus here on how multivariate morphometric data can be rendered graphically and explored visually.

## 8.2.4 Principal Components Analysis and Dimension Reduction

Let us extend feature space from two or three measurement variables to an arbitrary number of $K$ variables, and proceed from bi- or trivariate to multivariate analysis. The resulting multidimensional space has $K$ axes representing $K$ variables $v_1, \ldots, v_i \ldots, v_K$, such that specimens are represented as $K$-dimensional points. Although statistical analytical techniques devised for two or three dimensions can be extended to $K$ dimensions, it is no longer possible to provide an immediate visual representation of multidimensional spaces with more than three dimensions.

Various techniques of *dimension reduction* have been devised to capture the most statistically significant aspects of data scatter with two or three new variables that can be readily visualized with two- and three-dimensional graphs. Earlier in this book (see Figs. 5-9 and 5-10), we considered how a virtual object's position and orientation could be described in space. The basic approach was to shift and rotate an object's intrinsic system of coordinates relative to world coordinates. Here, we consider an opposite scenario: We attempt to recover the intrinsic system of coordinates of an object situated anywhere in world coordinates. For example, consider yourself sitting in front of this book anywhere in a room. Obviously, the midsagittal plane and long axis of your body represent your natural system of coordinates, whereas the axes of the room represent an extrinsic system of coordinates with no specific meaning to your body.

These considerations can be applied to multivariate statistics, where we deal with "data objects" located in abstract spaces of high dimensionality. Our aim is to find an intrinsic system of coordinates that describes "natural axes" through a cloud of data points. Figure 8-8A and B show examples. For visibility's sake, we first restrict ourselves to two dimensions. Cranial length, height, and breadth are strongly correlated in our sample, which is most likely an effect of overall cranial size. It is therefore sensible to define a new system of coordinates with axes "along" and "across" the two-dimensional data cloud. These axes can be expressed mathematically as the two principal axes of an ellipse circumscribing the data scatter, where the length of the axes measures the amount of variation along each direction. The new axes, termed *principal components* (PC), are at a right angle to each other, implying that they describe statistically independent (uncorrelated) patterns of variation in the data set. Note that this decomposition is a purely statistical procedure, so there is no guarantee that principal components have specific biological significance. In the present case, however, it is sensible to consider components PC1 and PC2 as capturing size-related and size-

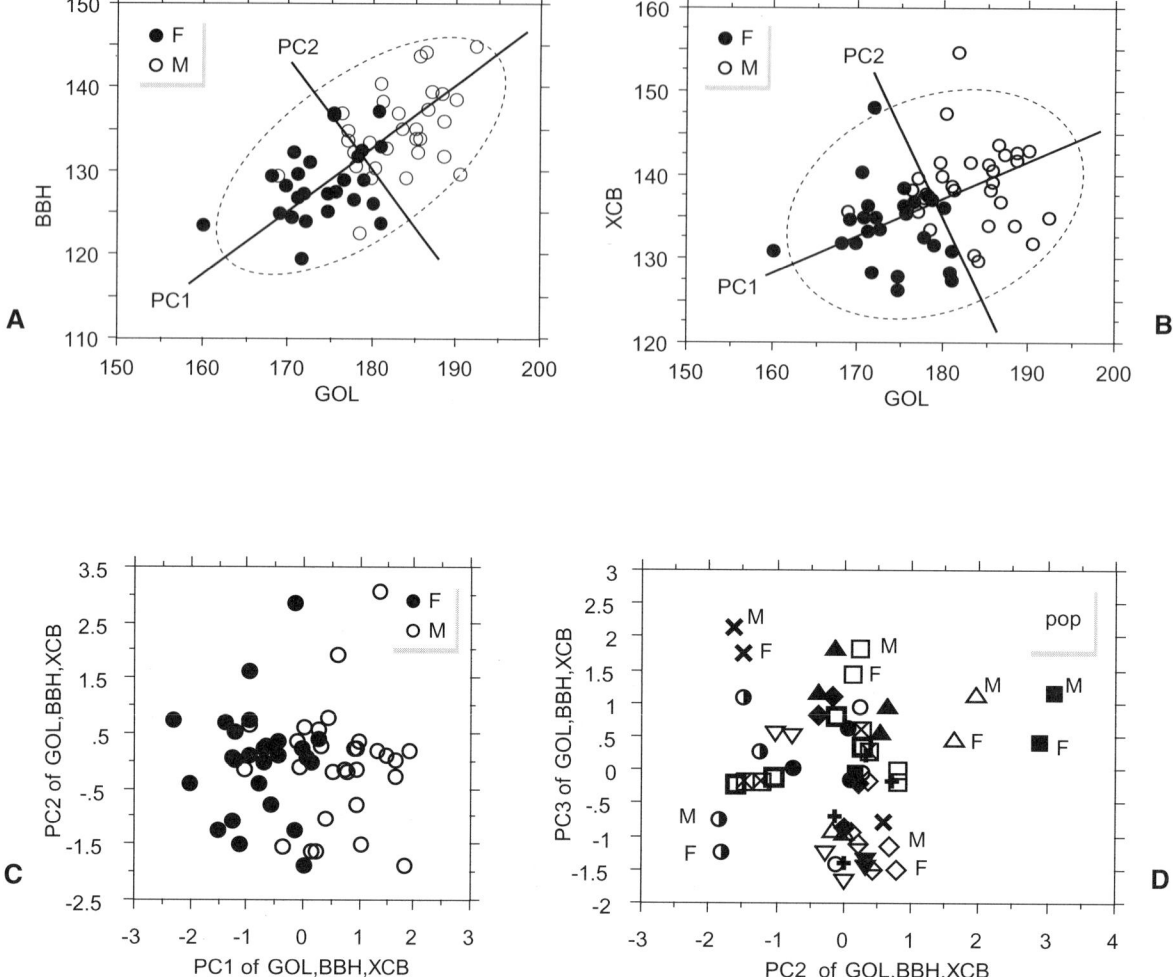

**FIGURE 8-8**  Determining natural axes through a cloud of data points in multivariate space. **A, B**: Plots of cranial length (GOL, glabella-opisthion length) against height (BBH, basion-bregma height) and breadth (XCB, maximum cranial breadth). The original system of coordinates is replaced by a system whose origin coincides with the centroid of the data scatter and whose axes PC1 and PC2 are along directions of largest and smallest variance of the data set (note separation of sexes along PC1). **C, D**: Principal components analysis of GOL, BBH, XCB together. PC1 accounts for most of the sample variance (62.5%) and can be regarded as a proxy of cranial size; PC2 (25.6%) separates populations from each other; and PC3 (11.9%) separates sexes among each population (the symbols in **D** represent populations). Data source: Howells (1989, 1996).

independent variations in cranial form (we may say that PC1 and PC2 are proxies for cranial size and shape, respectively). Similar procedures can be applied to a three-dimensional data set of cranial length, height, and breadth (Fig. 8-8C,D). We obtain three PCs, which can tentatively be interpreted as expressing three factors of form variation: The first accounts for variation in size, the second for variation in population, and the third for sex-related interpopulation differences. Again, we need to point out that these PCs do not automatically express biological entities;

rather, they represent a useful means to visualize and explore patterns of variation in a sample.

For the general case of $K$ multivariate dimensions, *principal components analysis* (PCA) (Jolliffe, 1986) decomposes data variation into $K$ statistically independent principal components (PCs), corresponding to $K$ orthogonal axes through multivariate space that account for the largest, second largest, and successively smaller proportions of the total sample variance. In most biological applications, the first few PCs account for most of the variation, such that subsequent PCs can be omitted without loss of significant information. In morphometric studies comprising more than three variables, therefore, the method of extracting PCs is often used to reduce dimensions of the original multivariate data set.

The general idea of PCA is to provide a look at multidimensional data from the perspective of statistical relevance. Disregarding the information contained in less significant PC axes does not mean that this information is lost. The original $K$-dimensional data scatter is simply expressed in a new system of coordinates with an equal number of dimensions, such that it is always possible to switch forth and back between original measurements and their statistically ordered representation.

## 8.2.5 Classical Multivariate Morphometry: Geometry Lost

When we apply multivariate techniques in morphometry, it is fundamental to distinguish the physical space in which organisms exist from the multivariate feature space in which the morphometric analysis takes place. Figure 8-7, for example, depicts the physical space from which morphometric data are acquired by using two dimensions, whereas the feature space in which these data are represented has three dimensions. In both spaces, all axes are calibrated in millimeters, yet the dimensions of physical and feature space express fundamentally different aspects of physical and statistical reality, respectively. In physical space we measure distances between landmarks on specimens, whereas in multivariate space we measure distances between specimens themselves. Hence, a multivariate analysis may comprise physically quite diverse variables that all characterize a given specimen—for example, individual age (measured in years), body temperature (measured in degrees centigrade), body mass (measured in kilograms), stature (measured in centimeters), and cross-sectional femoral area (measured in square centimeters). These measurements constitute the axes of one and the same abstract multivariate space, which by itself is completely devoid of any physical meaning.

The fact that this space can unite axes denoting diverse physical quantities is a strength of multivariate analysis (consider that physicists still wonder about the actual number of dimensions of our universe). In morphometric applications of multivariate analysis, however, this strength also is a weakness (Fig. 8-9). Using

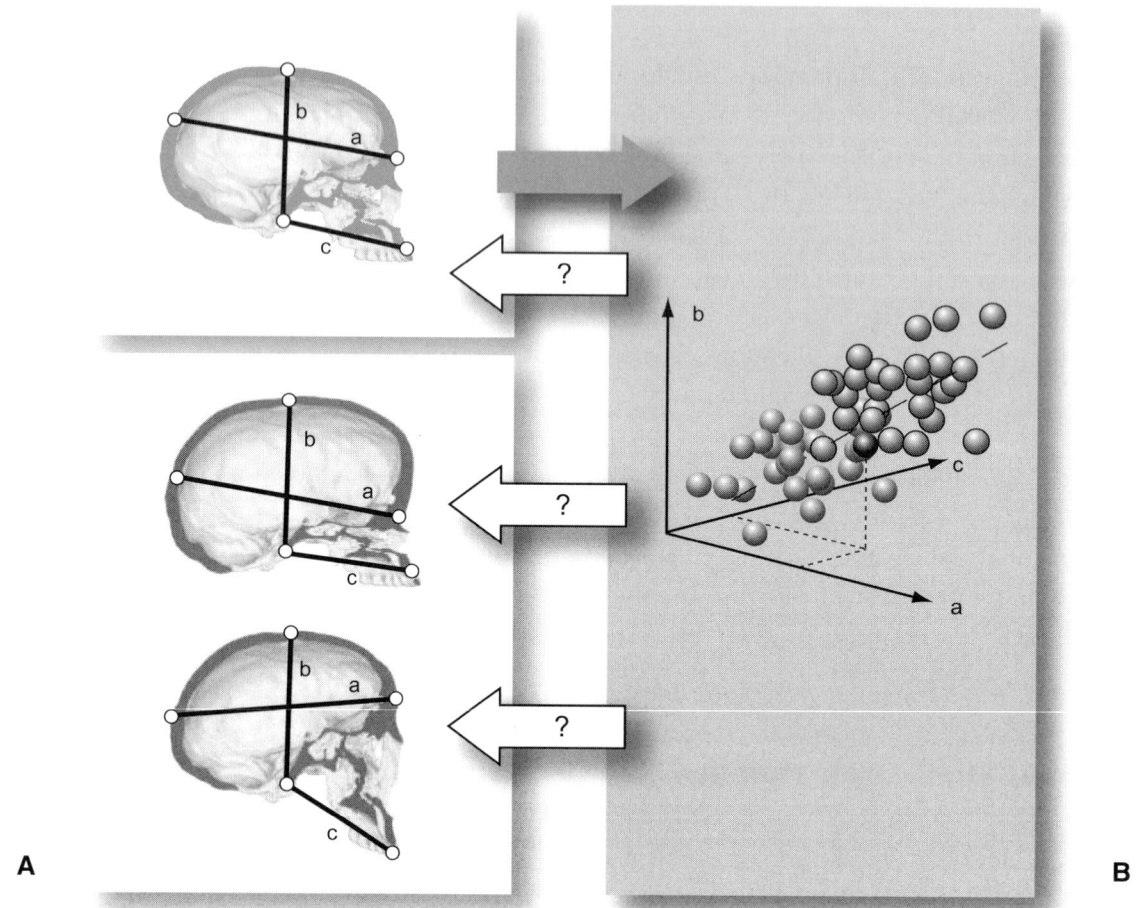

**FIGURE 8-9** Classic multivariate analysis of morphometric data. Physical measurements of neurocranial length (glabella-lambda; a), neurocranial height (basion-bregma; b), and facial length (basion-prosthion; c) (**A**) are used to construct an abstract multivariate space, in which specimens appear as points (**B**) and patterns of form variability appear as correlations, clusters, and other statistical features. However, because spatial relationships between these measurements are not specified, it is no longer possible to convert a point in multivariate space back into one single cranial form. This ambiguity makes it difficult to express and interpret abstract patterns of form variability in **B** in terms of actual cranial form variability in **A**.

interlandmark distances to characterize cranial form leads to a singular situation: The only geometric property that bears explicitly upon evaluation of statistical parameters is the relative position of specimens in abstract multivariate space, whereas physical positions of landmarks relative to each other do not. In fact, most of the geometric information contained in the original three-dimensional landmark configuration (painstakingly recovered from CT-scanned fossil fragments that we reconstructed on the computer screen . . .) is discarded before the analysis commences. As a consequence, it is no longer possible to reexpress multivariate patterns of form variability that are detected in feature space in terms of positional rearrangements of landmarks relative to each other (Fig. 8-9).

To summarize, in classic multivariate morphometric analysis based on inter-landmark distances, we discard a complex network of geometric data in favor of a simple list of extent-based data. The elimination of potentially significant geometric information during the transition from physical to feature space turns out to be a major drawback of classic multivariate morphometrics. This effect parallels a fossil specimen that is submitted to morphometric analysis before and after reconstruction. Before reconstruction, isolated features can be measured on single fragments; however, only after reconstruction does the specimen display the full picture of spatial morphometric relationships. Classic morphometric analysis thus involves unnecessary fragmentation of the integrated morphologic ensemble into isolated parts that cannot easily be recomposed into a whole.

## 8.2.6 Geometric Morphometrics: Geometry Recovered

Here we study how to integrate the full picture of physical spatial relationships between morphometric units into morphometric analysis. Three major issues deserve consideration:

- *Data acquisition*: How can we achieve exhaustive sampling of geometric information contained in organismic structures?
- *Analysis*: How can we achieve a lossless transfer of geometric information from physical space to feature space, to permit multivariate statistical analysis of shape variability?
- *Visualization*: How can we achieve lossless transfer of statistical information from feature space back to physical space, to express multivariate patterns of shape variability in terms of real space patterns of shape variability?

Over several decades, new morphometric tools have been developed to tackle these issues. These methods, known as *geometric morphometrics*, aim to integrate real space geometric properties of the form of organisms into multivariate analyses (Rohlf and Marcus, 1993). As we discussed above, there is no single best solution to the preceding issues. Rather, the method of choice depends on the biological hypotheses formulated at the outset of the analysis, and on the type of information that we expect to obtain.

We will consider three major current developments. The first type of analysis combines concepts of shape space with shape deformation (Bookstein, 1991), the second type investigates exhaustive sets of interlandmark distance measurements establishing a form space (Lele and Richtsmeier, 2001), and the third type is based on outlines rather than isolated landmarks (Ferson et al., 1985). Figure 8-10 gives an overview of the logic behind each type of analysis.

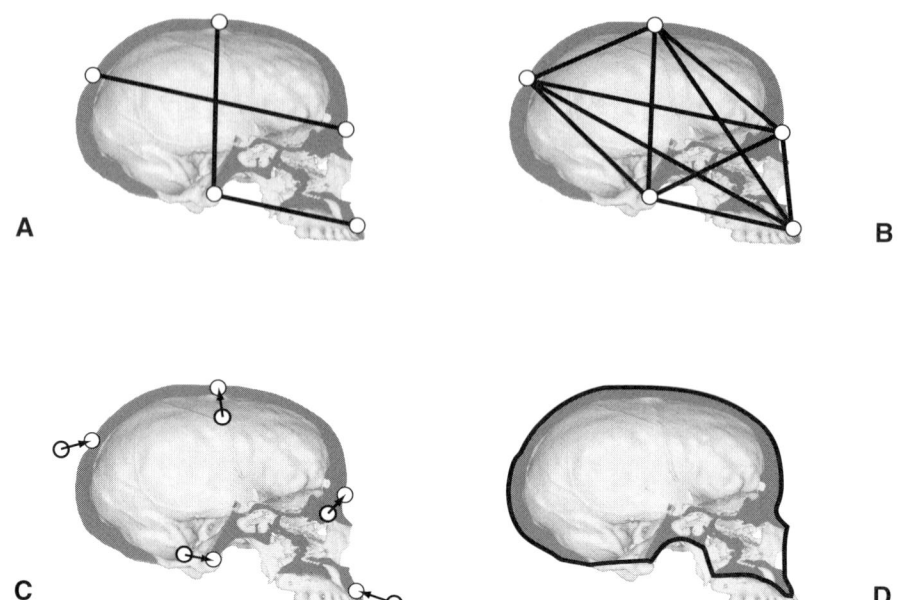

**FIGURE 8-10**  Data sources for methods of morphometric analysis. **A–C**: Landmark-based methods take anatomic points of reference to establish correspondence relationships between specimens in a sample. **A**: Classic multivariate analysis is based on sets of selected distance and angular measurements between landmarks. **B**: Euclidean distance matrix analysis (EDMA) is based on exhaustive sets of interlandmark distances, thus incorporating all spatial relationships between landmarks. **C**: Shape space analysis explicitly refers to the geometry of the landmark configurations, expressing a specimen's shape as deviation or deformation of its configuration with respect to a reference configuration. **D**: Fourier analysis establishes correspondence between specimens through outline data.

- *Shape space analysis* (Procrustes analysis and deformation analysis) is based on the idea that the shape of a specimen can be quantified as its deviation from a reference shape (as we define geographic locations in terms of "deviation" from the Greenwich meridian and the Equator).[3] Expressing shape as deviation from a norm has several advantages. It is visually intuitive because this is the way we perceive shapes, and it is biologically sensible because we consider shape transformation as an important aspect of ontogenetic and evolutionary change (Bookstein, 1991; Rohlf, 2001).
- *Euclidean distance matrix analysis* (EDMA) (Lele and Richtsmeier, 2001). The basic idea is to sample *all* possible distances that can be measured within a given set of landmarks. This results in a complete triangulation in the landmark configuration, whereas in classic morphometric analysis,

---

[3] In the current literature, *shape analysis* is used in a wider sense to differentiate geometric morphometric approaches from classic multivariate morphometry (Dryden and Mardia, 1998). We use here the term *shape space analysis* to denote analyses making reference to Kendall's definition of shape space (Kendall, 1981).

only a subsample of distances is considered. A major advantage of EDMA is that issues related to the definition of a reference shape can be elegantly circumvented.

- *Outline analysis.* In many cases, we are interested in overall comparative analyses of corresponding forms, notably where biological point-to-point correspondence between specimens is not known. Outline analysis is the method of choice for the morphometric investigation of nonlocalized correspondence between forms. Using Fourier analysis, it is possible to transform real-space geometric properties into the multivariate space of harmonic functions (Ferson et al., 1985; Lestrel, 1989). Metaphorically, organismic shape is decomposed into a sound spectrum, such that, in principle, we could "hear" the difference between a human and a chimp skull in the same way that we could hear the difference between the sound of a violin and a cello.

In the following sections, the foundations of each method are explored in more detail with illustrated examples. To permit quantitative comparisons between the different approaches, we apply each method to a reference data set consisting of midsagittal sections through human, chimp, and gorilla crania (Fig. 8-11).

## 8.3 SHAPE SPACE ANALYSIS

### 8.3.1 From D'Arcy Thompson to Kendall

In his book *On Growth and Form* (1917), D'Arcy Wentworth Thompson devised a highly suggestive graphical solution to visualize the geometric relations between two organismic forms: by overlaying one form with a rectangular grid, "attaching" the grid to selected anatomic landmarks, and deforming it to fit homologous landmarks in the second form. In this manner, a comprehensive picture of shape transformation based on corresponding landmarks emerges.[4]

In essence, the method establishes an explicit link between the geometric transformation of one form into another form and the biological notion of a correspondence relation (e.g., homology). Over the past two decades, Thompson's ideas have been developed into an explicit mathematical framework (to mention just a few: Kendall, 1981, 1984; Rohlf, 1990; Bookstein, 1991; Goodall, 1991; Small, 1996; Dryden and Mardia, 1998), and it is now possible to render his intuitive drawings with computer graphics tools (Fig. 8-12).

---

[4]Thompson was inspired by earlier transformation schemes proposed by Renaissance artists, notably Leonardo da Vinci and Albrecht Dürer, who studied methods to formalize the description of human proportions and their variability.

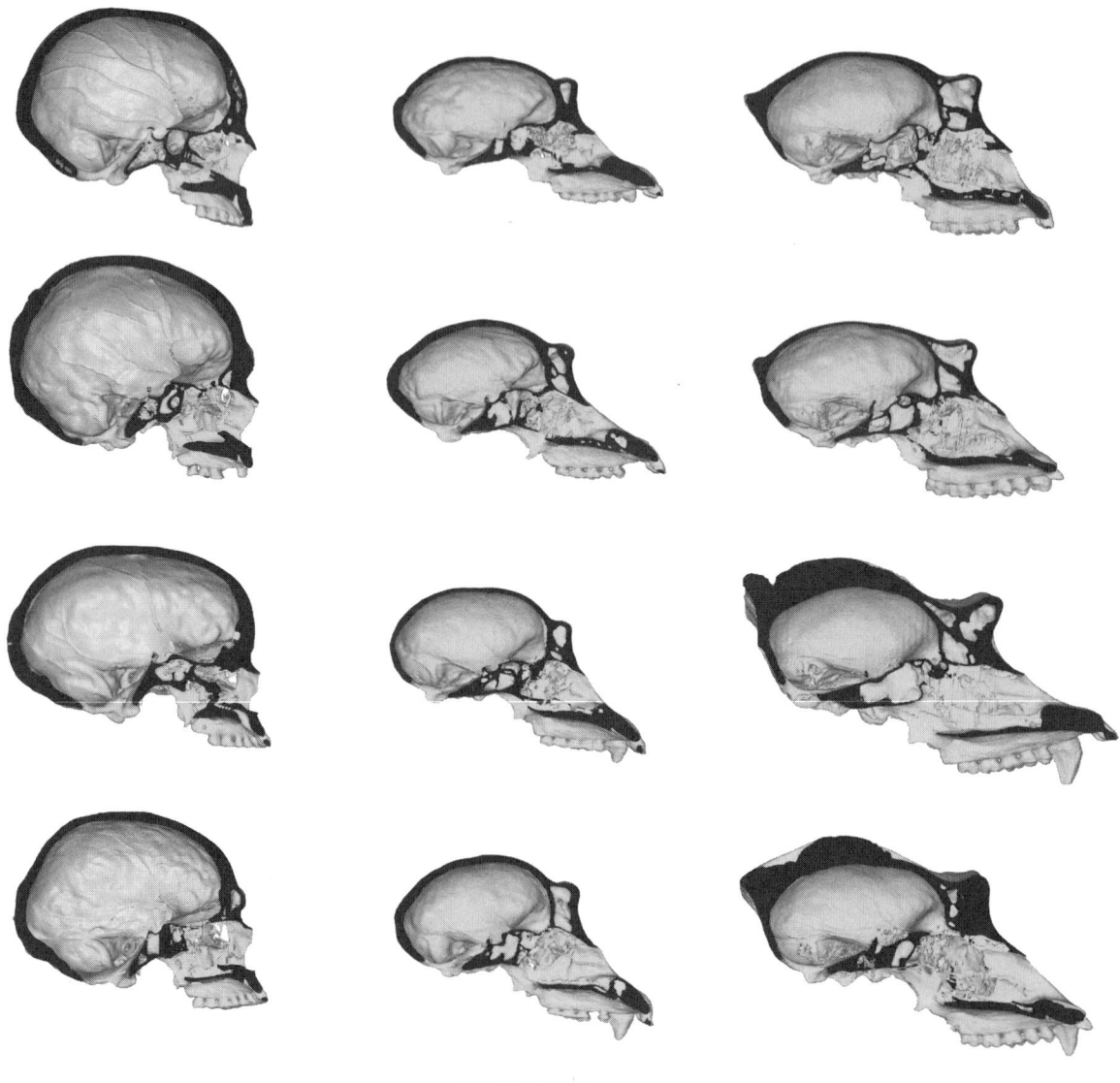

**FIGURE 8-11**  The human-chimp-gorilla sample used to compare the performance of different morphometric methods of analysis. Each taxon (from left to right: *Homo sapiens*, *Pan troglodytes troglodytes*, *Gorilla gorilla*) is represented by two female (top) and two male (bottom) skulls.

Although Thompson's approach was explicitly directed toward the comparison of organismic shape, quantification of shape variability was also investigated from a purely mathematical perspective. We owe major developments in this field to David Kendall (1981, 1984, 1985). He explored the mathematical and geometric properties of shape spaces in which each conceivable configuration consisting of $K$ landmarks is represented as a single multidimensional point and in which the similarity of shapes is measured as a distance between points (see Box 8-3).

## BOX 8-3

### KENDALL'S SHAPE SPACE

The full diversity of triangular forms can be generated by positioning vertices A, B, and C of a triangle at any location in the plane. Because each edge can be moved independently along $x$ and $y$ axes, we have $3 \times 2 = 6$ degrees of freedom (DF) of movement. Let us focus on triangular shapes, bearing in mind that *shape* is what remains of *form* after differences in size (a scaling factor), position (translation of the entire triangle along $x$ and/or $y$), and orientation (rotation of the triangle about an axis perpendicular to the plane) have been removed. This procedure reduces the 6 original DF by 4 (1 for scaling, 2 for translation, 1 for rotation); the remaining 2 DF are the two dimensions of shape space.

Kendall showed that this space is by no means a plane, but a sphere. Spherical geometry differs from planar geometry in various respects (see Fig. 8-2). Most notably, distances between points (which, in our case, represent triangular shapes) are measured on great circle segments rather than along straight lines (Fig. 8-13). This is in accordance with our intuition that dissimilarity between triangular shapes cannot grow indefinitely and corresponds to the fact that distances between any two points on our planet cannot be greater than half its circumference.

How about spaces representing landmark configurations consisting of four and more vertices, or representing three-dimensional configurations? For quadrilaterals in three-dimensional physical space, form space has $4 \times 3 = 12$ dimensions, corresponding to arbitrary positions of the four vertices along axes $x$, $y$, and $z$. Shape information is extracted by elimination of differences in scaling (i.e., size; 1 DF), translations along $x$, $y$, $z$ (3 DF), and rotations about $x$, $y$, $z$ (3 DF), yielding $12 - 7 = 5$ dimensions of shape space. The resulting five-dimensional hypersphere is hard to imagine; however, distances between quadrilateral shapes are still measured on great hypercircle segments. Generalizing these arguments, the hyperspherical shape space of two and three-dimensional $K$-landmark configurations has $2K - 4$ and $3K - 7$ dimensions, respectively.

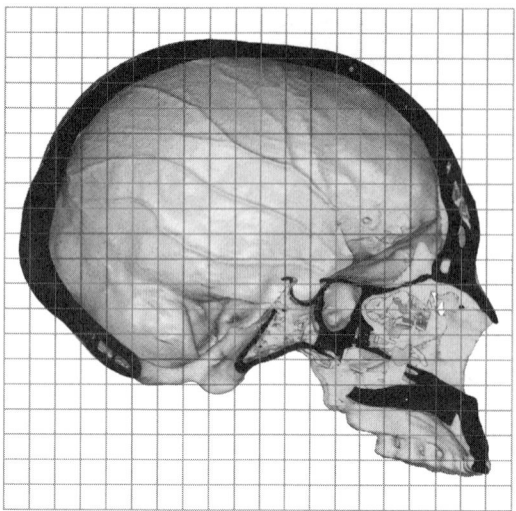
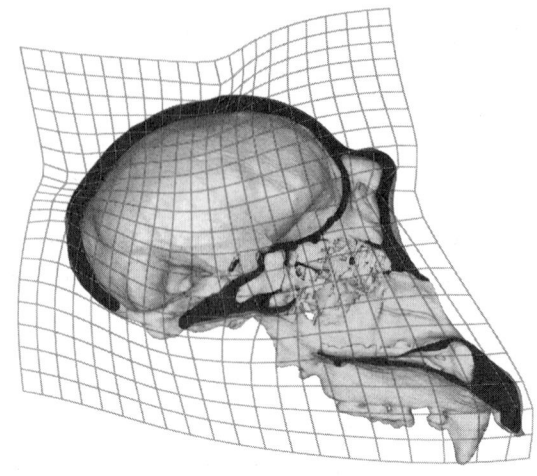

**FIGURE 8-12** The spirit of D'Arcy W. Thompson. Transformation grid adapted to a human and a chimpanzee skull with a thin plate spline interpolation function (see below). The grid renders local differences in cranial shape in a quantitative and visually comprehensive manner.

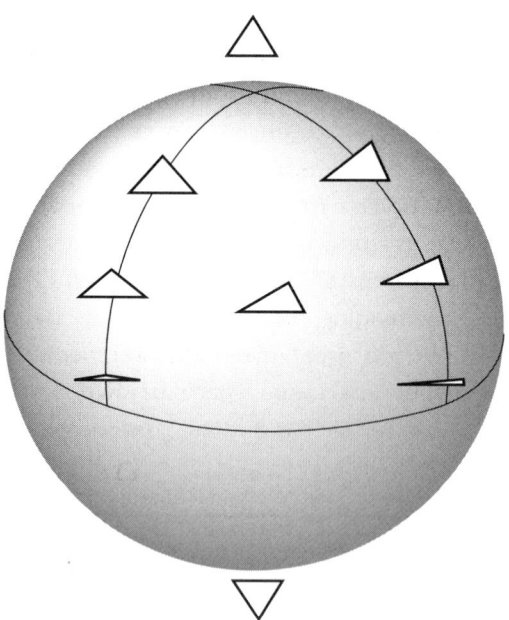

**FIGURE 8-13** Kendall's shape space of triangles. The space of all possible triangular shapes is a sphere. Poles represent equilaterals; points on the equator represent triangles collapsed to a line; points along the lateral boundaries of an octant represent isosceles triangles. The northern and southern hemispheres comprise triangles with counterclockwise and clockwise orientation, respectively. Distances between triangular shapes are measured along great circle segments.

From the perspective of morphometric applications, Kendall's most significant finding is that shape spaces have the geometric properties of a sphere, or its multidimensional extension, a hypersphere (see Fig. 8-13). As a consequence, distances (i.e., dissimilarities) between shapes are measured along great circle segments in the same way as we measure distances between geographic locations on the earth. Shape spaces thus have nonlinear geometric properties, which bring us into considerable practical difficulties. Most critically, in Kendall's shape space, it is not possible to analyze patterns of shape variability with linear multivariate statistical procedures, because these methods are based on the evaluation of linear distances in a euclidean space.

Fortunately, various methods of approximation are at hand to tackle these issues. To be able to perform multivariate analyses, it is necessary to construct a linearized, euclidean version of Kendall's shape space. As a familiar instance of the procedure of linearization of a spherical surface, consider the construction of a geographic map where a restricted area of the spherical surface of the Earth is projected onto a tangential plane through a given point of reference (Fig. 8-14). In the vicinity of this reference, linear distances measured on the planar map match spherical distances closely. However, as soon as the map incorporates more extended areas, deviations from linearity become more problematic. Extending these concepts to shape hyperplanes being tangential to shape hyperspheres (see Appendix F.7), we arrive at the concept of a *linearized shape space*, or *tangent shape space* (Dryden and Mardia, 1998).

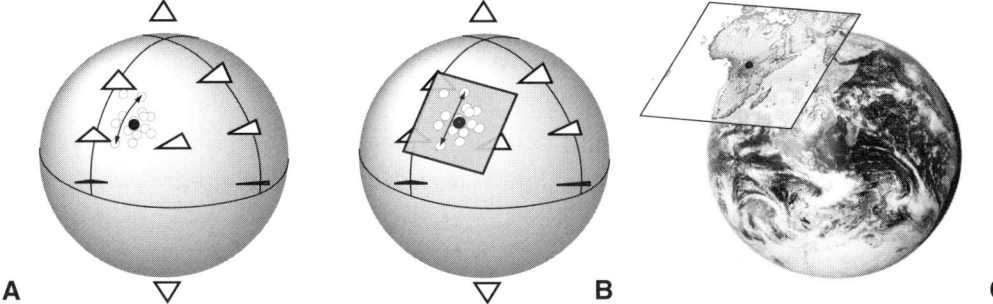

**FIGURE 8-14** Linearization of spherical spaces. Construction of a linearized version of Kendall's spherical shape space (**A**, **B**; also see **FIGURE 8-13**) follows the same logics as construction of a geographic map (**C**). To perform multivariate analyses, it is necessary to project shape data (**A**) onto a tangential plane (**B**) leading through a "mean" shape (black dot). Distances between shapes can now be measured along straight lines. As long as shapes of a sample (white dots) are clustered within a sufficiently small surface patch of the sphere, linearization entails negligible amounts of distortion.

## 8.3.2 The Workflow of Shape Space Analysis

We have now established the necessary conceptual framework to study how shape analysis in linearized shape space works. Here are the main steps:

- Eliminate size information by scaling each specimen's landmark configuration to centroid size = 1.
- Superimpose all configurations and determine a consensus configuration, which represents a sample average shape.
- Use the consensus shape as the coordinate origin of linearized multidimensional shape space and express all specimens as deviations from the consensus.
- Use classic linear multivariate techniques to analyze patterns of shape variability in linearized shape space.
- Visualize the results by transforming statistical patterns of shape variability back into real space patterns of shape variability.

Figure 8-15 is a more formal diagram of the workflow, whereas Box 8-4 presents a pictorial guide to the logics of shape analysis.

## 8.3.3 Determining a Reference Shape

Whereas details of the formalism of shape space analysis are given in Appendix F.6 and F.7, we focus here on the biologically sensitive issues. Above in this chapter (Fig. 8-3), we stated that biological reasoning comes into play at two stages of morphometric analysis, at the outset when biological concepts are cast into computational terms and at the end when results are interpreted in terms of biology. It appears that the pivotal point in shape space analysis literally occurs in the definition of a point, namely the selection of a sample's reference shape that serves as the coordinate origin of linearized shape space.

This reference must represent a "natural" geometry defined on biological considerations because, as we express shape as deviation from a reference, we expect it to reflect at least some aspects of an underlying biologically relevant process. How can the optimum reference be evaluated? There is no single best solution to do so, neither on biological nor on statistical/computational grounds.[5]

For example, in a phyletic analysis it might be appropriate to use an ancestral shape as a reference, whereas in a developmental analysis a juvenile form may

---

[5] Almost every culture has or had its geographic "navel of the world," whose definition, seen from an outside cultural perspective, seems rather arbitrary. Likewise, the definition of the "navel of shape analysis" is a rather arbitrary procedure.

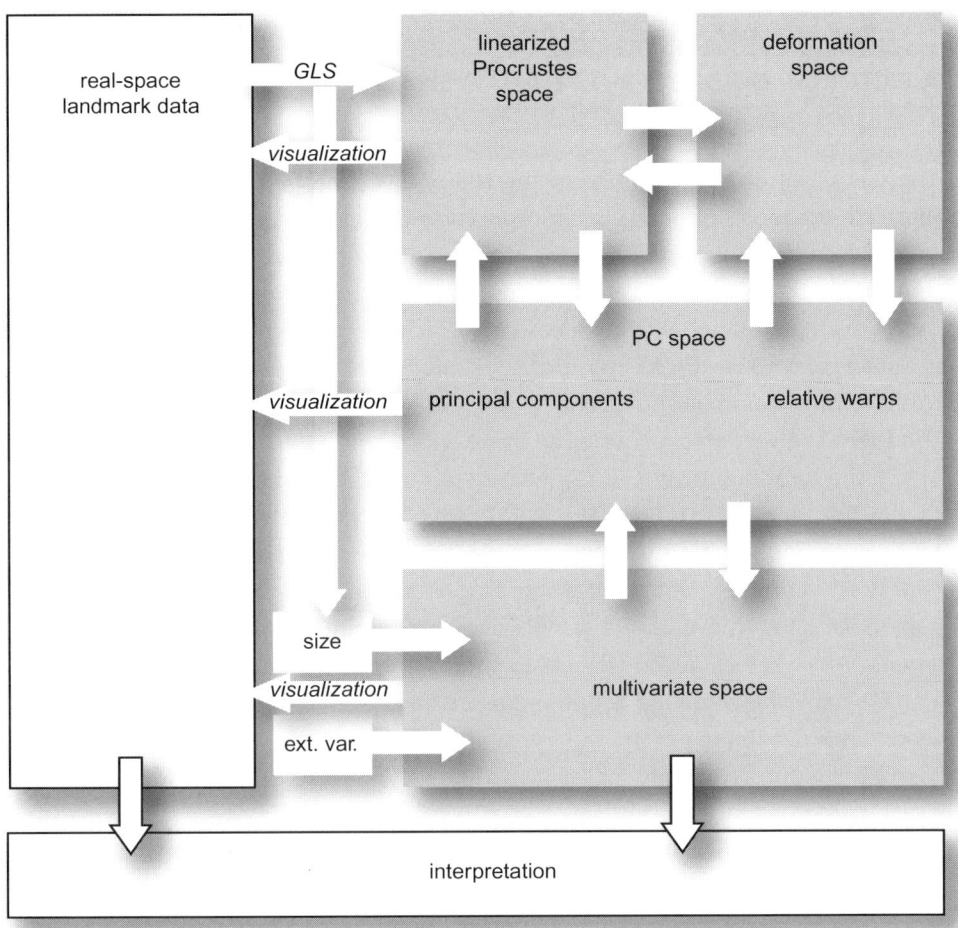

**FIGURE 8-15** The workflow of shape space analysis. Landmark configurations characterizing the form of specimens are normalized to unit size and superimposed with generalized least-squares fitting (GLS) procedures to obtain a consensus shape. The consensus serves as a common reference for all further analyses of shape variability (gray areas). Shapes are expressed as deviations from the consensus (in linearized Procrustes space) and/or as deformations of the consensus (in thin plate spline deformation space). Techniques of principal components analysis are used to reduce the dimensionality of these abstract multivariate spaces and provide a visual grasp of multivariate patterns of shape variability. Because all transformations from physical space to the various types of shape space are reversible, it is possible to reexpress results at any stage of the analysis in terms of real-space patterns of shape change. The final interpretation of the results in terms of underlying biological processes typically requires consideration of external variables, such as size, individual and geological age, sex, taxon specificity, and various biogeographical parameters.

serve this purpose. As we deal with individual specimens rather than abstract shapes, an averaging procedure is needed to remove interindividual variation from the reference shape. It is often sensible to use a sample mean configuration, or so-called consensus, as a reference against which the shape of all specimens is

## BOX 8-4

### A PICTORIAL GUIDE TO SHAPE ANALYSIS

The sample of animal forms from the Niaux cave (see Fig. 8-4) is used here to illustrate the logics of geometric-morphometric analysis of shape (Fig. 8-16A, B). In a first step, we discard size information by scaling specimens up or down to unit centroid size, and by scaling mirror images with –1, if necessary (Fig. 8-16C). In this example, the quantitative basis on which these operations are carried out is a set of $K = 10$ landmarks positioned at biologically relevant points along the outlines of the animal figures (Fig. 8-16D; for methods of analyzing outlines see the following section on Fourier analysis).

In the subsequent step, specimens are superimposed to eliminate differences in position and orientation (Fig. 8-17). This is achieved through appropriate translation and rotation until a best-fit criterion is reached (which is similar to that used during evaluation of a regression line; see Appendix F.6). The aim of superposition is to evaluate a sample average configuration, the so-called consensus, which serves as a reference to specify the shapes of all specimens.

Each specimen's shape is defined as its deviation—landmark coordinate by landmark coordinate—from the consensus configuration (Fig. 8-18). A specimen's deviation from the consensus is equivalent to its position in linearized shape space.

Linearized shape space typically has a large number of dimensions, such that it cannot be visualized immediately. Yet it is possible to apply dimension reduction techniques such as principal components analysis to represent the statistically significant portion of shape variability in a low-dimensional space (Fig. 8-19). Because each point in dimension-reduced shape space corresponds to the shape of one specific landmark configuration, it is possible to express arbitrary points as hypothetical real-space objects and to switch forth and back between statistical and real-space representations of the results. Furthermore, each specimen can be expressed as a deformation of the consensus and visualized with deformation grids. Generalizing the concept of shape deformation or transformation, any direction in shape space can be visualized as a real-space pattern of deformation of the consensus.

**FIGURE 8-16**  **A**, **B**: The sample. **C**: Normalizing specimens to centroid size = 1 and optional mirror-imaging. **D**: Anatomic landmarks are used to quantify size and shape information (note a close analogy to the representation of animals in celestial constellations).

**FIGURE 8-17**  Superimposition of specimens permits calculation of a sample mean (consensus) configuration (**A**) as a reference shape (**B**).

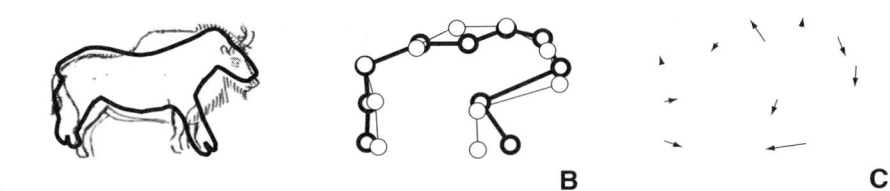

**FIGURE 8-18**  The shape of a specimen (**A**, **B**, thin lines) is defined as its deviation from the consensus (**A**, **B**, bold lines). Vectors (**C**) denote direction and magnitude of deviation at each landmark, defining the specimen's position in linearized multidimensional shape space.

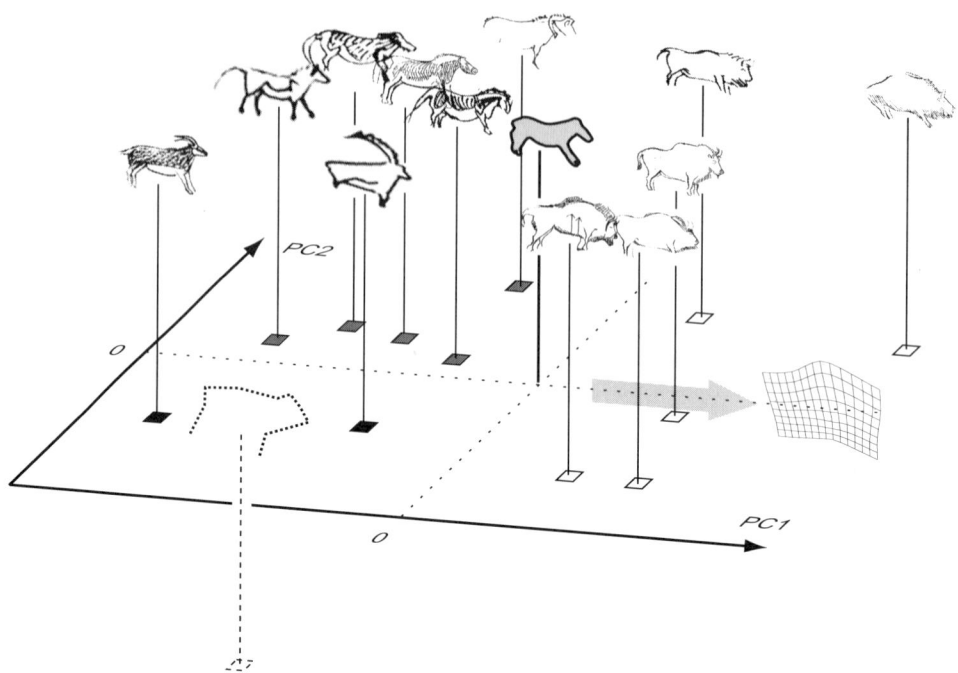

**FIGURE 8-19**   Applying dimension reduction techniques such as PCA to the data scatter in multidimensional linearized shape space permits visual inspection of patterns of shape variability in a low-dimensional space (PC1, PC2). Real-space specimens map to points in shape space; its coordinate origin represents the consensus shape. Arbitrary points in shape space can be expressed as hypothetical specimens (dotted outline), and directions in shape space (gray arrow) correspond to physical patterns of shape deformation (grid).

expressed. The procedure used to evaluate a consensus shape is known as *Procrustes superimposition* (Goodall, 1991), or generalized least-squares fitting (GLS) (Rohlf, 1990).[6]

In close analogy to finding a best-fit regression line through a cloud of data points, GLS methods align specimens until a best-fit criterion is met (see Appendix F.6). To illustrate the interplay between biological reasoning and computation during evaluation of a consensus shape, let us consider the example in Figure 8-20. From an evolutionary perspective, the major difference in shape between a giraffe and a specimen from our sample of prehistoric animals (Fig. 8-17) concerns vertical elongation. The associated changes in overall body structure make it difficult to find a best-fit superimposition. Straightforward GLS fitting tends to spread shape difference equally over the entire landmark configuration, whereas biologically motivated "resistant-fit" procedures (Rohlf and Slice, 1990b) may concentrate on superimposition of the trunk area.

---

[6] In allusion to the mythical highway robber Procrustes who, good host as he was, stretched out or cut short his guests to fit his unique guest bed.

A                                                                                           B

**FIGURE 8-20**   Issues of measuring shape. The shape of a giraffe represents an outlier in the relatively homogeneous sample of prehistoric animals (see Fig. 8-17). **A:** Procrustes superposition distributes shape difference evenly over all landmarks. **B:** Taking into account biological arguments, it is more sensible to locate most of the shape difference in the neck and legs.

No best biological solution exists to the problem of how to determine a reference shape. As we will see in Section 8.4 on euclidean distance matrix analysis, the same is also true for statistical/computational methods. However, even principal statistical limitations shall not prevent us from seeking solutions to the *biological* task of quantifying and visualizing shape variation. This approach is legitimate because, as we have seen in Fig. 8-10, biological reasoning determines the choice of morphometric/statistical procedures and the interpretation of statistical results, whereas methods themselves do not provide biological meaning. It is vital to examine the outcome of morphometric analyses not only for biologically relevant results, but also for potential biases intrinsic to the chosen method. Verification with alternative approaches is essential, for example, through application of different methods of fitting and subsequent comparison of results, or through applying alternative geometric-morphometric methods such as EDMA and outline analysis to the same data set.

### 8.3.4 Analyzing Data in Shape Space

As soon as we have settled on suitable methods to superimpose specimens and to calculate a reference shape, the geometry of linearized shape space, in which all further analyses are carried out, is fully determined. As we have seen in Box 8-3, shape space typically has a large number of dimensions, such that it is impossible to obtain an immediate visual representation of the data scatter in that space. However, its linearization permits application of standard multivariate procedures such as PCA to reduce dimensions (see Appendix F.4 and F.7).

We now return to the human-ape sample (see Fig. 8-11) in order to illustrate these procedures (Fig. 8-21A–D). The shape of each specimen is characterized by

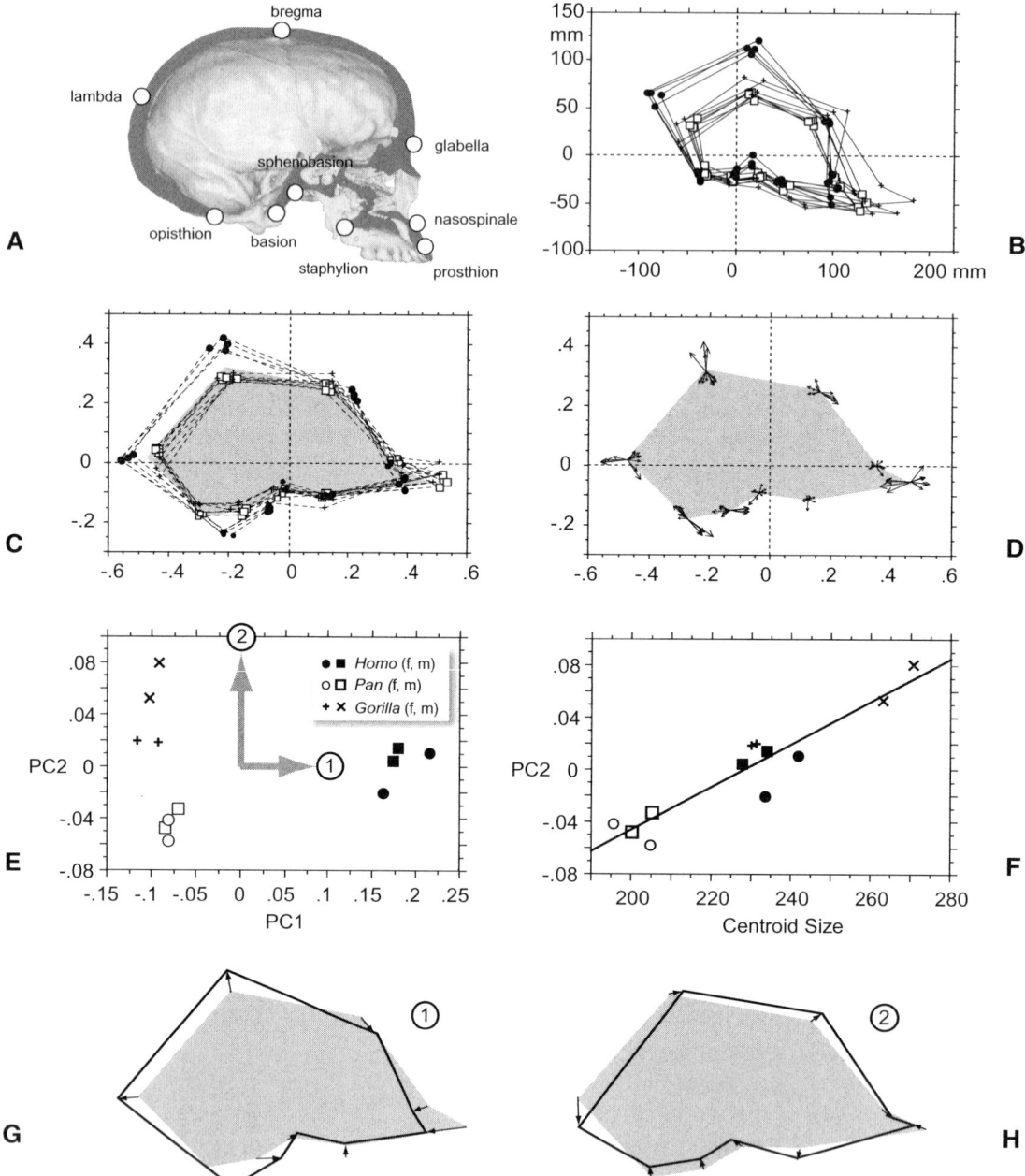

**FIGURE 8-21** Shape space analysis of a human-chimp-gorilla sample. **A**: The nine landmarks used to define cranial midsagittal shape. **B**, **C**: Original landmark data before (**B**) and after (**C**) superimposition (the gray area is the consensus shape, centroids of all specimens are located in the coordinate origin). **D**: Deviation of specimens from the consensus define a linearized shape space. **E**: Principal components analysis is used to reduce the dimensionality of shape space; this permits visual inspection of patterns of shape variability. The first two PCs account for 81.2% and 7.5% of the total sample variance, respectively. **F**: PC2 is strongly correlated with centroid size, indicating that it captures allometric trends of shape variation in the data set. **G**, **H**: Locations in PC space (circled numbers 1 and 2 in **E**) correspond to specific landmark configurations and directions (gray arrows in **E**) to specific patterns of landmark displacement of the consensus.

the deviation of $K = 9$ two-dimensional landmarks from the consensus. Taking into account reduction in degrees of freedom due to scaling, translation, and rotation, linearized shape space has $(2K) - 4 = 14$ dimensions. Applying PCA to the data scatter in this 14-dimensional space yields 14 principal components. We restrict our considerations to the first two PCs that account for 81.2% and 7.5% of the total sample variance, respectively.[7]

At this point in the analysis, we may consider the potential biological meaning of these components of shape variation. Plotting PC2 versus PC1 (Fig. 8-21E) reveals that apes and humans are at opposite ends in the graph along PC1, whereas chimps and gorillas are at opposite ends along PC2. Correlation with centroid size further shows that PC2 captures an allometric trend in shape variability ($r^2 = 0.89$; Fig. 8-21F). PC1 is not correlated with size, but because it accounts for most of the morphologic variability in our data set and permits differentiation between apes and humans, it is important to examine the real-space patterns of shape transformation that it expresses.

## 8.3.5 Visualizing Patterns of Shape Difference and Shape Change

We now return to the initial question of D'Arcy W. Thompson: How can differences in shape between two specimens be rendered visually? With the full toolkit of morphometric analysis and computer visualization methods at our disposal, we may generalize this question: How can patterns of shape variation in a sample be transformed from shape space into real space?

We make use of the links that exist between the geometry of an original landmark configuration and its representation as a single point in multidimensional shape space. During transformation from real space to linearized shape space to PCA space, the original geometric information is fully preserved, so that it is possible to reverse the procedure and transform points in PCA space back into real-space geometries. Any given point in PCA space is equivalent to a specific pattern of displacement of the consensus landmarks (see Appendix F.7). This pattern can be rendered readily as a set of vectors connecting landmark positions before and after displacement (Fig. 8-21G,H).

To obtain a more comprehensive visual impression, it is sensible to interpolate the effects of landmark displacement over the entire plane or space (see Fig. 8-23). This task can be realized with spline functions. Recall that spline functions

---

[7] As we will see below, the data scatter revealed here is similar to that obtained with PCA on interlandmark distances (see Fig. 8-29). The fact that quite different methods of analysis yield convergent results represents an important validation of the consistency of the current analysis.

are based on so-called nodes that act as anchor points of a smooth deformation (Fig. 2-13). Here, we use landmark positions before and after displacement as nodes, and we generate smooth deformations with thin plate spline (TPS) functions. As their name expresses, TPS functions draw on the metaphor of a thin metal plate, whose minimum-energy shape under deformation is considered (see Box 8-5).

Traditionally, TPS functions are visualized with deformation grids (Fig. 8-12, Fig. 8-23B,C), which provide an immediate quantitative account of how subregions of the organismic structure under investigation are rearranged relative to each other. However, although grids are visually appealing, their use in a biological context is problematic for several reasons. The first is a practical one. Although grids represent a suitable means to visualize shape change in two dimensions, substantial problems arise during visualization of three-dimensional shape transformations. Three-dimensional cuboid grids tend to be confusing, because our visual attention is inevitably directed toward undesired boundary effects at the edges of the cuboid, which no longer have any relation with the biological object itself, while more significant internal alterations remain unnoticed.

The second reason concerns virtual reconstruction, specifically our endeavor to infer processes of shape change from patterns of shape difference. Deformation grids represent an external frame of reference without explicit biological meaning. This makes it difficult to interpret geometric "deformation" in terms of potential biological mechanisms of shape variation. Indeed, having devoted great care to choosing appropriate reference structures at every step of shape space analysis, displaying the results in an arbitrary system of coordinates would constitute a leap backward.

To find a "natural" system of reference for the visualization of TPS deformation, we make use of the fact that TPS interpolation functions specify the amount and direction of displacement for every point in space, that is, they define a *displacement field*. This information can be visualized in various ways (Fig. 8-23). A straightforward strategy consists of computing a sequence of intermediate stages of deformation of a consensus object and rendering it as an animation. However, to obtain detailed quantitative insights into the effects of shape transformation, we need more sophisticated strategies. Let us first consider the two-dimensional case. Recall the concept of a gradient field (Fig. 4-11), which indicates local magnitudes and directions of intensity change in an image. Similarly, the TPS displacement field indicates local magnitudes and directions of shape change that can be graphed as a vector field (Fig. 8-23F,G). Alternatively, we may visualize the direction-free amount of local expansion and/or contraction, thus expressing regions of positive versus negative local allometry. Ideally, a two-color scheme is used to differentiate between opposite trends, but grayscale images work as well (Fig. 8-23D,E).

## VISUALIZING SHAPE TRANSFORMATION WITH THE THIN PLATE SPLINE

In the 1960s, the automobile industry used this technique to generate smooth and flowing forms. This was a major incentive to study deformation functions formally and to exploit them for graphical design. Let us consider one specific engineering problem illustrated in Figure 8-22A, the deformation of an infinitely thin, flat metal plate (Duchon, 1976; Meinguet, 1984). Imagine such a plate in the $x$-$y$ plane (at height $h = 0$) that is attached to $N$ surface points $p_i$ (the nodes). The plate is deformed by vertical displacement of the points $p_i$ by individual amounts $h_i$. The TPS function describes the shape of the deformed plate (i.e., height $h$ at any point of the surface) that is attained with a minimum amount of bending energy.

$$h(x,y) = TPS(x,y) \qquad (8\text{-}3)$$

In TPS-based analysis of real organisms, the spline functions are used in the following way: anatomic landmarks represent the nodes of the TPS function, which deforms the space such that a specimen **A** is transformed into a specimen **B**. Contrasting with the metal sheet, however, deformations are metaphorical and are no longer modeled as displacements perpendicular to a flat surface, but in any direction of space (for example, along the $y$-axis of the plane itself; Fig. 8-22B). This implies the use of two $(x,y)$ and three $(x,y,z)$ spline functions for two- and three-dimensional landmark configurations, respectively. These functions specify a *displacement field*, that is, they describe how any point $(x,y)$ and $(x,y,z)$ is transformed into $(x',y')$ and $(x',y',z')$, respectively.

$$2D:(x,y) \mapsto (x',y'); \quad x' = TPS_x(x,y), \quad y' = TPS_y(x,y)$$
$$3D:(x,y,z) \mapsto (x',y',z'); \quad x' = TPS_x(x,y,z),$$
$$y' = TPS_y(x,y,z), \quad z' = TPS_z(x,y,z), \qquad (8\text{-}4)$$

These functions have no immediate biological significance, but they provide the basis to visualize shape transformations comprehensively.

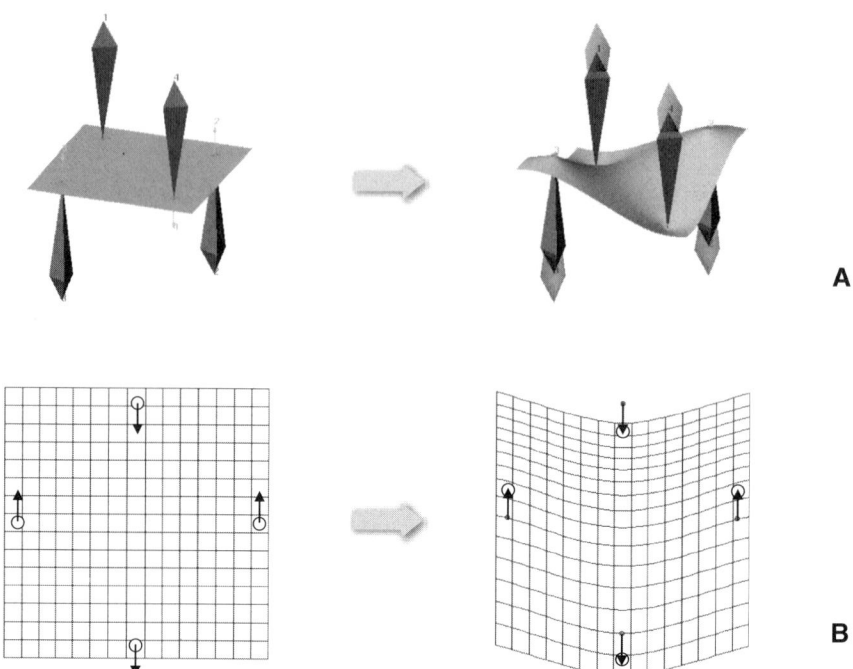

**FIGURE 8-22** The thin plate spline (TPS) interpolant function. **A**: The spline function is used to evaluate deformation of an infinitely thin metal sheet that is displaced in perpendicular direction to its surface (at four points in this example). **B**: Exactly the same mathematical formalism can be applied to deform a planar grid along one direction in the plane (here, the $y$-direction). The TPS formalism can be extended to model deformation of two- and three-dimensional landmark configurations in any direction of space, by using two and three TPS functions, respectively.

For the visualization of three-dimensional patterns of deformation, let us recall that the landmarks that form the basis of shape space analysis (and of the TPS function) are typically derived from surface structures of the specimens. During visualization, it is therefore sensible to make explicit reference to object surfaces rather than to the entire displacement field. The primary biological motivation of this approach is to obtain a process-oriented visualization of patterns of shape variation (see Box 8-6). Applied to a developmental series of human crania (Zollikofer and Ponce de León, 2002), these graphs provide a hint of the patterns in growth and development that generate adult cranial morphologies (Fig. 8-24).

To conclude this section, we return to a specific implementation of D'Arcy W. Thompson's idea of representing shape variation by means of shape transformation. In fact, TPS methods were originally proposed as an analytical rather than a visualization tool (Bookstein, 1991). As it is possible to study principal components of shape *variability*, it is also possible to evaluate principal components of shape *deformation*—so-called *principal warps*. Figure 8-19 demonstrates the complementary ways in which these methods look at shape variation. Whereas PCA

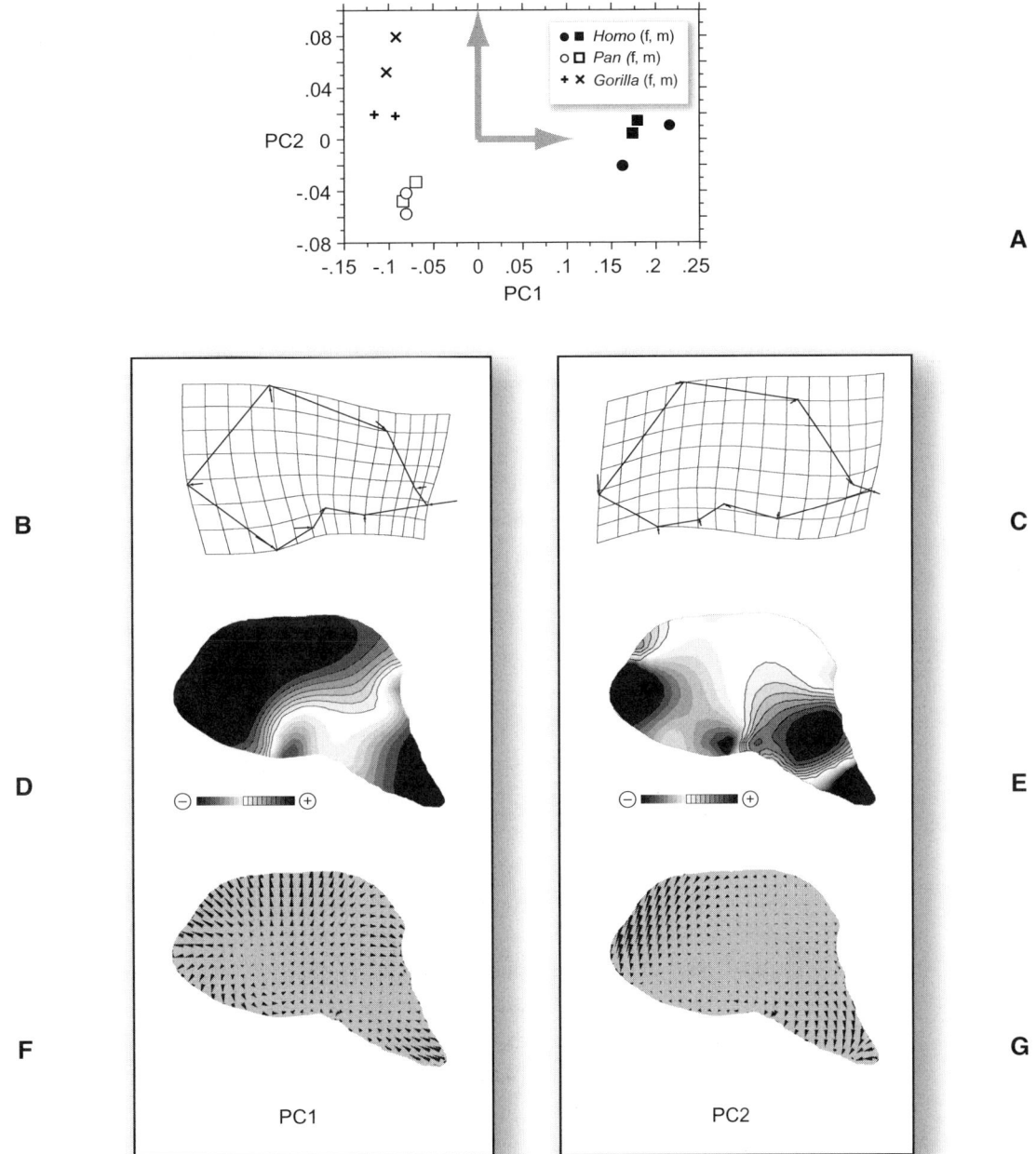

**FIGURE 8-23**  Visualization of shape transformation in the human-ape sample. **A**: Directions along PC1 and PC2 in linearized shape space correspond to hypothetical transitions from ape to human and from chimp to gorilla crania, respectively (same data as in Fig. 8-21). **B, C**: With thin plate spline functions, patterns of shape transformation can be visualized as deformations of a grid superimposed onto the consensus shape. **D–G**: Grid-free visualization of shape transformation. **D, E**: The amount of relative expansion/contraction (+/−) reveals areas of local positive/negative allometric differences between crania. **F, G**: The direction and magnitude of shape transformation is visualized with a vector field.

## BOX 8-6

### VISUALIZING THREE-DIMENSIONAL PATTERNS OF SHAPE VARIATION

Three-dimensional patterns of shape variation are difficult to visualize on a sheet of paper. The loss of one dimension, however, can be compensated with computational techniques and specific methods of visualization. As a general strategy, we restrict visualization to object surfaces and do not consider the entire volume because the original landmark data of shape space analysis represent surface locations (Zollikofer and Ponce de León, 2002).

We focus on (infinitesimally) small surface patches (Fig. 8-24) and consider their behavior under TPS shape transformation. Surface patches may expand or contract relative to the total surface area. Expansion/contraction can be measured by a scaling factor $a$ and visualized with opposite colors. In growth studies, $a$ indicates whether a local area grows with positive ($a > 1$) or negative ($a < 1$) allometry or isometrically ($a = 1$), relative to the entire object surface under consideration.

To obtain directional information about shape transformation, the three-dimensional vectors **d** specifying local displacement are decomposed into components normal (**n**) and tangential (**t**) to the surface. Tangential components are visualized as a vector field, whereas normal components are best visualized with opposite colors (see Web Companion) that indicate whether the vector points inward or outward (intensity codes for vector length).

A central biological motivation of this method of visualization is interpretation of normal/tangential components and local allometries as hypothetical relative directions and relative magnitudes of growth and development. For example, subregions of the vertebrate skull grow by expansion, drift, and/or passive displacement relative to neighboring structures (Moss and Young, 1960; Enlow, 1990). Visualizing normal and tangential components of shape change as well as allometric trends represents a first step toward the interpretation of patterns of shape change in terms of underlying processes.

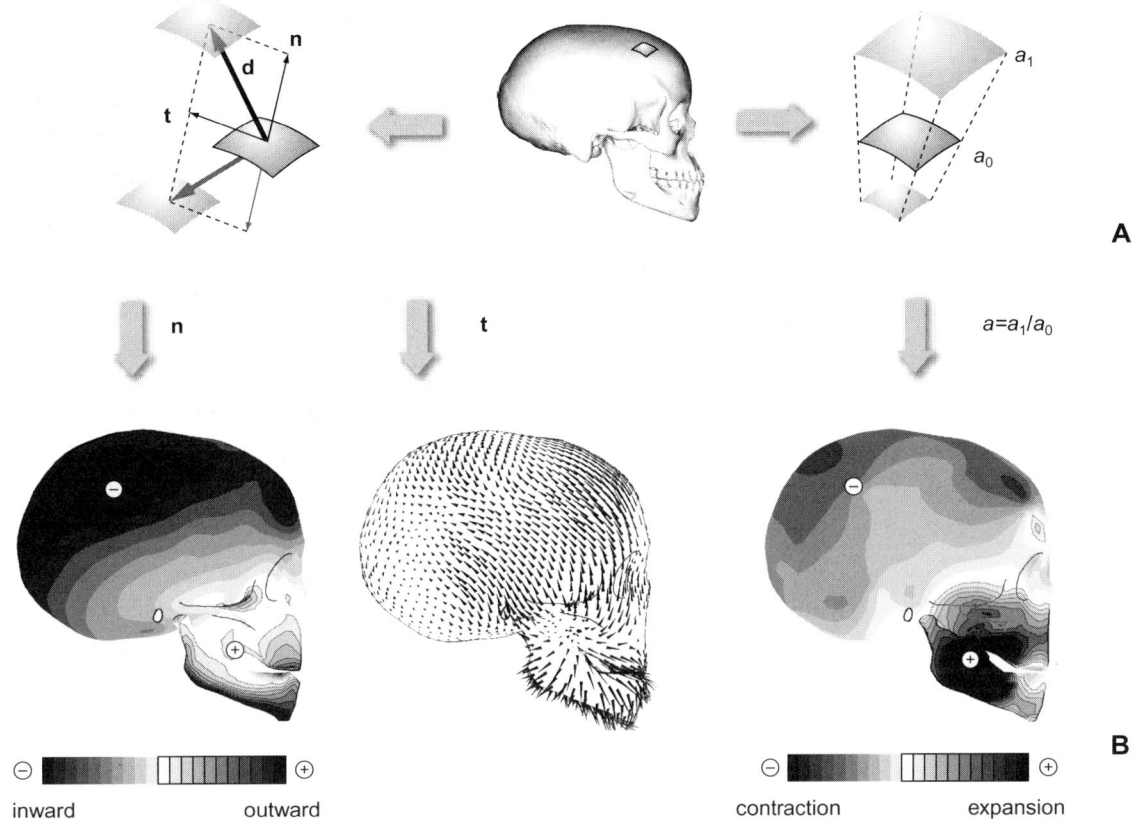

**FIGURE 8-24** Visualizing three-dimensional shape transformation. **A**: Displacement (left side) and relative expansion/contraction (right side) of a small surface patch. **B**: Visualization of normal (**n**) and tangential (**t**) components of displacement and of relative expansion/contraction (*a*) of a human skull during growth (shaded/outlined-shaded areas indicate inward/outward orientation of the normal vectors, and contraction/expansion, respectively). Note contrasting growth characteristics of the braincase and the face (data from Zollikofer and Ponce de León, 2002).

decomposes shape variation into components accounting for large to small proportions of the total variability, TPS analysis decomposes shape variation into components accounting for global to local modes of deformation. An account of these techniques and their potential biological relevance is given in Appendix F.8.

## 8.4 EUCLIDEAN DISTANCE MATRIX ANALYSIS

### 8.4.1 In Search of the Golden Mean

In previous sections, we devoted considerable effort to establishing criteria to find a consensus shape that represents a biologically and statistically suitable reference against which individual shapes can be measured. We ultimately arrived at the conclusion that no best estimate of a mean form exists. Let us consider the statistical side of this issue. When we estimate a mean value $\bar{x}_{est}$ from a series of $N$ measurements of $x$, we rely on the fact that, with increasing sample size $N$, $\bar{x}_{est}$ converges on the "true" (population) mean value $\bar{x}$ of a probability density function. Unfortunately, during Procrustes shape analysis, increasing the number $N$ of specimens in the analysis does not increase the consistency of the estimated mean shape. In fact, it has been shown that GLS fitting methods fail to yield a consistent reference shape (Dryden and Mardia, 1998; Lele and Richtsmeier, 2001), except for the very special case of spatially homogeneous (isotropic) data scatter around each landmark (Bookstein, 1991).

Euclidean distance matrix analysis (EDMA) (see Lele and Richtsmeier, 1991; Lele, 1993; Richtsmeier and Lele, 1993; Lele and Richtsmeier, 2001) circumvents inconsistency issues in an elegant manner. The basic idea is straightforward: analyze all possible euclidean (i.e., straight-line) distances that can be measured within a given set of landmarks on a specimen (these can be summarized in the *Euclidean distance matrix*). Note that, whereas classic multivariate morphometrics considers a small subsample of interlandmark distances (see Fig. 8-10), EDMA is based on a full triangulation of the landmark configuration. Compared to shape space analysis, the concept of a reference shape is explicitly abandoned in EDMA, because no consistent statistical or best biological estimator of such an entity exists. In effect, the distance matrix of each specimen is self-referential. However, the geometric correspondence relation between specimens is preserved because the same set of interlandmark distances—the euclidean distance matrix—is measured for each specimen.

EDMA essentially consists of analyzing patterns of variability within the high-dimensional distance matrix space termed *form space*. As indicated by its name, form space contains all possible *forms*, rather than *shapes* that can be built with a given number of landmarks. It is possible to analyze size and shape information as an ensemble, but it is also possible to decompose form into size and shape according to various user-specified criteria. The properties of form spaces are explored in Box 8-7.

## 8.4.2 Exploring Form Variability with EDMA

Recalling that we used PCA as a dimension reduction technique to explore shape variability in linearized Procrustes space, it seems tempting to apply similar procedures to explore form variability in euclidean distance space. However, because PCA involves evaluation of a reference form (the coordinate origin of PC axes), and we have seen that reference form estimates are statistically inconsistent, such an approach clearly contradicts a guiding principle of EDMA—to be reference-free. This problem can be circumvented by measuring distances between specimens in form space, rather than distances between a reference and each specimen.

## BOX 8-7

### FORM SPACE AND EUCLIDEAN DISTANCE MATRIX ANALYSIS (EDMA)

Contrasting with Kendall's two-dimensional spherical space of all triangular *shapes* (Box 8-3, Fig. 8-13), the euclidean space of all triangular *forms* has three dimensions, which measure the three sides *a*, *b*, and *c* (i.e., interlandmark distances) by which any triangle can be specified. Because the relationships between side lengths in a triangle obey geometric constraints—for example, the sum of any two sides is equal to or greater than the third side—triangular forms occupy only a subspace of the entire three-dimensional euclidean space (Fig. 8-25A).

Each triangle can be expressed by one specific point (*a*,*b*,*c*) in this subspace. Conversely, each point in the subspace corresponds to a set of triangles with edge lengths (*a*,*b*,*c*), whose sense (clockwise vs. counterclockwise), orientation, and position in real-space are not determined, accounting for the fact that these parameters are irrelevant to the definition of form itself.

Forms having the same shape but different size lie along lines emanating from the origin. Separation of form into size and shape can be specified according to user-defined hypotheses about the biological significance of these terms. In Figure 8-25B, for example, centroid size (see Fig. 8-6) is used as a measure of size.

The three-dimensional euclidean distance space of triangular forms has its higher-dimensional counterparts. For *K*-landmark configurations, form space has $D_K = K(K - 1)/2$ dimensions, corresponding to the total number of interlandmark distances that can be measured between the *K* landmarks. Accordingly, each *K*-landmark configuration is represented by a $D_K$-dimensional point in form space, and each point in that space corresponds to one specific *K*-landmark configuration and all its mirror-imaged, rotated, and translated versions.

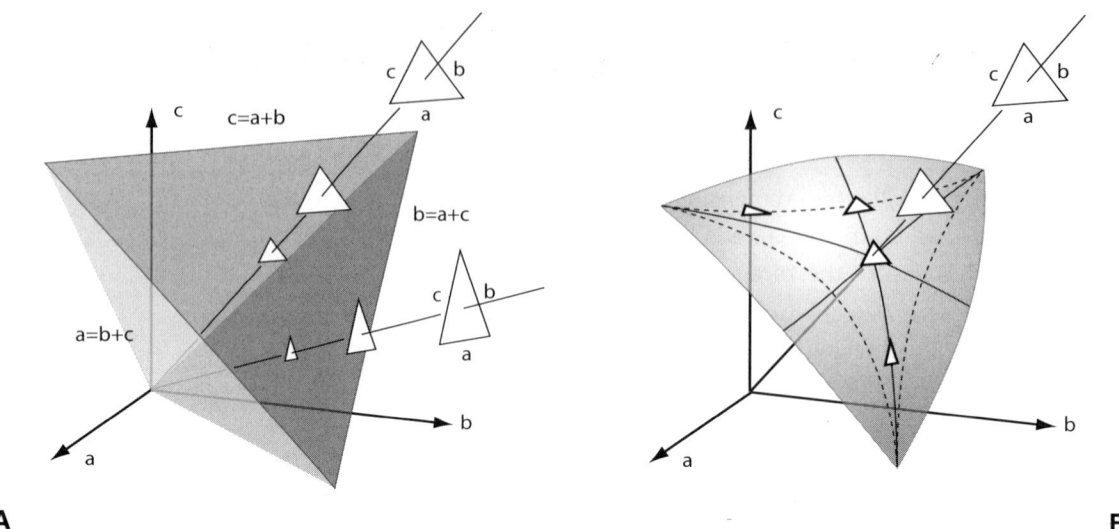

**A**                                                                                                                              **B**

**FIGURE 8-25** Euclidean distance space of all triangular forms. **A**: The form of any triangle with edge lengths $a$, $b$, and $c$ can be represented as a point $(a,b,c)$ in a subspace of three-dimensional space delimited by three planes $a = b + c$, $b = a + c$, and $c = a + b$. Triangles of similar form lie along lines emanating from the origin. **B**: Triangles of equal size lie on a surface, whose geometry depends on the definition of size. For example, normalization to centroid size $\sqrt{(a^2 + b^2 + c^2)} = 1$ projects all triangles on a spherical slice. Isosceles triangles lie on great circles (solid lines), right-angled triangles lie on parallel circles (dashed lines), and equilaterals are in the center. Compare this representation with Kendall's shape space representation of triangles in Figure 8-11.

These distances can be summarized in a *dissimilarity matrix* (Fig. 8-26C), from which patterns of shape variability (Fig. 8-26D) can be extracted by means of *principal coordinates analysis* (PCO) (Borg and Groenen, 1997).[8]

The entire procedure is illustrated with the human-ape sample in Figure 8-26.

EDMA has the advantage of being freed from the problems of statistical inconsistency that arise during the evaluation of a reference form. However, this advantage turns into a disadvantage when it comes to visualization. In fact, real-space patterns of shape variability can only be visualized if a reference shape located in a common coordinate system is available. As a consequence, the problems associated with selecting a reference that arose at the outset of shape space analysis and that were so elegantly circumvented by EDMA, reemerge during visualization (Table 8-1).

Nevertheless, special-purpose visualization tools are available to assess the role of each landmark in bringing about the observed (reference-free) pattern of shape variability (Lele and Richtsmeier, 1992; Cole and Richtsmeier, 1998). The *influential landmarks* detection method is illustrated in Figure 8-27. We have

---

[8] Principal coordinates analysis (PCO) is similar to PCA but applied to distance measures between specimens of a sample. Given a full set of interspecimen distances or dissimilarities (measured by any metric), PCO evaluates the most statistically significant patterns of association between specimens.

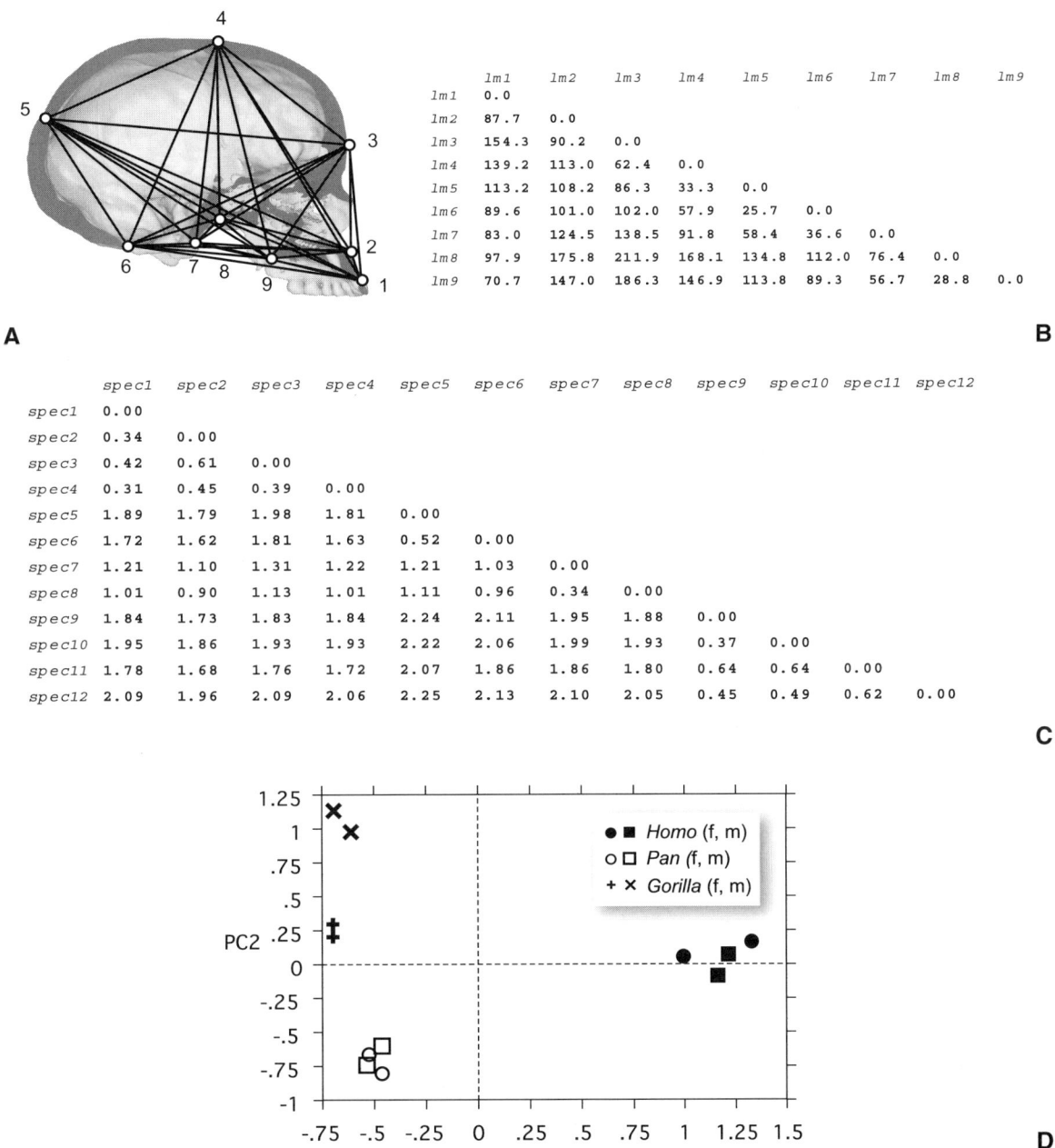

**A**

| | lm1 | lm2 | lm3 | lm4 | lm5 | lm6 | lm7 | lm8 | lm9 |
|---|---|---|---|---|---|---|---|---|---|
| lm1 | 0.0 | | | | | | | | |
| lm2 | 87.7 | 0.0 | | | | | | | |
| lm3 | 154.3 | 90.2 | 0.0 | | | | | | |
| lm4 | 139.2 | 113.0 | 62.4 | 0.0 | | | | | |
| lm5 | 113.2 | 108.2 | 86.3 | 33.3 | 0.0 | | | | |
| lm6 | 89.6 | 101.0 | 102.0 | 57.9 | 25.7 | 0.0 | | | |
| lm7 | 83.0 | 124.5 | 138.5 | 91.8 | 58.4 | 36.6 | 0.0 | | |
| lm8 | 97.9 | 175.8 | 211.9 | 168.1 | 134.8 | 112.0 | 76.4 | 0.0 | |
| lm9 | 70.7 | 147.0 | 186.3 | 146.9 | 113.8 | 89.3 | 56.7 | 28.8 | 0.0 |

**B**

| | spec1 | spec2 | spec3 | spec4 | spec5 | spec6 | spec7 | spec8 | spec9 | spec10 | spec11 | spec12 |
|---|---|---|---|---|---|---|---|---|---|---|---|---|
| spec1 | 0.00 | | | | | | | | | | | |
| spec2 | 0.34 | 0.00 | | | | | | | | | | |
| spec3 | 0.42 | 0.61 | 0.00 | | | | | | | | | |
| spec4 | 0.31 | 0.45 | 0.39 | 0.00 | | | | | | | | |
| spec5 | 1.89 | 1.79 | 1.98 | 1.81 | 0.00 | | | | | | | |
| spec6 | 1.72 | 1.62 | 1.81 | 1.63 | 0.52 | 0.00 | | | | | | |
| spec7 | 1.21 | 1.10 | 1.31 | 1.22 | 1.21 | 1.03 | 0.00 | | | | | |
| spec8 | 1.01 | 0.90 | 1.13 | 1.01 | 1.11 | 0.96 | 0.34 | 0.00 | | | | |
| spec9 | 1.84 | 1.73 | 1.83 | 1.84 | 2.24 | 2.11 | 1.95 | 1.88 | 0.00 | | | |
| spec10 | 1.95 | 1.86 | 1.93 | 1.93 | 2.22 | 2.06 | 1.99 | 1.93 | 0.37 | 0.00 | | |
| spec11 | 1.78 | 1.68 | 1.76 | 1.72 | 2.07 | 1.86 | 1.86 | 1.80 | 0.64 | 0.64 | 0.00 | |
| spec12 | 2.09 | 1.96 | 2.09 | 2.06 | 2.25 | 2.13 | 2.10 | 2.05 | 0.45 | 0.49 | 0.62 | 0.00 |

**C**

**D**

**FIGURE 8-26** Euclidean distance matrix analysis of the ape-human sample data set. **A**: The full set of interlandmark distances measured in one of the human specimens. **B**: Representation of the interlandmark distances as a euclidean distance matrix (the matrix is symmetrical along its main diagonal; its entries define the position of the specimen in form space). **C**: The dissimilarity matrix denotes distances between specimens in form space. **D**: Principal coordinates analysis is used to extract statistically most relevant components of shape dissimilarity (note close similarity of this graph to the results of shape space analysis; Fig. 8-21).

**TABLE 8-1  A Comparison of Geometric-Morphometric Methods**

| | Shape Space Analysis | EDMA | Outline Analysis |
|---|---|---|---|
| Data | Landmark coordinates | Interlandmark distances | Outlines |
| Definition of shape | Deviation from reference shape | Full set of reference-free interlandmark distances | $\Delta x$- and $\Delta y$-path function of the outline |
| Multidimensional space | Linearized Procrustes space (a tangent space to Kendall's spherical shape space) | Form space (a subspace of euclidean distance space) | Fourier space |
| Parameters estimated | Reference shape, specimens' variation around reference shape, TPS deformation modes of reference shape (*Note*: Estimation of reference shape is statistically inconsistent) | Mean form and variance-covariance pattern of landmark scatter  Form/shape difference matrix  Specimen dissimilarity matrix | Coefficients of Fourier transform |
| Multivariate data visualization | Principal components analysis of linearized shape space | Principal coordinates analysis of dissimilarity matrix | Eigenshape analysis |
| Real-space visualization | TPS deformation-based rendering of patterns of shape variability | Graphical identification of influential landmarks characterizing between-group differences | Comparison of outlines (*Note*: Shape differences not localizable) |

➤

**FIGURE 8-27**  Detecting influential landmarks bringing about major differences between ape and human skulls. **A:** The form difference matrix expresses, for each interlandmark distance, the ratio between human and ape mean forms (a *ratio* of 1 indicates identity). **B:** The shape difference matrix expresses, for each interlandmark distance, the arithmetic difference between ape and human mean shapes (a *difference* of 0 indicates identity). In both matrices, diagonal entries are set to zero by convention. **C–H:** Visualizing form and shape difference matrices. **C, D:** Graphing matrix entries per column permits identification of landmarks exhibiting mainly positive (lm2 and lm3) or negative (lm8) deviations of interlandmark distances of the human from the ape average. **E, F:** Difference matrices can also be visualized by coding matrix entries with gray values. Note light and dark "stripes" indicating prevalence of positive and negative values for landmarks 2/3, and 8, respectively. **G, H:** Visualizing maximum/minimum matrix entries permits localization of major differences between ape and human crania (solid/dashed lines represent positive/negative values).

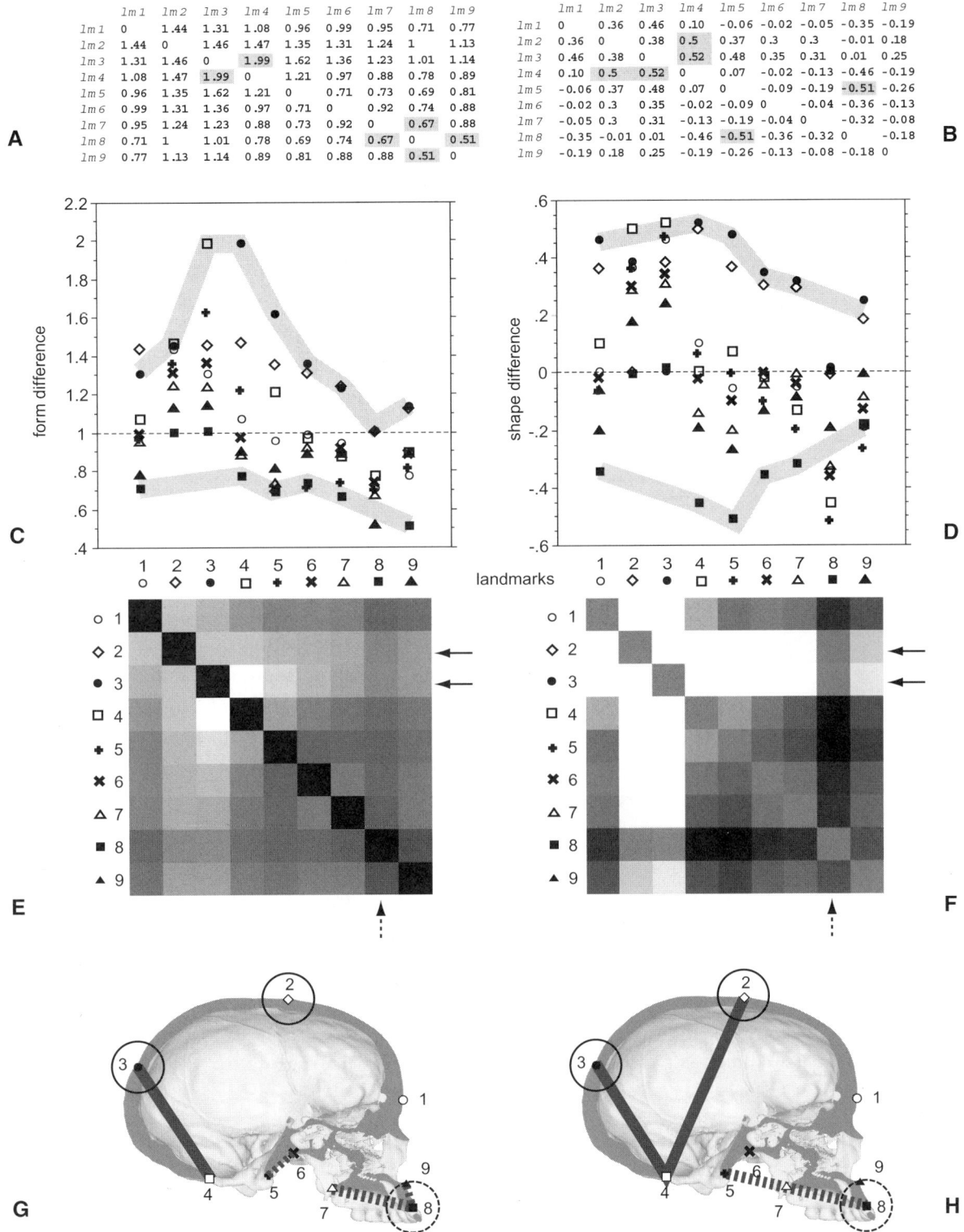

already observed that the most obvious difference in the human-chimp-gorilla data set occurs between apes and humans (Fig. 8-26D). Accordingly, we are interested in identifying those landmarks whose spatial rearrangement contributes most significantly to the ape-human difference. To do this, we calculate ape and human mean forms, respectively, and then express the ape mean form as a ratio (interlandmark distance by interlandmark distance) of the human mean form. As a result, we obtain a *form difference matrix* (FDM; Fig. 8-27A).[9]

Next, we normalize mean forms to size = 1 in order to express the difference between ape and human mean *shapes*. This results in a *shape difference matrix* (SDM; Fig. 8-27B). Both FDM and SDM can be rendered graphically to get a quick visual grasp of the contribution of each landmark to the overall difference between groups (Fig. 8-27C–F). As a result, it appears that, in humans relative to apes, landmarks of the cranial vault exhibit consistently larger distances to any other landmarks, whereas landmarks of the anterior dentition exhibit consistently smaller interlandmark distances.

## 8.5 OUTLINE ANALYSIS

Shape space analysis and euclidean distance matrix analysis are based on landmarks denoting locations of biological and geometric correspondence between specimens. Reconsidering our example of midsagittal cranial forms (Fig. 8-21), it appears that most landmarks are located in the face and cranial base, whereas few landmarks define the form of the cranial vault. This is because the braincase exhibits comparatively few boundary structures that permit establishment of biological point-to-point homologies between specimens.

How can we sample data more extensively in regions between landmarks? Let us first consider the biological side of this issue. For example, the *outline* connecting bregma and lambda (see Fig. 8-21A) corresponds to the interparietal suture (the joint between left and right parietal bones). It is reasonable to assume that interparietal sutures are homologous structures in humans, chimps, and gorillas. However, except for the sutural end points bregma and lambda, we cannot determine point-to-point biological homologies along the suture line (at least as long as no detailed knowledge about relative growth rates along a suture line is available).

As we are forced to abandon biological homologies, let us turn to the geometric side of the issue. Two strategies may be followed to sample additional landmark data along an outline. One strategy consists of spreading evenly a

---

[9]While consistent *reference forms* cannot be estimated, consistent *mean forms* can. This is because a mean form is specified as a euclidean distance matrix without the need to place it into a reference coordinate system.

predetermined number of so-called *semilandmarks* between end points (Bookstein, 1996b, 1997). During this process, semilandmarks are allowed to glide along the outline according to a geometric optimality criterion (e.g., minimum thin plate spline deformation energy) that guarantees their even distribution. Similar procedures can be used to spread semilandmarks over a surface (Andresen and Nielsen, 2001). In any case, the position of points *along* an outline or surface is determined geometrically, thus representing geometric rather than biological correspondence between specimens, whereas their position *on* the outline or surface represents biological homology. Accordingly, differences in semilandmark positions between specimens convey a confusing message, as they have a geometrically relevant (but biologically irrelevant) tangential component and a biologically relevant normal component. This odd situation is difficult to translate into localized, biologically meaningful patterns of shape difference. Nevertheless, semilandmarks remain a valuable tool for exploring shape variation in landmark-depleted regions of a morphology (Bookstein et al., 1999).

The second strategy tries to avoid the problem of specifying point-to-point homology. As a first step, only closed outlines—which can be represented as periodic functions—are considered. Because we live in a finite world, however, free-form outlines must be specified as a sequence of line segments (Fig. 8-28A; see also Fig. 2.12). For convenience, we subdivide the outline into segments of equal length $\Delta l$. Next, we walk along the path with step length $\Delta l$ and measure increments $\Delta x$ and $\Delta y$ from step to step. Because we may walk around over and over again, we obtain two periodic functions $g$ and $h$ with period length $L$ ($L$ is the total length of the perimeter):

$$\Delta x = g(\Delta l) \text{ and } \Delta y = h(\Delta l). \tag{8-5}$$

If we pace at constant speed, we may measure step lengths $\Delta l$ in time units, $\Delta t$. Most readers will be familiar with periodic functions, such as sine and cosine functions. We have already encountered spatial sine patterns in Figure 3-2, and we alluded to temporal sine patterns (tones, whose frequency $f$ is inversely proportional to wave length $L$) during the discussion of MRI signals in Chapter 3. In Chapter 4, we used the *Fourier transform* to convert an image into a spectrum of sine waves with various frequencies and to filter out specific frequencies.

Let us have a look at what the Fourier transform actually does. The Fourier theorem states that it is possible to generate any periodic function $g$ with base frequency $f$ (and corresponding greatest wavelength $L$) by superposition of the 0th, 1st, 2nd, and so on, harmonics (meaning multiples of frequency $f$) of sine and cosine functions with characteristic amplitudes $a_i$ and $b_i$:

$$g(x) = \sum_{i=0}^{\infty} (a_i \cos(ix) + b_i \sin(ix))$$
$$= a_0 + (a_1 \cos(x) + b_1 \sin(x)) + (a_2 \cos(2x) + b_2 \sin(2x)) + \ldots \tag{8-6}$$

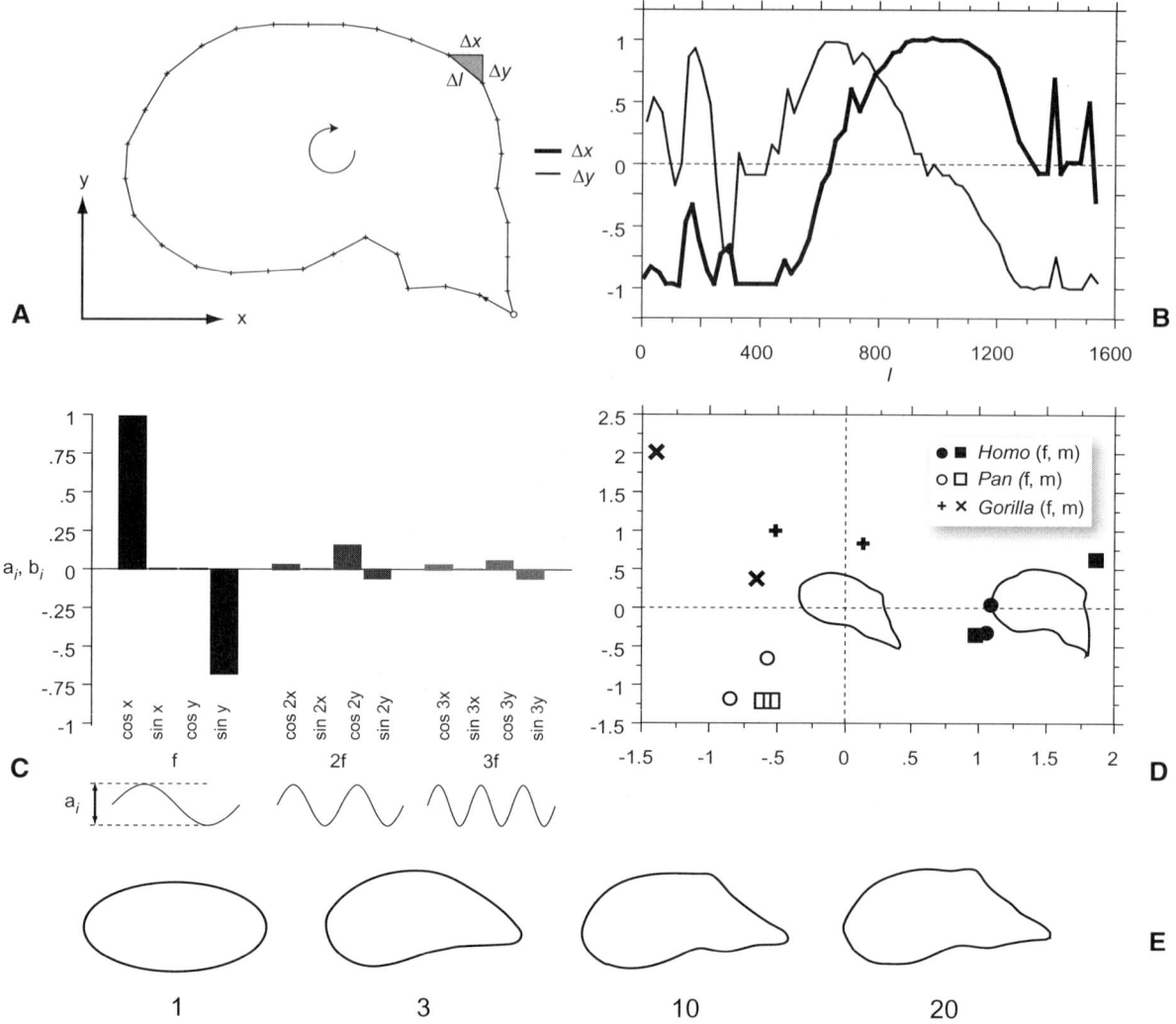

**FIGURE 8-28**   Elliptic Fourier analysis (EFA). **A**, **B**: The midsagittal section through a skull is quantified by two functions describing deviations in $x$ and $y$ directions, as one steps around its outline. **C**: In Fourier space, each of these functions is expressed as a spectrum of sine and cosine functions of frequency $f$ and its multiples ($2f$, $3f$, . . . $if$), each with characteristic amplitudes $a_i$ and $b_i$. **D**: Application of EFA to the human-chimp-gorilla data set followed by principal components analysis of coefficients $a_i$ and $b_i$ reveals statistically relevant patterns of shape variation in the sample. **E**: Sample mean shape calculated with 1, 3, 10, and 20 harmonics.

Here, we make use of the Fourier theorem to transform the periodic functions $g(\Delta l)$ and $h(\Delta l)$ (see Eq. 8-5) describing a morphologic outline into a spectrum of wave functions. This procedure is called *elliptic Fourier transform* (EFA) (Ferson et al., 1985; Lestrel, 1989). Its application to our human-ape data set is illustrated in Figure 8-28. It is important to note at this point that, although EFA circumvents the problem of defining homology between specimens along outlines, it is necessary to define homology at the outset of the analysis. As a minimum requirement, outlines of all specimens in a sample must have homologous orientations, and outline functions must have homologous starting points. In the cranial sample, specimens were oriented along a standard anatomic plane (the Frankfurt plane), and outline functions started at the anteriormost point (i.e., prosthion; see Fig. 8-21A).

The goal of EFA is to express the real space geometric properties of outlines in Fourier space. Note that the principal thrust of this analysis is similar to GLS/TPS and EDMA methods, where we express real-space geometric properties of landmark configurations in linearized Procrustes space/deformation space and form space, respectively.

What are the computational consequences of EFA? We evaluate coefficients $a_i$ and $b_i$ for a limited number $I$ of harmonics $i$, such that we obtain a $2I$-dimensional space of Fourier coefficients. Making use of our multivariate skills, we may apply dimension reduction techniques (e.g., PCA to extract statistically relevant patterns of outline shape variation from high-dimensional Fourier space; Fig. 8-28C,D). This method of extraction is known as *eigenshape analysis* (MacLeod, 1999).

This permits us to reexpress points in Fourier coefficient space as real-space outlines. However, because we abandoned the concept of point-to-point homology along outlines at the beginning of our analysis, it is not possible to localize in the real space regions of greatest or least shape difference along the outline. Nevertheless, by exploiting decomposition into various spatial frequencies, it is possible to explore the level of detail at which significant patterns of shape variation occur. In our ape-human data set, for example, the first three harmonics are sufficient to distinguish between species-specific cranial outlines (Fig. 8-28C–E).

## 8.6 A COMPARISON OF GEOMETRIC MORPHOMETRIC METHODS

### 8.6.1 Criteria for Comparison

The reader is now acquainted with three important geometric-morphometric approaches and with a working example that demonstrates their performance. This permits a comparative look at how close these methods come to the stated goals formulated at the outset of this section:

- *Data acquisition*: quantitative representation of organismal geometry
- *Data analysis*: lossless transfer of geometric information from physical to multivariate space
- *Data visualization*: lossless transfer of multivariate data back to real-space geometry

To facilitate comparisons, a pictorial overview of methods and their application to the human-ape sample is given in Figure 8-29, with the main points summarized in Table 8-1. We first compare patterns of shape variation as revealed after

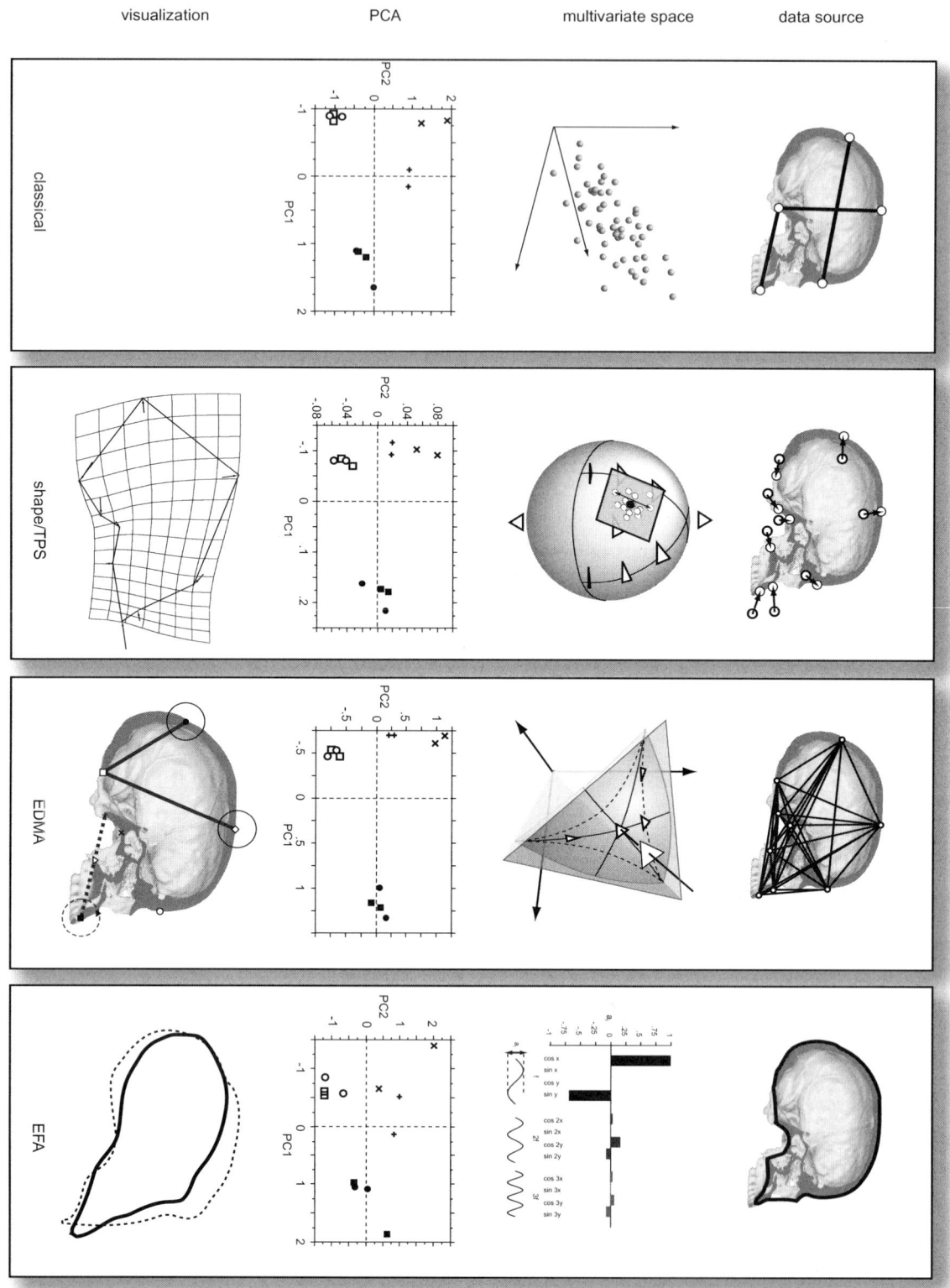

**FIGURE 8-29** Comparative summary of classic multivariate, shape space, EDMA, and outline-based geometric-morphometric analysis. Note convergence of results in dimension-reduced multivariate space.

applying dimension reduction techniques (PCA, PCO) to multivariate space. Interestingly, the three types of geometric-morphometric analysis yield broadly similar results, as does even a classic analysis of only three cranial measurements. This might seem surprising, given that the four methods considered here use quite diverse concepts of shape and shape comparison. Nevertheless, convergent results indicate that biologically relevant patterns of shape variability in the sample can be extracted by any of these methods. Furthermore, they provide verification of each method, showing that potential methodological biases do not obscure biological signals.

Beside these commonalities, however, obvious differences exist in the ability of geometric-morphometric methods to retransform multivariate patterns of shape variability into real-space patterns of shape variability. Shape space and TPS-based analyses yield the most comprehensive picture of taxon-specific spatial rearrangements in cranial morphology. EDMA permits identification of influential landmarks, but restriction to interlandmark distances makes it more difficult to interpret the results in terms of actual morphologic modifications. Fourier-based outline analysis yields a comprehensive picture of overall cranial shape variability, but it cannot pinpoint these effects within the morphology. Finally, classic morphometric analysis cannot reexpress its multivariate results in terms of real-space morphology.

The strengths and weaknesses of each method are a direct consequence of their analytical preconditions. The comprehensiveness of shape space/TPS-based visualizations results from the definition of a reference shape, whose estimation lacks rigorous statistical consistency. Less intuitive visualization of influential landmarks is a consequence of the reference-free approach of EDMA, but this method yields statistically consistent estimators of various form parameters. The fact that Fourier analysis cannot localize shape differences reflects the initial decision to refrain from defining point-to-point homologies in the sample.

### 8.6.2 From Pattern to Process

At the beginning of this chapter, we stated that the ultimate goal of morphometrics is the reconstruction of biological processes that bring about the observed patterns of form variation. This is certainly one of the most demanding aspects of virtual reconstruction, because organismal structure results from a complex interplay of genetic, developmental, environmental, and evolutionary processes. During inference of process from morphologic pattern, therefore, we need to interpret our morphometric results not only by visualizing patterns of shape transformation but also in light of what is already known about underlying processes. No fixed recipes of inference exist. Here, we consider how several morphometric methods operate on our human-ape sample.

Classic morphometric analysis based on selected interlandmark distances (Fig. 8-29, left) shows that human and ape crania differ in the length of the face relative to the height of the cranial vault. This relates to one of the principal themes in hominid evolution—reduction of the viscerocranial elements of the skull and concomitant increase of the brain and its bony case. Shape space analysis yields a detailed spatial picture of how the face and the braincase are related to each other. The TPS deformation grid in Figure 8-29 shows that reduction of the human face occurs concomitantly with changes in its orientation relative to the braincase (Lieberman et al., 2002). Because visualizing patterns depends on the initial choice of the reference shape, it is a good idea to verify these results with the reference-free approach provided by EDMA. Visual representation of the human-ape shape difference matrix shows that bregma, lambda, and prosthion are influential landmarks. Distances measured between bregma/lambda and most other landmarks are larger in humans than in apes, whereas for prosthion, the situation is exactly the reverse. However, it is difficult within the context of the entire landmark configuration to assess the direction into which these landmarks are displaced in apes relative to humans. Finally, eigenshape analysis of cranial outlines confirms the correlation of a low braincase and a projecting, large face in apes, or of a high braincase and a vertically oriented, small face in humans. The quantitative description of these patterns of shape transformation merely represents the outset of process analysis. Further hypotheses must specify how genetic and developmental modifications might influence the observed patterns of shape variation in the hominoid cranium.

## 8.7 EXPLORING MORPHOMETRIC PATTERNS

The methods of morphometric analysis described so far use predefined sets of quantifiable morphologic characters to express geometric and biological correspondence between specimens. Patterns of shape variation revealed by these methods are first expressed in abstract multidimensional space and then reverted to real-space patterns of shape variation, which allow tentative interpretations of underlying processes. Typically, this analytical workflow is iterative: Preliminary results lead to refinement of hypotheses, leading to new morphometric data that can be defined and acquired, which ultimately permits more detailed insights into patterns of shape variability, and so on.

Here, we investigate from a more general perspective how the exploration of morphometric data can be used to generate new morphometric hypotheses, always keeping in mind Darwin's statement at the beginning of this chapter that the acquisition of data is always hypothesis-driven. The practical question is, How can we specify morphometric properties beyond landmarks, outlines, and surfaces? Recalling that these features represent relational properties of our original data volume (see Chapter 2), a promising strategy is to explore a wide spectrum of relational data in search for new morphometrically relevant properties. This open field of research is illustrated here with two examples.

The term *brain mapping* subsumes a wide range of techniques used to localize and visualize the physiological correlates of behavioral/cognitive processes within the three-dimensional structure of the brain. Brain mapping thus provides a new functional morphometry of the brain that can be correlated with classic morphometry based on anatomic structures (Friston, 1998; Sereno, 1998; Van Essen, 2002). Let us focus here on a slightly different aspect of "brain mapping," which is particularly relevant to paleoanthropologists, namely, the quantitative relationship between the brain and the braincase. Endocranial capacity has long been used as a proxy of brain size in human evolutionary studies and the surface structure of endocranial casts (so-called endocasts) as an indicator of brain structure (Holloway, 1975; Falk, 1993). The fundamental question of how endocranial structures are related to actual brain structures was considered in one important monograph (Connolly, 1950) but has received comparatively little attention in subsequent evolutionary studies. Medical imaging techniques now permit a new approach to this question, as it is possible to quantify the brain-bone interface in living subjects. A preliminary study using "morphologic brain mapping" (Fig. 8-30) shows that this relation is highly ambiguous (Zollikofer and Ponce de León,

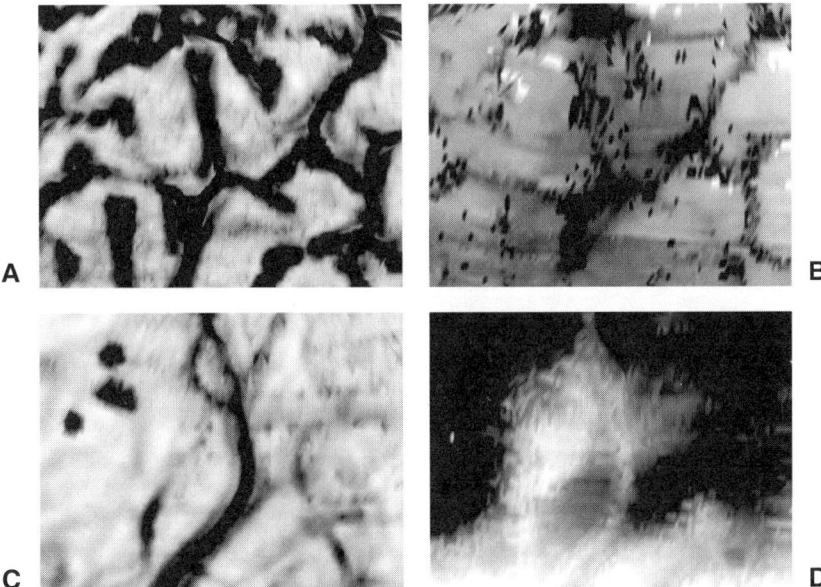

**FIGURE 8-30** Morphologic brain mapping of a 5 × 4-cm area of the left hemisphere around the central fissure (Visible Human data set). Grayscale maps represent cortical surface curvature (**A**), distance between the cortical surface and the inner surface (endocast) of the braincase (**B**), the curvature of the endocast (**C**), and bone thickness (**D**). Brain structures are difficult to identify in morphometric maps derived exclusively from hard tissue.

2001a). Small-scale sulci and gyri that exhibit large interindividual variability are well-represented, whereas large-scale features such as the Sylvian fissure that characterize interindividual homologies tend to be underrepresented. As a consequence, the interpretation of endocranial fine structure in terms of cortical surface structure and, ultimately, in terms of behavioral/cognitive performance must be carried out with caution.

As another application of morphometric mapping techniques, we consider bone thickness. Cranial bone thickness has clearly diminished over the course of human evolution, yet this feature exhibits considerable environmental and activity-related variation (Lieberman, 1996). Bone thickness is difficult to quantify conventionally, as it fluctuates over the cranial surface. Nevertheless, it is possible to obtain overall thickness measures (Zollikofer and Ponce de León, 2001b) and to use bone thickness maps as a clinical diagnostic tool (Fig. 8-31). Thus, using the concept of morphometric mapping, it is possible to provide comprehensive pic-

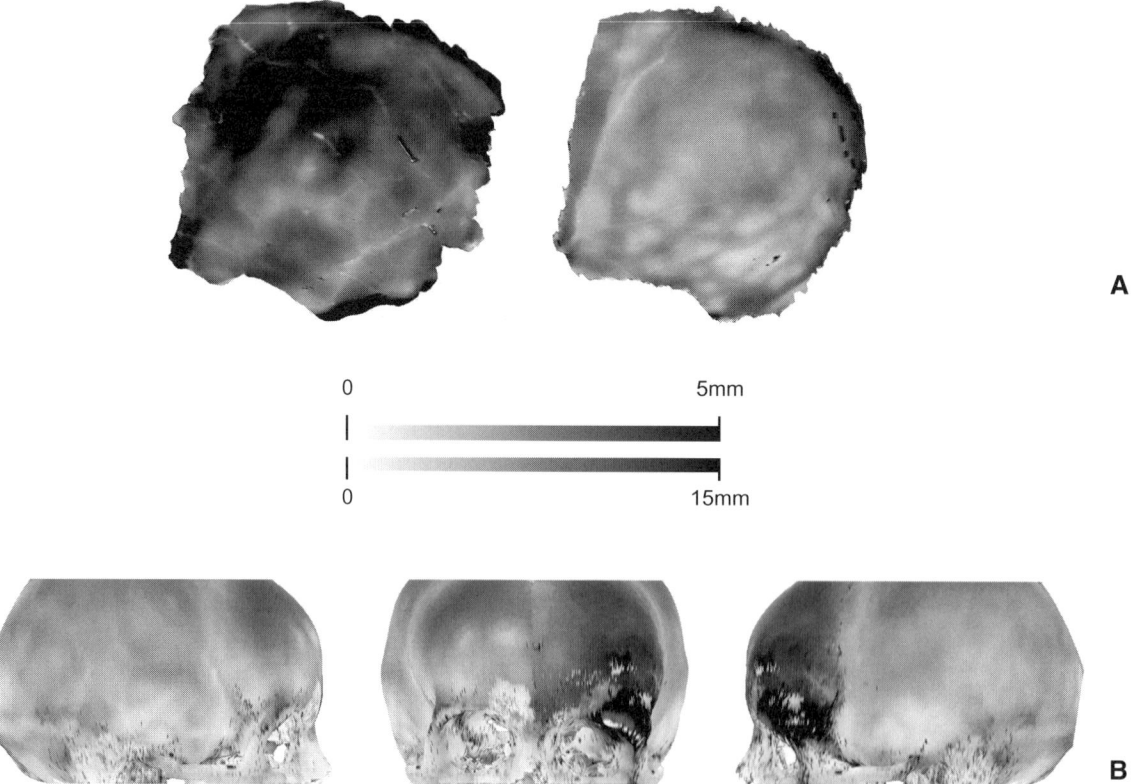

**FIGURE 8-31** Morphometric mapping of bone thickness. **A**: Comparative thickness maps of a Neanderthal and a modern human left parietal bone. **B**: Diagnostic mapping of bone thickness in a patient with a tumor in the left frontal bone. The thickness map reveals that the tumor is confined to the left frontal bone, which, in this patient, is separated from the right side by a patent metopic suture (see also Fig. 7-9).

tures of bone thickness distribution over entire bones. Quantitative comparative investigation of these patterns is still an open field of research.

Let us sum up this chapter about the complexities of morphometric analysis: Geometric-morphometric methods are tools that facilitate the detection and interpretation of complex patterns of morphologic variability, both between and within specimens. Morphometric tools in the hands of scientists are like music instruments in the hands of musicians. The real art not only consists in finding an adequate instrument to interpret a given tune, but also in understanding the score and identifying and conveying its relevant message.

## 8.8 FURTHER READING

A classic textbook on multivariate analysis of biological data is Sokal and Rohlf's *Biometry* (1995). Morrison (1990) is a thorough compendium of multivariate methods. Principal components analysis is treated exhaustively in Jolliffe (1986), and Jackson (1991) is an excellent hands-on guide. For those interested in a wider perspective on linear algebra and matrices, we recommend Janich (1994), Axler (1997), Harville (1997), and Serre (2002). Multivariate analysis and its application to morphometry is treated in Reyment's textbooks (1984, 1991). Classics on size, scaling, and allometry are Gould (1966), McMahon and Tyler Bonner (1983), and Schmidt-Nielsen (1984). Examples of multivariate allometric analysis are provided by Klingenberg (1996) and Klingenberg et al. (1996).

The basics as well as current topics of geometric morphometrics are now nicely covered in a series of textbooks. Bookstein (1991) is a classic, providing a sound formal basis of shape space and TPS analysis as well as thoughtful insights into the connections that can be established between geometry and biology. Dryden and Mardia (1998) is a comprehensive volume, considering a wide range of methods of shape analysis. It presupposes familiarity with mathematical concepts and mathematical notation. Small (1996) provides an overview of the concepts of shape space and their applications in morphometrics. Lele and Richtsmeier's book on euclidean distance matrix analysis (2001) treats the subject in a highly instructive manner. Giving equal weights to biological and statistical issues, it provides critical insights into the potential and limitations resulting from the combination of both. The proceedings volumes of three workshops in morphometrics yield a plethora of information in the form of comprehensive reviews, original contributions, and hands-on instructions (Rohlf and Bookstein, 1990; Marcus et al., 1993, 1996). As an introductory lecture, we recommend the following review articles: Bookstein (1996a), Rohlf (1998), and Richtsmeier et al. (2002).

Fourier-based shape analysis is illustrated in a series of original contributions (Rohlf and Archie, 1984; Ferson et al., 1985; O' Higgins and Williams, 1987; Lestrel,

1989; Renaud et al., 1996; MacLeod, 1999). Further applications are presented in an edited book (Lestrel, 1997). General textbooks on Fourier analysis abound; Morrison (1994) and Gray and Goodman (1995) proved very valuable.

Various studies are dedicated to the question of how geometric morphometric methods—notably shape space analysis and EDMA—can be compared. Rohlf's study (2000) is very useful as it not only compares methods but also provides a succinct overview of each approach and supplies specific software that allows a user to explore the characteristics of various shape spaces (see the tpsPower software at `http://life.bio.sunysb.edu/morph`). Further references include Lele and Richtsmeier (1990), Lele and Cole (1996), and Coward and Conathy (1996), as well as sections in the textbooks mentioned above (Dryden and Mardia, 1998; Lele and Richtsmeier, 2001). A practical example documenting how the same biological question can be tackled with different morphometric methods is provided by a series of studies dedicated to the investigation of growth and development in modern humans, Neanderthals, and Great Apes (Williams, 2000; Ponce de León and Zollikofer, 2001; Rogers, Ackermann, and Krovitz, 2002).

Problems of overlaying various morphometric and functional maps of the brain are reviewed in a fascinating article by Van Essen (2002). And, after all, we should not forget about origins. Galileo's *Discorsi* (1638) and D'Arcy W. Thompson's *On Growth and Form* (1917) are still highly inspiring and show that the basic questions of morphology and morphometry were recognized and formulated well before our time.

Software for morphometric data acquisition, analysis, and visualization is available as freeware or as commercial packages (a link list is provided on the Web Companion). Here, we draw the reader's attention to special-purpose freeware that is constantly updated. Shape space- and TPS-based morphometric analysis tools are provided by F. James Rohlf's "tps suite" of programs at `http://life.bio.sunysb.edu/morph`. The tps suite further provides modules for acquisition of two-dimensional morphometric data and various test modules for the exploration of the properties of shape space. EDMA-based analysis is best performed with WinEDMA by Theodore M. Cole III (`http://c.faculty.umkc.edu/colet`). The software PAST (PAleontological Statistics; `http://folk.uio.no/ohammer/past`), written by Oyvind Hammer and colleagues, is outstanding, as it permits analysis and visualization of morphometric data with a variety of methods (classic multivariate, TPS, EDMA, Fourier, and much more) within a single, compact, and intuitive application.

# IMAGE DATA ACQUISITION SYSTEMS: PERFORMANCE CONSIDERATIONS

The performance of an image data acquisition system, such as the human eye or the photographic camera, is limited by its optical properties and the characteristics of its light receptors. The optical system behaves as a light collector that focuses light waves originating from a point in space onto a point in the focal plane. Because of the wave characteristics of light, a sharp line pattern is transformed into a blurred diffraction pattern. In the focal plane, photographic film or its digital counterpart (a charge-coupled device, CCD) acts as receptors in analogy to retinal rods and cones; they collect data by "counting" the number and location of incident photons (which represent the particle aspect of light). This is a stochastic sampling process that ensures that the "true," discrete intensity distribution of the line pattern is mapped into a bell-shaped (Gaussian) distribution with mean value $m(x)$ and standard deviation $s(x)$. Therefore, accuracy, precision and resolution are defined in statistical terms (Fig. A-1). *Accuracy* is the degree of coincidence between the measured value $m(x)$ and its "true" value $\mu(x)$ that is known from the position of the standard line pattern. *Precision*, on the other hand, is the estimated standard deviation, $s_x$.

From these definitions, it becomes apparent that image data can be precise but also inaccurate, for example, exhibiting a small standard deviation and systematic measurement errors. In practice, a camera may yield sharp but optically distorted pictures (Fig. A-2A). Conversely, a distortion-free, blurred image is accurate but lacks precision (Fig. A-2B).

*Spatial resolution* is the minimum distance at which the intensity distributions of two neighboring dots compose statistically different samples (as evaluated by a $t$-test of sample mean difference). Accordingly, spatial resolution depends not only on the distance $m(x_2) - m(x_1)$ between dots but also on the widths $s(x_1)$ and $s(x_2)$ (Fig. A-1B). Similar arguments are applicable in quantifying the resolution of intensity differences in an image. Consider a setting in which two adjacent structures have different overall light intensities (Fig. A-3).

*Virtual Reconstruction: A Primer in Computer-Assisted Paleontology and Biomedicine.*
By Christoph P. E. Zollikofer and Marcia S. Ponce de León.
Copyright © 2005 John Wiley & Sons, Inc.

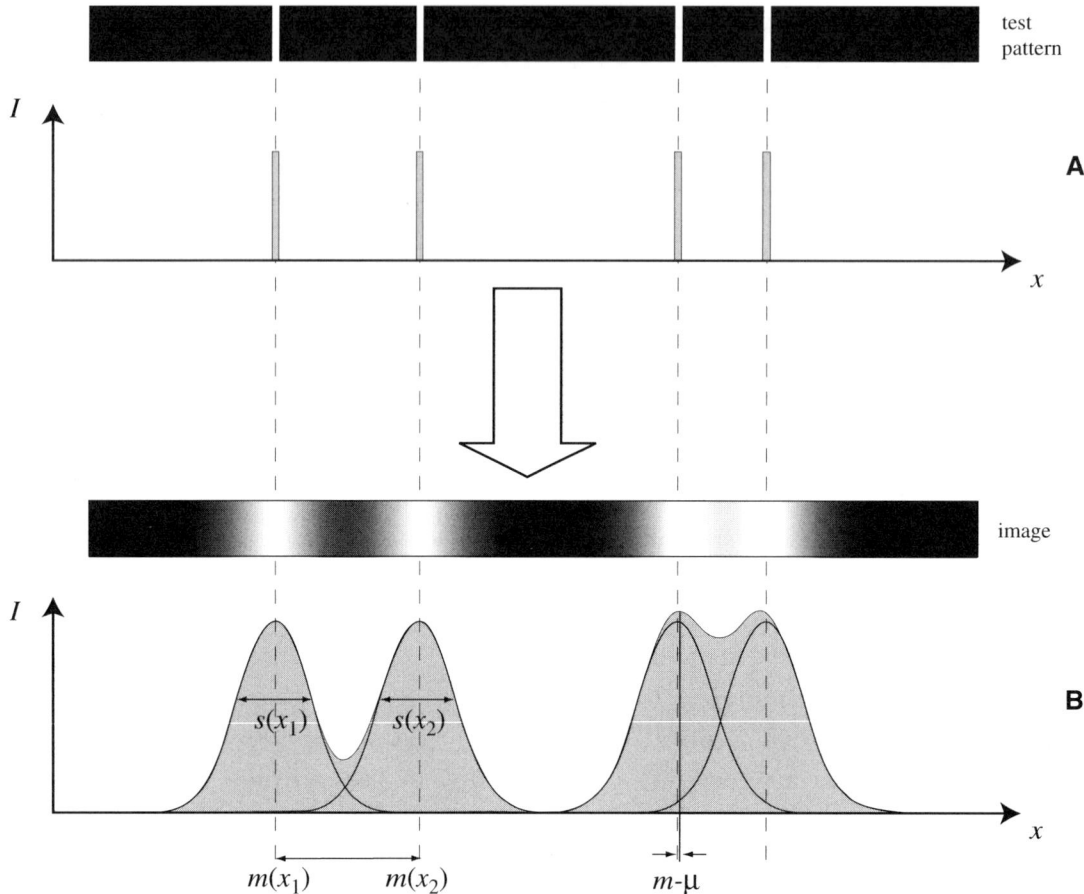

**FIGURE A-1** Accuracy, precision, and spatial resolution of an imaging device. **A** shows a test pattern (white bars on black background) and its intensity distribution. **B** shows the image of the test pattern and its intensity distribution. Precision is the standard deviation $s(x_i)$ of the "blurred" intensity distribution in the image. Accuracy is the degree of coincidence between estimated ($m$) and "true" ($\mu$) locations of the bars in the image. The limits of spatial resolution are reached when intensity distributions of neighboring bars overlap by one standard deviation (right side of the graph).

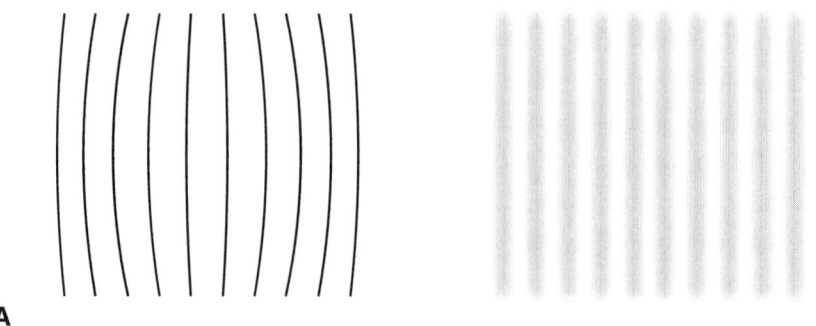

**FIGURE A-2** **A**: A precise (sharp) but inaccurate (distorted) image. **B**: An accurate (distortion-free) but imprecise (blurred) image.

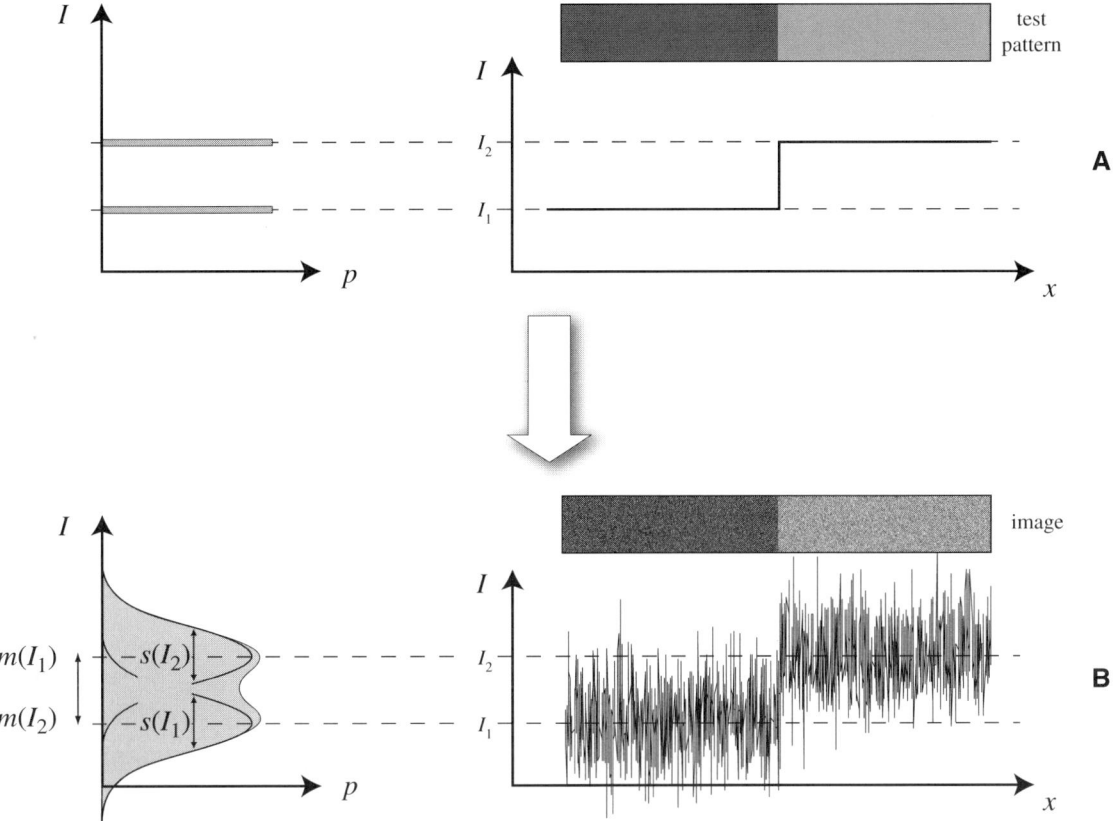

**FIGURE A-3**  Contrast resolution. **A**: Test pattern exhibiting two intensity levels ($I_1$ and $I_2$). **B**: The "noisy" image exhibits intensity distributions with mean values $m(I_{1,2})$ and standard deviations $s(I_{1,2})$ ($p$ is the probability density). Contrast resolution is the minimum separable intensity interval when intensity distributions overlap one standard deviation.

Again, measuring light intensities by counting photons is a stochastic process. Accordingly, "true" signal intensity is superimposed with "noise" originating at various stages in the data acquisition process, such that we expect a bell-shaped (Gaussian) distribution of intensity fluctuations in the image. Accordingly, the empirical value $m(I)$ is an estimate of the "true" intensity value $\mu(I)$, and its precision is its standard deviation $s(I)$. A particularly relevant parameter for imaging devices is *contrast resolution*, which is the minimum intensity interval that a measuring device can detect. This parameter is measured similarly to spatial resolution, namely, it is the minimum statistically significant difference between two mean intensities $m(I_1)$ and $m(I_2)$ with associated standard deviations $s(I_1)$ and $s(I_2)$.

Spatial and contrast resolution depend critically on respective values of $s$ relative to $m$. This dependence is expressed in the *signal-to-noise ratio (SNR)* of the system that measures the proportion of the signal intensity relative to the stochastic imprecision superimposed on it.

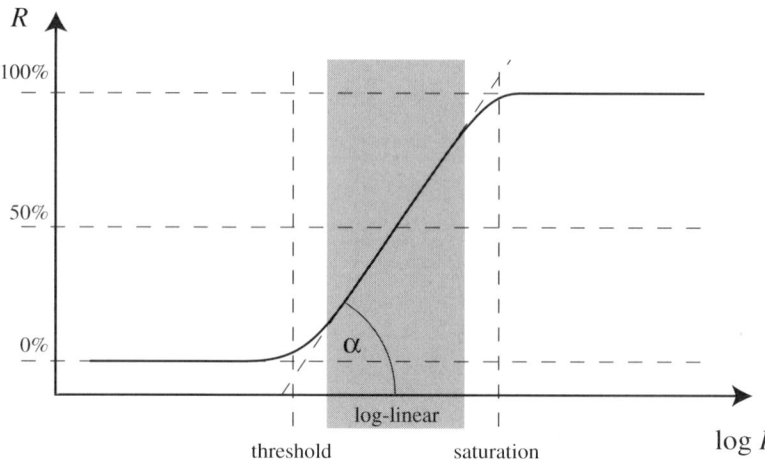

**FIGURE A-4**  Response characteristics of biological and technical light detectors.

$$\mathrm{SNR} = \frac{\text{signal}}{\text{noise}} = \frac{m}{s} \tag{A-1}$$

Considerations of resolution and SNR depend fundamentally on response characteristics of light-collecting structures, the grains of photographic film, the CCDs, or the receptor cells of the human retina. In all three instances, "response" (output) of light detectors depends on the input (Fig. A-4).

Detectors typically respond with log-linear characteristics (note that the input axis in Fig. A-4 is scaled in orders of magnitude) within a given range of input values. Below the threshold value, the system generates no output (0%); at the upper end, it attains saturation (100%). Shifting the curve to the left increases sensitivity; a steeper slope $\alpha$ results in a narrower response range with improved contrast resolution. To achieve optimum sensitivity and contrast resolution, the visual system constantly adjusts slope and position of neuronal response curves to current lighting conditions. In classic photography, different film types provide a wide range of sensitivity and contrast resolution.

# PARAMETERS INFLUENCING THE QUALITY OF CT IMAGE DATA

The following four parameters are primarily relevant for resolution in sinograms, but each ultimately impinges upon the spatial resolution of resulting CT images:

- *The size and number of detectors.* The more detectors in an array of given angle, the better the spatial resolution of the projection data along that array. Medical CT scanners have between 500 and 1000 detectors in an array that fits within a fan angle of 40 to 50°.
- *The diameter of the examination area that is projected onto the detector array,* typically called the *scanned area.* As can be seen in Figure 3-13, a small area projected onto a large detector array yields higher spatial resolution.
- *The number of projections that can be recorded in one full rotation of the gantry.* When more projections are available, that is, the increment of $\alpha$ is smaller, the results from filtered backprojection are better. Medical scanners use between 1000 and 2000 projections per rotation.
- *The thickness of the CT section and the distance between consecutive sections.* The notion of a CT "slice" adequately characterizes the fact that, although one generally thinks of a CT section as a two-dimensional image consisting of pixels, it actually represents a plane of voxels. The depth of voxels in the *z*-direction is referred to as the *slice thickness* or *collimation* and is determined by the width of the X-ray beam and the optical aperture of the detectors. The distance between consecutive slices is called *slice distance, slice increment,* or *slice index.*

Contrast resolution is often subsumed under the notion of sensitivity. Contrast resolution involves the signal-to-noise ratio of a CT device. The generation, absorption, and detection of X rays are inherently stochastic processes. Subtle local differences in density $\mu$ in the reconstructed images may be obscured by noise in the image. The following parameters and trade-offs help reduce image noise (Zonneveld and Vijerberg, 1984):

---

*Virtual Reconstruction: A Primer in Computer-Assisted Paleontology and Biomedicine.*
By Christoph P. E. Zollikofer and Marcia S. Ponce de León.
Copyright © 2005 John Wiley & Sons, Inc.

- *The response characteristics*, or dynamic range, of detectors, that is, how many different X-ray attenuation values they can resolve (see Fig. A-4).
- *The integration time*, that is, how much time the detectors are given to perform (repeated) intensity measurements at a given location. The longer this time, the better the signal-to-noise ratio. However, long acquisition times are impractical when large object volumes are scanned, when time is at premium (e.g., to minimize involuntary movement of the patient), and when the X-ray dose must be minimized.
- *The aperture of detectors*. The aperture of a detector can be imagined as its visual field. Decreasing the aperture restricts the spatial volume from which a detector can sample data. This results in a more focused sampling area, less between-detector overlap, and thus higher contrast resolution. On the other hand, small apertures reduce signal intensity and thus lower the signal-to-noise ratio.—As we have seen before, in the $z$-direction, the aperture determines slice thickness. Increasing the aperture results in shorter data acquisition times, but because object properties may vary along $z$, a large $z$-aperture usually tends to blur image details, similar to a large aperture in photography.
- *The X-ray tube voltage and current*. A "bright" X-ray source can penetrate larger or denser objects, or produce a stronger signal that permits sampling at smaller detector apertures. As mentioned previously, energy of cathode electrons hitting the anode determines energy of X-ray photons. Total radiation energy increases with the square of tube voltage and linearly with tube current.[1] Increasing the voltage has the additional effect of generating a larger proportion of short-waved X rays, whereas increasing the current maintains the spectral composition of X-ray radiation.

There are three primary factors that establish upper limits to radiation energy. First, as in photography, an image may be "overexposed" when a large proportion of the incident X-ray energy passes through the object unabsorbed. Second, in medical applications, X-ray energy must be minimized to avoid tissue damage. Third, with higher radiation energies, the tube overheats more easily, which requires extended cool-down pauses until data acquisition can resume.

Adjusting the energy spectrum of an X-ray source by changing the tube voltage has interesting side-effects during CT scanning. X-ray attenuation coefficients have a spectral bias, that is, low-energy radiation is slightly more attenuated than high-energy radiation. While the X-ray beam passes through the object, the proportion of high-energy radiation increases, such that attenuation coeffi-

---

[1] This is like in traffic, where kinetic energy increases with the square of traveling speed of a car and linearly with the number of cars per time unit.

cients measured at the detectors become lower. This effect, known as *beam hardening*, may introduce artifacts into image reconstruction (Fig. 3-15). All medical CT scanners are calibrated to compensate for beam hardening effects caused by the human body. However, scanning nonstandard objects such as fossil or recent skeletal material requires special measures to correct for beam hardening (see Appendix C).

Medical CT scanners typically provide preadjusted suites of parameters for specific diagnostic situations, so-called *protocols*, such as imaging different tissues (e.g., bone vs. soft tissues) or structures of different size (e.g., children vs. adults, or the thorax vs. the head). During fossil scanning, these protocols often need modification because they incorporate beam-hardening corrections that are adapted to the properties and dimensions of the human body.

A parameter that is vital during patient scanning, but of less importance in other situations, is the temporal resolution of the system. This has two critical aspects, including the time necessary to acquire raw data and the time required for image reconstruction. Typically, the duration of the former is shorter than the duration of the latter, such that raw data are acquired from entire volumes in consecutive slices or along continuous spirals, whereas cross-sectional images are reconstructed subsequently in a more time-consuming step.

# CT SCANNING OF FOSSIL SPECIMENS AND RECENT SKELETAL SPECIMENS: HOW TO PROCEED?

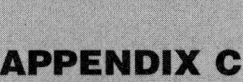

The following is a cookbook-style recipe for CT scanning fossil specimens. It is intended to assist potential users in considering all points that warrant attention for producing optimal results.

## C.1 PREPARATION

### C.1.1 Mounting the Specimens

To position the specimen, or an entire set of fragments, in the CT scanner, embed them in sheets (boxes work as well) of Styrofoam, the white foamlike rigid material used in packing and insulation (Fig. C-1). By virtue of its low X-ray density, any disruptive effects it may have during scanning are minimized. Completely immobilize material with the Styrofoam, because specimens may be displaced when the CT table changes position during scanning. While mounting specimens, anticipate a desirable estimate of field of view for the reconstructed image. When necessary, several smaller items can be placed in close proximity to one another, or Styrofoam boxes can be stacked on top of each other and scanned simultaneously.

### C.1.2 Materials Used for Fixation

Do not bring any material other than Styrofoam into direct contact with the fossil. Avoid using metal pins, modeling clay/putty, but also rubber bands when positioning the boxes or specimens because they are materials exhibiting high X-ray density (Velcro strips and medical tape may be used as long as they do not come into direct contact with the specimen). White glues should not be used to fix Sty-

*Virtual Reconstruction: A Primer in Computer-Assisted Paleontology and Biomedicine.*
By Christoph P. E. Zollikofer and Marcia S. Ponce de León.

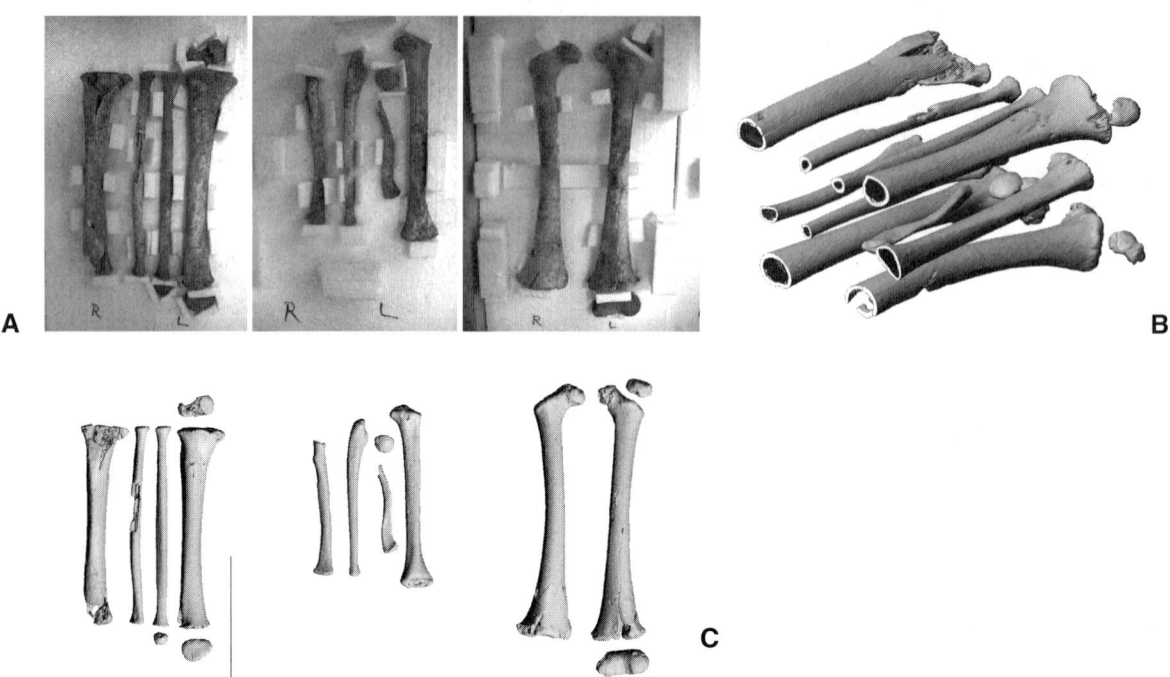

**FIGURE C-1** Preparation of specimens for CT scanning. **A**: Long bone remains of the Lagar Velho Gravettian child skeleton placed in Styrofoam boxes to be stacked on top of each other for CT scanning. **B**: 3D reconstruction of the scanned set. **C**: Separation of individual pieces is straightforward, because Styrofoam has a Hounsfield density resembling that of the surrounding air (scale bar is 10 cm).

rofoam because these often contain metallic oxides, which have high Hounsfield values. Glues based on organic solvents, however, work well. It is not advisable to introduce metal objects such as rulers and wires. Even though these are sometimes useful for creating a known-distance scale in conventional radiographs, the metal causes image reconstruction artifacts that degrade the entire data set, even if they are placed outside the scanned area.

### C.1.3 Placement

Place the boxes containing the specimens directly onto the CT table (first remove any cushions and soft mats, which create instability). As specimens may shift during table positioning, secure everything tightly with medical tape. Document the situation with photographs. Scanning specimens in two orthogonal directions may be beneficial, especially during data acquisition from spherical structures. This provides denser data sampling in regions that are nearly parallel to the scanning plane (Fig. C-2).

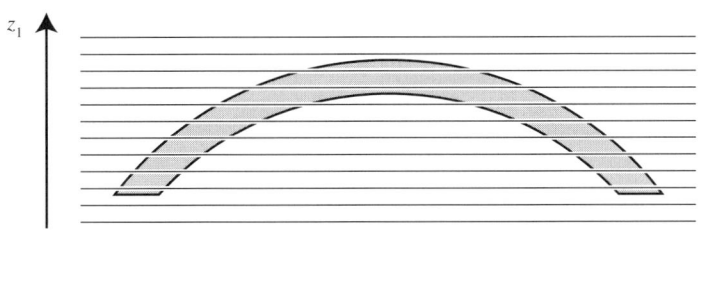

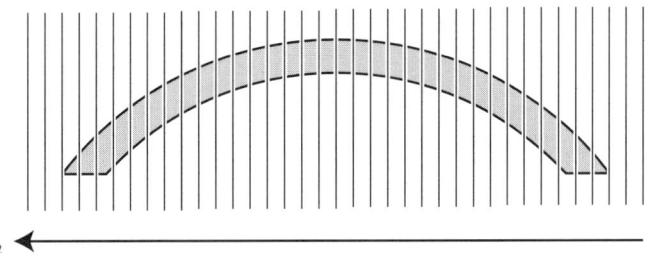

**FIGURE C-2** Principle of homogeneous CT data acquisition. In regions of the specimen where object surfaces are nearly parallel to the scanning plane, performing an additional orthogonal scan along ($z_2$) can compensate for undersampling along the scanning direction ($z_1$).

## C.2 PARAMETERS FOR CT DATA ACQUISITION

### C.2.1 Scanned Area

The scanned area, or scan angle, is the diameter of the examination area that is projected onto the detector array (see Fig. 3-9; note that "field of view" is the size of the reconstructed image; see below). Most medical scanners permit two settings, typically ~25 cm (for head scans) and ~50 cm (for thorax and abdomen scans), where in the first case the active detector array is obviously shorter than in the second. Some CT manufacturers (typically older devices of Philips and Toshiba) offer geometric enlargement; the distance between the X-ray source and the scanned object, and thus the angle along which it is projected onto the full detector array, can be adjusted for better resolution in small objects.

### C.2.2 X-Ray Tube Current and Voltage

Proper adjustment of the X-ray beam that irradiates the fossil is critical for generating artifact-free and undistorted images. Tube voltage is normally around 120 kV (kilovolts), whereas the current is normally between 50 and 150 mA (milliamperes). These values apply for slice thicknesses of ~1 mm; thicker slices require an X-ray beam with larger diameter (focal spot size), and thus larger currents (200–500 mA). Although fossils might be considerably more dense than living bone tissue, an absence of surrounding soft tissues often makes it possible to use relatively low milliampere values when scanning fossils.

### C.2.3  Gantry Tilt

The CT gantry (i.e., the $x$-$y$ plane) can be tilted relative to the scanning axis ($z$). This option is useful to obtain sections at predefined diagnostically relevant planes, particularly when the patient cannot be moved accordingly (example: measurement of intervertebral spaces). However, it is strongly recommended that this option not be used during fossil scanning, because recovering three-dimensional data from the consequent skewed voxel volume is difficult. Moreover, in conjunction with helical scanning, a gantry tilt causes unnecessary image reconstruction artifacts.

### C.2.4  Scanning Direction and Object Orientation

During data acquisition, the patient table can be moved outward or inward. The object can be oriented head-first or feet-first and placed on its dorsal or ventral side. Images can be reconstructed with the anatomic right side on the image's left side (standard) and vice versa. It is highly recommended that the choice of these parameters be documented.

### C.2.5  Object Positioning

Place specimens in the center of the gantry. This can be ascertained by raising/lowering the patient table and using optical guides (typically red laser beams) provided by most devices to position the patients or specimens. Once the specimens are positioned appropriately, the radiologist should perform a *scanogram*[1] of the object. Scanograms are similar to conventional radiographic projections, except that the scanogram projection is distortion-free (see Chapter 5 for orthographic vs. perspective projection). Projection data are recorded at a fixed angular position, for example, from above the object or from the side of the object, during movement through the gantry (Fig. C-3).

- *Scanning modality*: Choose between sequential and helical scanning. Sequential scanning is recommended when deriving optimal image quality. Helical scanning is recommended when it is necessary to acquire large amounts of

---

[1]The name *scanogram* was introduced on EMI, Philips, and Elscint scanners. Subsequent synonyms are *scout view* (General Electrics, Toshiba), or *topogram* (Siemens).

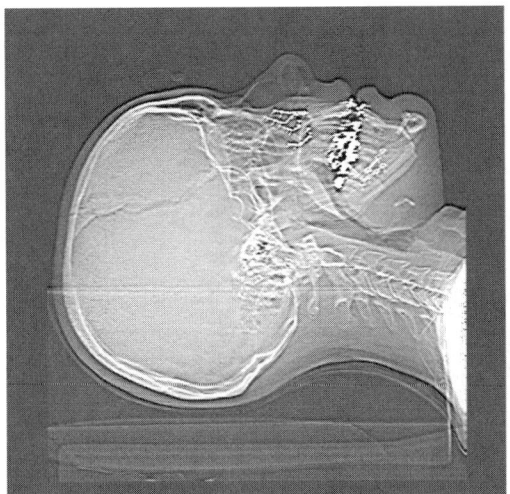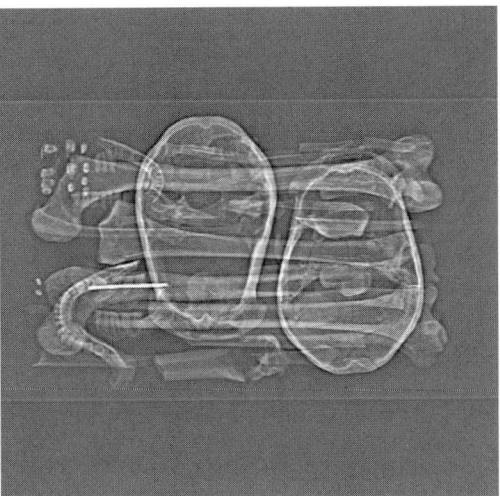

A                                                                     B

**FIGURE C-3** The topogram (synonyms: scanogram, scout view), analogous to a conventional radiograph, assists in determining initial and final table positions during volume data acquisition, and facilitates reconstruction of subvolumes at higher magnifications. Data are acquired with a fixed X-ray tube position, while the patient table proceeds through the gantry. **A** shows the lateral topogram of a patient, whereas **B** depicts the vertical topogram of the Spy Neanderthal remains.

volume data, for example, extended recent samples from museums or collections.

- *Sequential scanning*: Determine *slice thickness* (also known as scan collimation) and *table feed*, which determines the *interslice distance*. To obtain true volume data for subsequent three-dimensional object reconstruction, acquire contiguous, or slightly overlapping, slice data. Minimum slice thickness for current medical CT scanners is around 1 mm, whereas their minimum table feed is between 0.1 and 1 mm. Consider that increasing the slice thickness lowers contrast resolution but simultaneously reduces image noise.

- *Helical scanning*: Contrary to sequential scanning, in helical scans interslice distance and in multislice scanners slice thickness can be determined retrospectively, such as during image reconstruction from the raw data volume. Accordingly, *scan collimation*, that is, the thickness of the X-ray fan-beam in the z-direction, is used as a scanning parameter rather than slice thickness. To avoid creating helical artifacts during image reconstruction (see Fig. 3-18), table feed values should be optimized with respect to the required interslice distance. Table feed is measured as the advancement of the object along the z-axis per gantry revolution. The relationship between table feed and scan collimation, known as *pitch*, describes the degree of contiguity between subsequent turns of the helix (see Fig. 3-17). Never use pitch values greater than 1 because this results in incomplete coverage of the data volume. Ideally, pitch should be ≤1, whereas the table feed should have the same value as the required interslice distance. Satisfactory results are still possible with pitch = 1 and interslice distance = 1/2 of table feed (Kalender et al., 1994). Other combinations of pitch and interslice distance create noticeably suboptimal data. To summarize: for high-resolution data, set pitch = 1, slice thickness = 1 mm, table feed = 0.5 mm, or 1.0 mm and reconstruct images at a slice increment of 0.5 mm.

- *Multislice scanners*: Because multislice CT devices evolve rapidly, nomenclature and parameter settings may differ among device manufacturers. However, fundamental constraints imposed on helical CT scanning remain essentially the same (see Fig. 3-17). Currently, many hospitals replace their 2- or 4-array scanners with 16-array devices. Typically, 2- or 4-array scanners yield more helical artifacts than 16-array scanners, so that it is advisable to look out for availability of the latter.

- *Limitations*: Depending on the combination of scanning parameters, CT devices might be unable to acquire data volumes along the full z-length in a single step. This occurs partly because of X-ray tube overheating and partly because of limited data storage capacities. Under these circumstances, multiple, contiguous data sets are advisable. Units should overlap by a few millimeters to secure continuity.

## C.3 IMAGE RECONSTRUCTION

### C.3.1 Reconstruction Kernels

Image reconstruction uses several potential filter functions, also known as *kernels*. Medical scanners offer a series of predefined filter functions that are adjusted to specific tissues and organs. For fossils, the entire stack of cross-sectional images should be reconstructed with two different filters. The ramp filter, which yields a mathematically "objective" solution to the reconstruction problem, is the *standard kernel* on most CT consoles. This filter yields relatively smooth images that facilitate taking measurements and 3D reconstruction. The second filter should be a *bone* or *hard* filter, which will enhance small details and edge structures, which facilitates visual assessment of the data. However, be aware that bone filters alter a reconstructed image in several ways. Most notably, image intensities no longer represent "true" Hounsfield density values when a bone filter is used.

### C.3.2 Image Reconstruction

Reconstruct images at the smallest possible field of view (FOV) so that specimens fill the entire image. However, a field of view smaller than approximately 7.5 cm does not yield additional information, because pixel size reaches the limit of spatial resolution of the device. Note that image reconstruction differs fundamentally from image enlargement. For example, if you want to obtain a zoom reconstruction (small FOV) of the inner ear region within a fossil skull, make sure that you generate the images applying a reconstruction kernel to the raw data, *not* applying an enlargement procedure to already existing images.

## C.4 CT DATA STORAGE

### C.4.1 Raw Data Storage

CT attenuation profiles are known to radiologists as *raw data*. These data allow subsequent reconstruction of cross-sectional images according to various reconstruction protocols. Because raw data occupy 5- to 10-fold the storage space of image data, sufficient short-term memory and storage space should be made available. If storage space is filled prematurely, medical scanners tend to erase earlier data sets without advance warning. This is logical in a clinical environment, where emergencies have absolute priority, but this can be very disheartening during fossil scanning! Although raw data are discarded typically when a new data acquisition session is started, it is often sensible to save them on permanent storage media so that they are available for retrospective image reconstructions with various filters and different magnifications.

### C.4.2 Image Data Storage

It is possible to print cross-sectional image series on radiographic film, as is routine in medical diagnostics. However, it is more important to verify that image data are stored on a digital medium and in a format that can be recognized in future years. Currently, most CT consoles support the DICOM 3 standard format, and permit export of data to a standard medium, such as CD or DVD, or to a server. This is vital for postprocessing of the image data. DICOM-formatted image stacks can be read on PCs with one of the many DICOM reading and visualizing applications available as shareware (see links on the Web Companion).

## C.5 CALIBRATION

### C.5.1 Test Scans

If permitted, several test scans to assess image quality and adjust parameters, if necessary, should be performed. Whereas industrial scanners permit full control over a wide variety of parameters, modern medical scanners offer relatively fewer possibilities for optimizing parameters for data acquisition from fossils with potentially elevated X-ray density. Density overflow, which regularly appears in fossil tooth enamel, may be counteracted on some devices by extending the Hounsfield scale. Beam hardening, which creates underestimation and/or overestimation of the Hounsfield density toward the center of objects (Figs. 3-14, 3-15), is another issue that should be addressed. Medical scanners are designed to compensate for beam hardening in the human body, but not in fossils. Nevertheless, by placing small specimens into PVC or pewter tubes, which absorb the low-energy part of the X-ray spectrum that is responsible for the beam-hardening effect, these artifacts can be removed.

## C.5.2 Calibration

It is often advisable to perform test scans on a series of standard objects exhibiting a variety of known densities (e.g., Perspex [PMMA], PVC, POM, aluminum), and known dimensions (e.g., tubes or blocks with drilled holes). This facilitates subsequent calibration of images, especially when distance measurements at the subpixel level are desired (see Chapter 4).

# OBJECT MANIPULATION IN VIRTUAL SPACE

**APPENDIX D**

## D.1 MATRICES

Matrices are two-dimensional mathematical structures whose meaning depends on the context in which they are used. In the context of computer graphics, matrices are primarily used as tools to transform objects in virtual space, to navigate in space, or to generate viewing projections. Let us consider the following matrix **A**

$$\mathbf{A} = \begin{pmatrix} a_{11} & a_{12} \\ a_{21} & a_{22} \end{pmatrix} \tag{D-1}$$

Each matrix element $a_{ij}$ is identified by two index entries, indicating its row number ($i$) and its column number ($j$). Matrix **A** consists of two rows and two columns and is called a $2 \times 2$ matrix (in a generalized matrix, the number of rows and columns need not be the same). Let us define a second matrix, **B**:

$$\mathbf{B} = \begin{pmatrix} b_{11} & b_{12} \\ b_{21} & b_{22} \end{pmatrix} \tag{D-2}$$

**A** and **B** are added in the following way:

$$\mathbf{S} = \mathbf{A} + \mathbf{B} = \begin{pmatrix} a_{11} + b_{11} & a_{12} + b_{12} \\ a_{21} + b_{21} & a_{22} + b_{22} \end{pmatrix} = \begin{pmatrix} s_{11} & s_{12} \\ s_{21} & s_{22} \end{pmatrix} \tag{D-3}$$

Although matrix addition is straightforward, matrix multiplication is slightly more complex:

$$\mathbf{P} = \mathbf{A} \cdot \mathbf{B} = \begin{pmatrix} a_{11}b_{11} + a_{12}b_{21} & a_{11}b_{12} + a_{12}b_{22} \\ a_{21}b_{11} + a_{22}b_{21} & a_{21}b_{12} + a_{22}b_{22} \end{pmatrix} = \begin{pmatrix} p_{11} & p_{12} \\ p_{21} & p_{22} \end{pmatrix} \tag{D-4}$$

*Virtual Reconstruction: A Primer in Computer-Assisted Paleontology and Biomedicine.*
By Christoph P. E. Zollikofer and Marcia S. Ponce de León.
Copyright © 2005 John Wiley & Sons, Inc.

Row elements of matrix **A** are combined with column elements of matrix **B** ("row-times-column" rule). Furthermore, **AB** and **BA** do not generally yield the same result, that is, matrix multiplication is noncommutative.

Matrices of differing dimensions can be multiplied, as long as the number of columns of the first matrix equals the number of rows of the second matrix. For example, consider the multiplication of matrix **A** (2 × 2) with a vector **v** (which is a 2 × 1 matrix):

$$\mathbf{A} \cdot \mathbf{v} = \begin{pmatrix} a_{11} & a_{12} \\ a_{21} & a_{22} \end{pmatrix} \cdot \begin{pmatrix} v_1 \\ v_2 \end{pmatrix} = \begin{pmatrix} a_{11}v_1 + a_{12}v_2 \\ a_{21}v_1 + a_{22}v_2 \end{pmatrix} = \mathbf{v}' \tag{D-5}$$

the result of which is again a vector, **v'**.

## D.2 RIGID TRANSFORMS

Rigid transforms translate and rotate objects in the plane or in space, while the form of objects remains unaffected. We consider a triangle being rotated about the coordinate origin by an angle of φ (Fig. D-1A) and evaluate the coordinates of a point **p** before $(x, y)$ and after $(x', y')$ transformation. Using trigonometry, you may verify that the new coordinates are as follows:

$$\begin{aligned} x' &= x\cos\phi - y\sin\phi \\ y' &= x\sin\phi + y\cos\phi \end{aligned} \tag{D-6}$$

With matrix-vector notation, we may rewrite these equations as follows:

$$\mathbf{p}' = \mathbf{R} \cdot \mathbf{p} = \begin{pmatrix} \cos\phi & -\sin\phi \\ \sin\phi & \cos\phi \end{pmatrix} \cdot \begin{pmatrix} x \\ y \end{pmatrix} \tag{D-7}$$

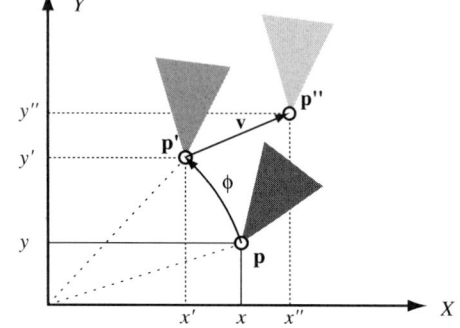 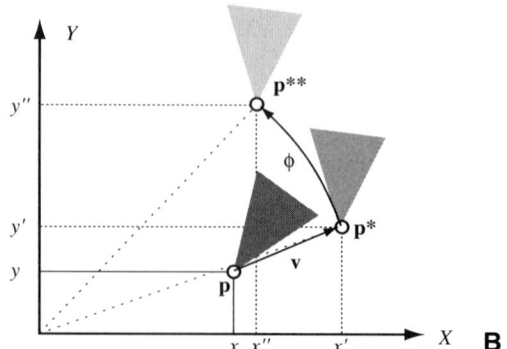

**FIGURE D-1** Rigid transforms of a triangle. **A**: Rotation about the origin by angle φ. followed by translation along vector **v** (exemplified for a point **p** representing a vertex of the triangle). **B**: Translation followed by rotation yields a different final position of the triangle.

that is, $\mathbf{p}'$, the new position of point $\mathbf{p}$, is evaluated by multiplication of rotation matrix $\mathbf{R}$ with vector $\mathbf{p}$. Likewise, translation of a point $\mathbf{p}'$ by a vector $\mathbf{v}$ into a point $\mathbf{p}'$ can be formulated as follows:

$$\mathbf{p}'' = \mathbf{p}' + \mathbf{v} = \begin{pmatrix} x \\ y \end{pmatrix} + \begin{pmatrix} x_v \\ y_v \end{pmatrix} \tag{D-8}$$

## D.3 HOMOGENEOUS MATRICES

In computer graphics, it is convenient to express all transforms (rigid transforms such as translation and rotation, as well as nonrigid transforms, some of which we will discuss here) in terms of matrix multiplications of so-called homogeneous matrices. In homogeneous matrix notations, the above operations (D-7) and (D-8) look as follows. Rotation around the origin by an angle $\phi$:

$$\mathbf{p}' = \mathbf{R} \cdot \mathbf{p} = \begin{pmatrix} \cos\phi & -\sin\phi & 0 \\ \sin\phi & \cos\phi & 0 \\ 0 & 0 & 1 \end{pmatrix} \cdot \begin{pmatrix} x \\ y \\ 1 \end{pmatrix} = \begin{pmatrix} x\cos\phi - y\sin\phi \\ x\sin\phi + y\cos\phi \\ 1 \end{pmatrix}. \tag{D-9}$$

Translation by a vector $\mathbf{v}$:

$$\mathbf{p}'' = \mathbf{V} \cdot \mathbf{p} = \begin{pmatrix} 1 & 0 & x_v \\ 0 & 1 & y_v \\ 0 & 0 & 1 \end{pmatrix} \cdot \begin{pmatrix} x \\ y \\ 1 \end{pmatrix} = \begin{pmatrix} x + x_v \\ y + y_v \\ 1 \end{pmatrix}. \tag{D-10}$$

Furthermore, scaling along $x$ and $y$ axes by factors $s_x$ and $s_y$ can be expressed as follows:

$$\mathbf{p}''' = \mathbf{S} \cdot \mathbf{p} = \begin{pmatrix} s_x & 0 & 0 \\ 0 & s_y & 0 \\ 0 & 0 & 1 \end{pmatrix} \cdot \begin{pmatrix} x \\ y \\ 1 \end{pmatrix} = \begin{pmatrix} x \cdot s_x \\ y \cdot s_y \\ 1 \end{pmatrix}. \tag{D-11}$$

Note that values of $-1$ for either $s_x$ and $s_y$ correspond to mirror-imaging point $\mathbf{p}$ at the $y$- and $x$-axes, respectively.

With homogeneous matrix notation, it is now straightforward to specify complex object manipulations by concatenating the matrices that describe consecutive transformations. For example, the position of point $\mathbf{p}'$ in Figure D-1 is evaluated as follows:

$$\mathbf{p}'' = (\mathbf{V} \cdot \mathbf{R}) \cdot \mathbf{p} = \mathbf{T} \cdot \mathbf{p} = \begin{pmatrix} \cos\phi & -\sin\phi & x_v \\ \sin\phi & \cos\phi & y_v \\ 0 & 0 & 1 \end{pmatrix} \cdot \mathbf{p} \tag{D-12}$$

where the sequence of matrices can be premultiplied into a single transformation matrix **T**. Changing the order of transformations in our example (translation first, rotation second; see Fig. D-1B) yields a different result:

$$\mathbf{p}** = (\mathbf{R} \cdot \mathbf{V}) \cdot \mathbf{p} = \mathbf{T}' \cdot \mathbf{p} = \begin{pmatrix} \cos\phi & -\sin\phi & x_v \cos\phi - y_v \sin\phi \\ \sin\phi & \cos\phi & x_v \sin\phi + y_v \cos\phi \\ 0 & 0 & 1 \end{pmatrix} \cdot \mathbf{p} \qquad \text{(D-13)}$$

You may gain more insight into homogeneous matrix-based object manipulations by experimenting with the Transformation Applet on the Web Companion (Chapter 5).

Object transformations in three dimensions can be described by $4 \times 4$ homogeneous matrices. Translation by a vector **v** $(x_v, y_v, z_v)$ corresponds to the following matrix **T**:

$$\mathbf{T} = \begin{pmatrix} 1 & 0 & 0 & x_v \\ 0 & 1 & 0 & y_v \\ 0 & 0 & 1 & z_v \\ 0 & 0 & 0 & 1 \end{pmatrix} \qquad \text{(D-14)}$$

Rotation around axes $x$, $y$, and $z$ by an angle of $\phi$ is described by the following matrices $\mathbf{R}_x(\varphi)$, $\mathbf{R}_y(\varphi)$, and $\mathbf{R}_z(\varphi)$, respectively:

$$R_x(\phi) = \begin{vmatrix} 1 & 0 & 0 & 0 \\ 0 & \cos\phi & -\sin\phi & 0 \\ 0 & \sin\phi & \cos\phi & 0 \\ 0 & 0 & 0 & 1 \end{vmatrix}$$

$$R_y(\phi) = \begin{vmatrix} \cos\phi & 0 & \sin\phi & 0 \\ 0 & 1 & 0 & 0 \\ -\sin\phi & 0 & \cos\phi & 0 \\ 0 & 0 & 0 & 1 \end{vmatrix} . \qquad \text{(D-15)}$$

$$R_z(\phi) = \begin{vmatrix} \cos\phi & -\sin\phi & 0 & 0 \\ \sin\phi & \cos\phi & 0 & 0 \\ 0 & 0 & 1 & 0 \\ 0 & 0 & 0 & 1 \end{vmatrix}$$

And finally, scaling an object by factors $s_x$, $s_y$, $s_z$ corresponds to matrix **S**

$$\mathbf{S} = \begin{pmatrix} s_x & 0 & 0 & 0 \\ 0 & s_y & 0 & 0 \\ 0 & 0 & s_z & 0 \\ 0 & 0 & 0 & 1 \end{pmatrix} . \qquad \text{(D-16)}$$

Mirror-images are created when one or three factors have a negative sign.

# A PARSIMONIOUS APPROACH TO CORRECTION OF TAPHONOMIC DEFORMATION

Many fossil hominid skulls underwent taphonomic compression, but the in situ orientation is unknown, be it due to reembedding and reorientation of fragments during diagenesis, or due to incomplete documentation at the site. Applying geometric considerations, it is possible to infer from a distorted fossil morphology a "most parsimonious" direction and amount of taphonomic compression and to realign the specimen accordingly.

Figure E-1 shows a bilaterally symmetrical object before and after compression. Compression results in skewing of the "natural" anatomic system of coordinates (**U** is the midsagittal plane and **T** a transversal direction; you may simulate this scenario with the transformation applet in the Web Companion). However, because the direction of compression is unknown, only that component of deformation can be observed that results in transversal slanting, whereas the amount of compression parallel and/or perpendicular to the midsagittal plane cannot be detected. The slanting component can be quantified by evaluating the average orientation $\bar{t}$ of vectors $\mathbf{t}_i$ connecting contralateral anatomic landmarks. The amount of slanting is then given by the deviation of $\bar{t}$ from the normal vector **n** to the midsagittal plane.

Among the many potential compression scenarios that may result in the observed slanting, we determine the direction vector **f** of taphonomic forces that explains the observed deviation with a minimum amount of compression. This direction can be evaluated as follows. Define a plane through $\bar{t}$ and **n**; in that plane, determine a unit vector **u** parallel to the midsagittal plane. The bisector **b** of $\bar{t}$ and **u** is the direction of minimum compression. The amount of decompression necessary to render $\bar{t}$ perpendicular to the midsagittal plane is $\tan(\beta)$, where $\beta$ is the bisecting angle.

In a virtual fossil, these parameters can be estimated as follows. Assuming that the midsagittal landmarks span a reasonably planar surface (i.e., assuming only linear deformation), the surface normal vector **n** is given by the eigenvector associated with the smallest eigenvalue of the landmark covariance matrix. The average slanting orientation $\bar{t}$ is obtained by normalizing the sum of all vectors $\mathbf{t}_i$.

*Virtual Reconstruction: A Primer in Computer-Assisted Paleontology and Biomedicine.*
By Christoph P. E. Zollikofer and Marcia S. Ponce de León.

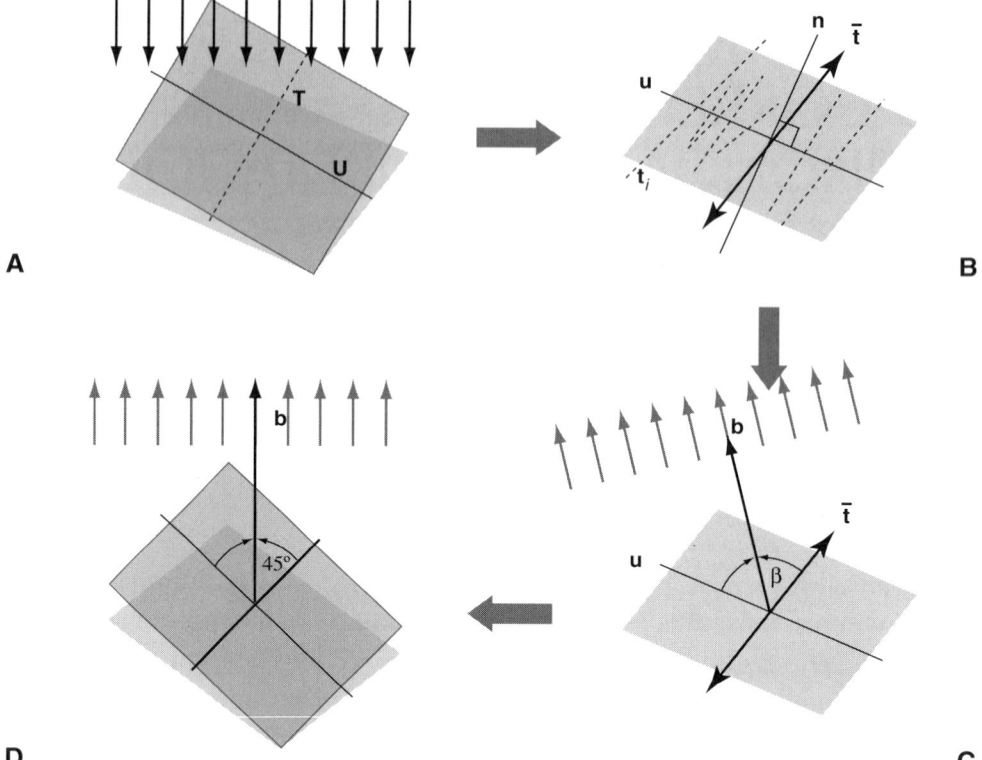

**FIGURE E-1**. Correcting compressive deformation. **A**: Compression of a bilaterally symmetrical fossil. **B**: The recovered fragmentary fossil exhibits a distorted morphology; transversal anatomic axes $t_i$ deviate from vector **n** that is perpendicular to the midsagittal plane. **C**: The bisector **b** of the average slanting direction $\bar{t}$ and the midsagittal plane **u** is the most parsimonious estimate of the original direction of compression. **D**: Decompression renders angle $\beta > 45°$ to $\beta = 45°$.

# MORPHOMETRY

## F.1 ANATOMIC AXES AND PLANES

Figure F-1 shows the current terminology for anatomic axes, rotations around axes, and principal planes (note that most terms have synonyms).

## F.2 ACCURACY AND PRECISION OF MEASUREMENT

How accurate and how precise can morphometric analyses be? We must discern between precision and accuracy of the measuring device (see Chapter 2) and precision and accuracy of the operator taking the measurements. In the latter case, precision is the amount of scatter of a measurement around an average value, and accuracy is the amount of coincidence between the average value and the true value of the measurement. Operator precision and accuracy can be improved by cross-checking repeated measurements and/or measurements obtained by different persons and by providing exact definitions of measurement procedures and landmark locations.

## F.3 ALLOMETRY

Huxley (1924; 1932) proposed a morphogenetic model of how size-related changes in shape are generated during growth. Given two morphometric quantities $x$ and $y$, we consider their specific growth rates $\dfrac{dx}{dt}\dfrac{1}{x}$ and $\dfrac{dy}{dt}\dfrac{1}{y}$, postulating that these rates are coupled to each other by a constant $k$:

$$\frac{dy}{dt}\frac{1}{y} = k\frac{dx}{dt}\frac{1}{x} \qquad \text{(F-1)}$$

Canceling $dt$ from both sides of the equation yields

*Virtual Reconstruction: A Primer in Computer-Assisted Paleontology and Biomedicine.*
By Christoph P. E. Zollikofer and Marcia S. Ponce de León.
Copyright © 2005 John Wiley & Sons, Inc.

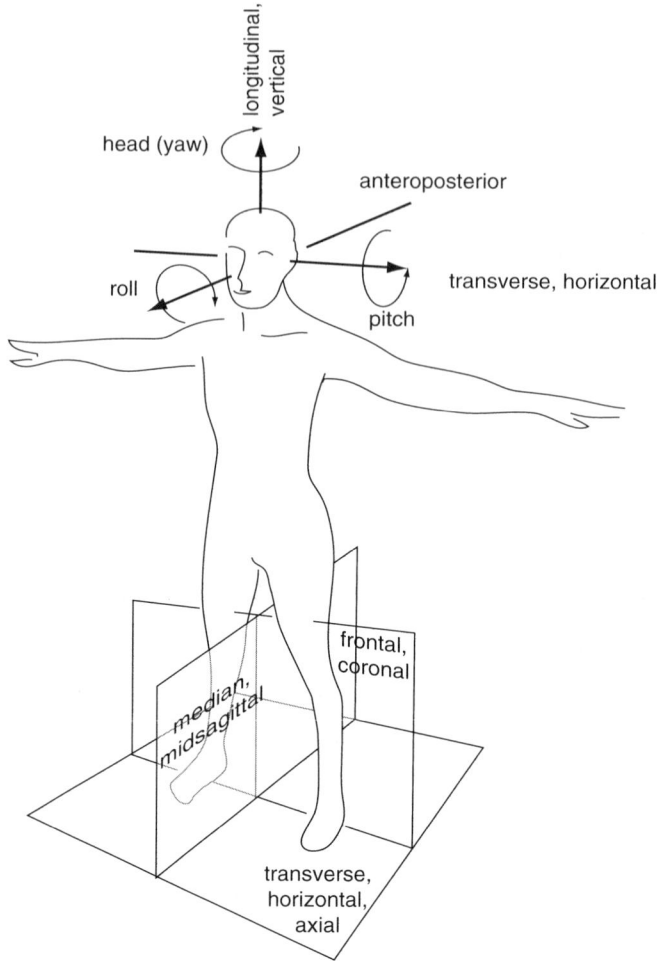

**FIGURE F-1** Principal anatomic axes and planes.

$$dy\,\frac{1}{y} = k \cdot dy\,\frac{1}{x} \qquad \text{(F-2)}$$

On integration, we obtain the time-independent allometric equation

$$\ln y = k \ln x + C \qquad \text{(F-3)}$$

where $C$ is an integration constant. Exponentiation yields

$$y = e^C x^k = a\,x^k \qquad \text{(F-4)}$$

which states that $y$ grows exponentially with increasing $x$. Values of $k > 1$ and $k < 1$ denote positive and negative allometry; $k = 1$ denotes isometry, that is, no change in shape with increasing size.

## F.4 MULTIVARIATE ANALYSIS AND DIMENSION REDUCTION

An important aim of multivariate analysis consists in exploration of patterns of variation in a sample, in which each specimen is represented by a series of measurements of $K$ variables $x_1, x_2, \ldots x_i, \ldots x_K$. In bioscientific samples, variables $x_i$ are typically correlated with each other to some degree. Each specimen is represented as a point in $K$-dimensional space. The idea behind dimension reduction is to replace the $K$ original axes $x_i$ by $K$ new axes $s_1, s_2, \ldots s_i, \ldots s_K$ that describe principal directions of variation within the sample. These *principal components* have two key properties: They are statistically independent of each other (i.e., uncorrelated, or, in geometric terms, orthogonal to each other) and they "explain," in a statistical sense, the largest, second largest, and subsequently smaller proportions of the total variation contained in the sample. Typically, the first few principal components account for an important part of the total variation, such that the remaining components can be omitted during visualization.

How can principal components be evaluated? Principal component analysis (PCA) is based on matrix algebra, notably the concept of eigenvectors and eigenvalues. Think of a $K \times K$ (i.e., square) matrix $\mathbf{S}$ (matrices are described in Appendix D) and find a vector $\mathbf{e}$ that, when multiplied by $\mathbf{S}$, is simply scaled by a factor $\lambda$:

$$\mathbf{Se} = \lambda \mathbf{e} \qquad \text{(F-5)}$$

When $\mathbf{S}$ is symmetric (i.e., elements $s_{ij} = s_{ji}$), $K$ solutions to this problem exist. Vectors $\mathbf{e}_i$ and associated factors $\lambda_i$ are the *eigenvectors* and corresponding *eigenvalues* of matrix $\mathbf{S}$.

Let us now consider an $N \times K$ data matrix, $\mathbf{X}$, in which each row corresponds to the measurements taken from one specimen (specimen numbers range from 1 to $N$, variable numbers from 1 to $K$)

$$\mathbf{X} = \begin{pmatrix} x_{11} & x_{12} & \ldots & x_{1K} \\ x_{21} & x_{22} & \ldots & x_{2K} \\ \ldots & \ldots & \ldots & \ldots \\ x_{N1} & x_{N2} & \ldots & x_{NK} \end{pmatrix} \qquad \text{(F-6)}$$

This matrix describes the specimens' positions in a $K$-dimensional space. For ease of further calculations, we "center" the entire data set such that each variable's mean value is situated at the coordinate origin. The resulting matrix $\mathbf{X}'$ looks as follows:

$$\mathbf{X}' = \begin{pmatrix} x_{11} - \overline{x}_1 & x_{12} - \overline{x}_2 & \dots & x_{1K} - \overline{x}_K \\ x_{21} - \overline{x}_1 & x_{22} - \overline{x}_2 & \dots & x_{2K} - \overline{x}_K \\ \dots & \dots & \dots & \dots \\ x_{N1} - \overline{x}_1 & x_{N2} - \overline{x}_2 & \dots & x_{NK} - \overline{x}_K \end{pmatrix} \tag{F-7}$$

We may now calculate the variance-covariance matrix $\mathbf{S}$ of matrix $\mathbf{X}'$ by multiplying $\mathbf{X}'$ with its transpose $\mathbf{X}'^{\mathrm{T}}$ (a matrix is transposed by writing rows as columns, and vice versa)

$$\mathbf{S} = \frac{1}{N}\mathbf{X}'\mathbf{X}'^{\mathrm{T}} = \begin{pmatrix} s_{11} & s_{12} & \dots & s_{1K} \\ s_{21} & s_{22} & \dots & s_{2K} \\ \dots & \dots & \dots & \dots \\ s_{K1} & s_{K2} & \dots & s_{KK} \end{pmatrix} \tag{F-8}$$

where each element $s_{ij}$ expresses the covariance between variables $X_i$ and $X_j$ ($s_{ii}$ is the variance of variable $X_i$). $\mathbf{S}$ is square-symmetric (i.e., covariances $s_{ij}$ are equal to covariances $s_{ji}$), such that we can evaluate its eigenvalues and eigenvectors. Various standard numerical procedures exist to perform these operations (see, e.g., Press et al., 1992). In compact form, we may annotate the result as follows:

$$\mathbf{E} = \begin{pmatrix} e_{11} & e_{12} & \dots & e_{1K} \\ e_{21} & e_{22} & \dots & e_{2K} \\ \dots & \dots & \dots & \dots \\ e_{K1} & e_{K2} & \dots & e_{KK} \end{pmatrix}; \quad \mathbf{L} = \begin{pmatrix} \lambda_1 & 0 & 0 & 0 \\ 0 & \lambda_2 & 0 & 0 \\ \dots & \dots & \dots & \dots \\ 0 & 0 & 0 & \lambda_K \end{pmatrix}, \tag{F-9}$$

where each row of $\mathbf{E}$ denotes an eigenvector, and each entry along the diagonal of $\mathbf{L}$ denotes its corresponding eigenvalue (note that the number of eigenvectors and the number of dimensions are both equal to $K$, the number of measurement variables in the original data set). Typically, matrix rows are ordered such that $\lambda_1 > \lambda_2 > \dots > \lambda_K$. Eigenvectors and eigenvalues have exactly the properties that we postulated for principal components. Eigenvectors are orthogonal to each other, that is, they represent statistically independent axes of covariation in the data set, and squared eigenvalues express the proportion of the total sample variance that is captured by each corresponding eigenvector.

As a final step of PCA, we need to express the original data set in terms of the eigenvectors of the covariance matrix. In other words, we calculate each specimen's position in the system of coordinates defined by the PCs. This can be achieved by the following matrix multiplication:

$$\mathbf{P} = \mathbf{X}' \cdot \mathbf{E}^{\mathrm{T}}, \tag{F-10}$$

where $\mathbf{E}^T$ is the transpose of $\mathbf{E}$. $\mathbf{P}$ is an $N \times K$ matrix, each row of which expresses the coordinates of a specimen in principal component space. These are the so-called PC scores of the specimens. Graphing PC scores for the first few principal components typically suffices to visually assess statistically relevant patterns of variation in a sample.

## F.5 CENTROID SIZE

Let us assume isotropic variation of a sample of $K$-landmark configurations around its consensus configuration, which implies that the scatter of the specimens' landmark positions around the mean position is the same in all spatial directions. Under these conditions, the only measure of size that is statistically independent from shape is *centroid size* (Bookstein, 1991). This is an isometric scaling factor, which is defined as follows:

$$S = \sqrt{\sum_{i=1}^{K}(\mathbf{p}_i - \mathbf{c})^2} \; ; \quad \mathbf{c} = \frac{1}{K}\sum_{i=1}^{K}\mathbf{p}_i, \tag{F-11}$$

where $\mathbf{c}$ denotes the centroid (i.e., the center of mass of the landmark configuration), and $(\mathbf{p}_i - \mathbf{c})^2$ the squared distance of landmark $\mathbf{p}_i$ from the centroid. Note that the formula for centroid size is similar to that of a statistical standard deviation.

Centroid size is not the only way to measure the size of a landmark configuration. Alternative size measures are, for example, the distance between two specific landmarks in the configuration (Bookstein, 1991; Dryden and Mardia, 1998), the arithmetic or geometric mean of all interlandmark distances (Lele and Richtsmeier, 2001), or the area or volume comprised by the landmark configuration (delimited by the so-called *convex hull* enclosing all landmarks; note that this area may be reduced to zero when all landmarks are aligned on a straight line). From a biological point of view, it is sensible to evaluate size in various ways and to assess the impact of size measures on the subsequent analysis of shape variation (Rohlf, 2000).

## F.6 PROCRUSTES SUPERIMPOSITION, GENERALIZED LEAST-SQUARES FITTING, AND LINEARIZED SHAPE SPACE

To compare the shape of specimens represented by $K$ landmarks, it is necessary to find a reference landmark configuration that serves as a "fixed point" against which all comparisons can be performed. As illustrated in Figure 8-14, Kendall's shape space containing all $K$-landmark configurations is a hypersphere, on which each point represents a specific shape. To apply methods of linear multivariate statistical analysis, it is necessary to linearize the spherical shape space. This is done by mapping shape data onto a plane that is tangential to the sphere. The point where the plane touches the sphere is the reference shape; because Kendall's shape sphere typically has more than three dimensions, we may state that the tangential *hyperplane* touches the shape *hypersphere* in one single point, the reference shape. The hyperplane itself is equivalent to linearized Procrustes space, in which data can be analyzed with classic multivariate methods such as principal components analysis (see Appendix F.4).

When no specific biological hypotheses govern the definition of a reference shape, it is sensible to evaluate a sample mean shape, or consensus, by means of geometry. The term *Procrustes superimposition* denotes various methods of finding a consensus. *Generalized least-squares fitting* (GLS) is the most straightforward method and comparable to regression analysis, where an regression line is laid through a data scatter such that the sum of squared deviations between data points and the regression line is minimized (thus the term least squares). During GLS, size differences are eliminated by normalizing all specimens to centroid size = 1, and positional differences are removed by superimposing the centroids (centers of mass) of all specimens' landmark configurations (Figs. 8-6 and 8-21C). Differences in orientation are eliminated by rotating specimens around their centroids until a best match is achieved according to a least-squares criterion (minimize the sum of squared distances between specimens, landmark by landmark).

Whereas positional differences can be eliminated in one step, removing rotational differences is typically performed by iteration. First, an arbitrary specimen is taken as a preliminary consensus, to which all other specimens are fitted. Second, a new consensus is evaluated as the arithmetic mean of all fitted specimens. Steps one and two are repeated until the sum of squares reaches a minimum. Rohlf and Slice (1990a) proposed variants of GLS, named *resistant fitting*, taking into account that "outlier landmarks" might lead to spurious results during straightforward GLS fitting (what these authors call the "Pinocchio effect" is illustrated with a giraffe in Fig. 8-20).

## F.7 SHAPE SPACE ANALYSIS

Shape space analysis can be defined as multivariate analysis of a sample of landmark configurations in linearized Procrustes space. Here, we give a brief overview of the logic of shape space analysis, using standard matrix algebra as introduced in Section F.4. Detailed accounts of methods of shape space analysis can be found in the literature (Dryden and Mardia, 1998; Rohlf, 1999, 2000).

Let vector $\mathbf{c}$ be the consensus configuration of a sample of $N$ three-dimensional $K$-landmark configurations

$$\mathbf{c} = (\bar{x}_1 \quad \bar{y}_1 \quad \bar{z}_1 \quad \bar{x}_2 \quad \bar{y}_2 \quad \bar{z}_2 \quad \ldots \quad \bar{x}_K \quad \bar{y}_K \quad \bar{z}_K) \tag{F-12}$$

where $\bar{x}_j$, $\bar{y}_j$, $\bar{z}_j$ are the coordinates of the $j$th landmark ($j = 1 \ldots K$). Furthermore, let vector $\mathbf{v}_i$ denote the landmark configuration of the $i$th GLS-fitted specimen ($i = 1 \ldots N$)

$$\mathbf{v}_i = (x_{i,1} \quad y_{i,1} \quad z_{i,1} \quad x_{i,2} \quad y_{i,2} \quad z_{i,2} \quad \ldots \quad x_{i,K} \quad y_{i,K} \quad z_{i,K}) \tag{F-13}$$

Note that vectors $\mathbf{c}$ and $\mathbf{v}_i$ have $3K$ elements, corresponding to the $3K$ landmark coordinates measured per specimen. Because the consensus configuration defines the origin of linearized Procrustes shape space, the location (i.e., the shape) of a specimen in that space is given by

$$\mathbf{x}'_i = \mathbf{x}_i - \mathbf{c} \tag{F-14}$$

In other words, the specimen's shape is measured as its deviation from the consensus configuration. We now combine the shape vectors $\mathbf{x}'_i$ of all aligned specimen into one single matrix

$$\mathbf{X}' = \begin{pmatrix} \mathbf{X}'_1 \\ \ldots \\ \mathbf{X}'_i \\ \ldots \\ \mathbf{X}'_N \end{pmatrix} = \begin{pmatrix} x'_{1,1} & y'_{1,1} & z'_{1,1} & \ldots & x'_{1,K} & y'_{1,K} & z'_{1,K} \\ \ldots & \ldots & \ldots & \ldots & \ldots & \ldots & \ldots \\ x'_{i,1} & y'_{i,1} & z'_{i,1} & \ldots & x'_{i,K} & y'_{i,K} & z'_{i,K} \\ \ldots & \ldots & \ldots & \ldots & \ldots & \ldots & \ldots \\ x'_{N,1} & y'_{N,1} & z'_{N,1} & \ldots & x'_{N,K} & y'_{N,K} & z'_{N,K} \end{pmatrix}. \tag{F-15}$$

Each specimen occupies one row in this matrix, such that the matrix has $N$ rows and $3K$ columns ($N \times 3K$ matrix). Next, we perform principal components analysis (PCA), that is, we calculate the variance-covariance matrix $\mathbf{S}$ of $\mathbf{X}'$

$$\mathbf{S} = \frac{1}{N} \mathbf{X}' \mathbf{X}'^T. \tag{F-16}$$

$\mathbf{S}$ has $3K \times 3K$ elements. Using eigenvectors $\mathbf{E}$ of the variance-covariance matrix $\mathbf{S}$ as principal components, and eigenvalues $\mathbf{L}$ to sort these components according to the variance proportion they account for (see Section F.4), each specimen's principal components score vector $\mathbf{p}_j$ ($j = 1 \ldots 3K$) can be evaluated as follows:

$$\mathbf{P} = \mathbf{X}' \mathbf{E}^T = \begin{pmatrix} \mathbf{p}_1 \\ \mathbf{p}_2 \\ \ldots \\ \mathbf{p}_N \end{pmatrix} = \begin{pmatrix} p_{1,1} & p_{1,2} & \ldots & p_{1,3K} \\ p_{2,1} & p_{2,2} & \ldots & p_{2,3K} \\ \ldots & \ldots & \ldots & \ldots \\ p_{N,1} & p_{N,2} & \ldots & p_{N,3K} \end{pmatrix} \tag{F-17}$$

Typically, only the first few components are considered, because they "explain," in a statistical sense, a large part of the total sample variance. An important property of the above transformation is that it can be inverted:

$$\mathbf{X}' = \mathbf{PE} \qquad\qquad (\text{F-18})$$

With Equations F-14 and F-18, the landmark configuration $\mathbf{x}_i$ of a specimen $i$ can be recovered from its principal component scores $\mathbf{p}_i$ and, more importantly in practical analyses, the landmark configuration of a hypothetical specimen, $\mathbf{x}^*$, can be evaluated from user-defined PC scores $\mathbf{p}^*$

$$\mathbf{x}^* = \mathbf{c} + \mathbf{p}^* \cdot \mathrm{E} \qquad\qquad (\text{F-19})$$

This procedure is mainly used to generate visual representations of modes of shape transformation associated with each principal component and to explore patterns of covariation in linearized Procrustes shape space.

## F.8 SHAPE VARIABILITY AS DEFORMATION: PRINCIPAL, PARTIAL, AND RELATIVE WARPS

In Chapter 8, thin plate spline (TPS) methods were primarily used as a visualization tool to facilitate the interpretation of complex spatial patterns of shape variability. At the same time, however, TPS can be used as an analytical tool to look at shape variability from the perspective of modes of deformation at various levels of spatial scale (Bookstein, 1991). Recall that principal components analysis (PCA) in shape space permits the decomposition of shape variability according to variance proportions. In close analogy, TPS analysis permits decomposition according to proportions of deformation energy. A given landmark configuration can be bent out of shape in many different ways, but there exists a set of bending modes by which its deformation can be described as a suite of least to most energy-demanding processes. These are the *principal warps* (note that they are a property of one single landmark configuration, not of a sample of landmark configurations!). An interesting aspect of this decomposition is that low-energy principal warps describe global features of deformation, whereas high-energy warps describe localized features of deformation.

When principal warps are evaluated for the consensus configuration of a sample, it is possible to express each specimen as a suite of global to local modes of deformation of the consensus. Formally, specimens are now expressed in *partial warp* space. The analysis of shape variability in partial warp space has attracted considerable attention, because it permits a scale-dependent approach. However, this appealing mathematical framework is of limited biological use. Although it might be informative to explore whether shape variation in a sample is due to global versus local modes, it must be kept in mind that the modes of deformation represented by partial warps are free of biological meaning (see discussion of this topic in Adams and Rosenberg, 1998; Rohlf, 1998). Hence, we are confronted with the recurrent problem of extrinsic versus intrinsic frames of reference: Partial warps do not represent a natural system of reference to characterize shape variation in a sample. Nevertheless, such a system can be superimposed onto partial warp space. The application of classic PCA to partial warps yields *relative warps*. Relative warps describe shape variation in the sample, optionally weighted according to global versus local scales at which shape variation takes place.

Before we ask what "optional weighting" means, we need to know what relative warps signify statistically and biologically. As shown in Figure 8-15, we first transformed our data from linearized shape space into deformation space and then applied PCA to extract patterns of shape variability. Statistically, it does not make any difference whether we extract principal components of shape variability directly from shape space or from deformation space. The results are the same. Accordingly, relative warps are equivalent to principal components in linearized shape space, as we encountered them in the section on analyzing data in shape space. Nevertheless, there is a difference between the two as soon as we weigh the modes of deformation according to the spatial scale at which they act. In an explorative study, for example, one might want to learn whether a pattern of shape variation revealed by PCA is associated with local or global changes in shape. Given the above caveats on extrinsic versus intrinsic reference systems, however, global versus local modes of shape change are not necessarily a biologically relevant issue, so it is wise to restrict weighting to preliminary and explorative studies. For almost all practical purposes, therefore, it suffices to perform PCA in shape space and use TPS functions as a visualization rather than as an analytical tool.

Another point needs consideration here. Principal warps of TPS functions account for nonlinear modes of deformation, whereas linear (i.e., affine, or shearing) modes must be added to describe the complete deformation from configuration **A** to **B**. Again, the decomposition of the deformation matrix into affine components and nonlinear components (principal warps) does not yield biological insights, because this matrix is a mathematical construct with no direct relation to underlying biological processes of shape transformation. However, one application scenario exists in which the separation of affine and nonlinear components is practical. In Chapter 6 and in Appendix E, we showed that overall compressive deformation of a fossil results in a skewed morphology. In a morphometric analysis comprising compressed specimens, it may therefore be sensible to disregard the affine component that accounts for left/right asymmetries.

An interesting new extension of warp analysis concerns *singular warps* (Bookstein et al., 2003). The idea is to decompose landmark configurations into two or more landmark subsets (which, for example, are hypothesized to represent developmental modules influencing each other during growth). Whereas relative warps represent principal components of shape variation of the entire landmark configuration, singular warps represent principal components of shape variation between subsets of landmarks.

# REFERENCES

Acharya, R., Wasserman, R., Stevens, J., and Hinojosa, C. (1995) Biomedical imaging modalities: a tutorial. *Comput. Med. Imaging Graph.* **19**, 3–25.

Adams, D. C., and Rosenberg, M. S. (1998) Partial warps, phylogeny, and ontogeny: a comment on Fink and Zelditch. *Syst. Biol.* **47**, 168–173.

Adams, L., Krybus, W., Meyer-Elbrecht, D., Rueger, R., Gilsbach, J. M., Moesges, R., and Schloendorff, G. (1990) Computer-assisted surgery. *IEEE Comput. Graph. Appl.* **10**, 43–51.

Akenine-Möller, T., and Haines, E. (2002) *Real-Time Rendering.* Natick MA: A.K. Peters.

Andresen, P. R., Bookstein, F. L., Conradsen, K., Ersboll, B. K., Marsh, J. L., and Kreiborg, S. (2000) Surface-bounded growth modeling applied to human mandibles. *IEEE Trans. Med. Imaging* **19**, 1053–1063.

Andresen, P. R., and Nielsen, M. (2001) Non-rigid registration by geometry-constrained diffusion. *Med. Image Anal.* **5**, 81–88.

Axler, S. (1997) *Linear Algebra Done Right.* Heidelberg: Springer.

Azuma, R., Baillot, Y., Behringer, R., Feiner, S., Julier, S., and MacIntyre, B. (2001) Recent advances in augmented reality. *IEEE Comp. Graph. Appl.* **21**, 34–47.

Bidgood, W. D., Jr., Horii, S. C., Prior, F. W., and Van Syckle, D. E. (1997) Understanding and using DICOM, the data interchange standard for biomedical imaging. *J. Am. Med. Inform. Assoc.* **4**, 199–212.

Boissonnat, J.-D. (1988) Shape reconstruction from planar cross-sections. *Comput. Vis. Graph. Image Proc.* **44**, 1–29.

Bookstein, F. L. (1989) "Size" and "shape": a comment on semantics. *Syst. Zool.* **38**, 173–180.

Bookstein, F. L. (1991) *Morphometric Tools for Landmark Data: Geometry and Biology.* Cambridge: Cambridge University Press.

Bookstein, F. L. (1996a) Combining the tools of geometric morphometrics. In *Advances in Morphometrics.* Marcus, L. F., Corti, M., Loy, A., Naylor, G. J. P., and Slice, D. E. (eds.) New York: Plenum Press, pp. 131–151.

Bookstein, F. L. (1996b) Applying landmark methods to biological outline data. In *Proceedings in Image Fusion and Shape Variability Techniques.* Mardia, K. V., Gill, C. A., and Dryden, I. L. (eds.) Leeds: University of Leeds Press, pp. 59–70.

Bookstein, F. L. (1997) Landmark methods for forms without landmarks: morphometrics of group differences in outline shape. *Med. Image. Anal.* **1**, 225–243.

Bookstein, F. L., Gunz, P., Mitterocker, P., Prossinger, H., Schäfer, K., and Seidler, H. (2003) Cranial integration in *Homo*: singular warps analysis of the midsagittal plane in ontogeny and evolution. *J. Hum. Evol.* **44**, 167–187.

Bookstein, F. L., Schäfer, K., Prossinger, H., Seidler, H., Fieder, M., Stringer, C., Weber, G. W., Arsuaga, J. L., Slice, D. E., Rohlf, F. J., Recheis, W., Mariam, A. J., and Marcus, L. F. (1999) Comparing frontal cranial profiles in archaic and modern *Homo* by morphometric analysis. *Anat. Rec.* **257**, 217–224.

Borah, B., Dufresne, T. E., Cockman, M. D., Gross, G. J., Sod, E. W., Myers, W. R., Combs, K. S., Higgins, R. E., Pierce, S. A., and Stevens, M. L. (2000) Evaluation of changes in trabecular bone architecture and mechanical properties of minipig vertebrae by three-dimensional magnetic resonance microimaging and finite element modeling. *J. Bone Miner. Res.* **15**, 1786–1797.

Borah, B., Gross, G. J., Dufresne, T. E., Smith, T. S., Cockman, M. D., Chmielewski, P. A., Lundy, M. W., Hartke, J. R., and Sod, E. W. (2001) Three-dimensional microimaging (MRmicroI and microCT), finite element modeling, and rapid prototyping provide unique insights into bone architecture in osteoporosis. *Anat. Rec.* **265**, 101–110.

Borg, I., and Groenen, P. (1997) *Modern Multidimensional Scaling. Theory and Applications.* Berlin: Springer.

Bowskill, J., and Downie, J. (1995) Extending the capabilities of the human visual system: an introduction to enhanced reality. *ACM Comput. Graph.* **29**, 61–65.

Bowskill, J., Morphett, J., and Downie, J. (1997) A taxonomy for enhanced reality systems. In *Digest of Papers. First International Symposium on Wearable Computers. IEEE Comput. Soc. 1997, pp.175–176. Los Alamitos, CA, USA.*

Bresenham, J., Jacobs, P., Sadler, L., and Stucki, P. (1993) Real Virtuality: StereoLithography—rapid prototyping in 3D. *Proc. SIGGRAPH 93*, 377–378.

Brochu, C. A., and Ketcham, R. A. (2003) Computed tomographic atlas of the skull of *Tyrannosaurus rex. J. Vertebr. Paleontol.* **22**.

Brodlie, K. W., Carpenter, L. A., and Earnshaw, R. A. (eds.) (1992) *Scientific Visualization. Techniques and Applications.* Berlin: Springer.

Buitrago-Téllez, C. H., Zollikofer, C. P. E., Gufler, H., Cantini, J. L., Kimmig, M., and Langer, M. (1997a) Complex craniofacial trauma: Spiral 3D-CT, 3D model manufacturing and virtual-reality prosthesis design for preoperative planning. *Eur. Radiol. (Suppl.)* **7**, 385.

Buitrago-Téllez, C. H., Zollikofer, C. P. E., Cantini, J. L., Gufler, H., Kimmig, M., Stoll, P., Langer, M., and Matter, P. (1997b) Preoperative planning of complex craniofacial fractures. *AO ASIF Dialogue* **10**, 18–19.

Caselles, V., Catté, F., Coll, T., and Dibos, F. (1993) A geometric model for active contours in image processing. *Num. Math.* **66**, 1–31.

Chan, T. F., and Vese, L. A. (2002) Active contour and segmentation models using geometric PDE's for medical imaging. In *Geometric Methods in Bio-Medical Image Processing.* Malladi, R. (ed.) Berlin Heidelberg New York: Springer.

Chen, M., Jones, M. W., and Townsend, P. (1996) Volume distortion and morphing using disk fields. *Comput. Graph.* **20**, 567–575.

Chung, R., and Ho, C. K. (2000) 3-D reconstruction from tomographic data using 2-D active contours. *Comput. Biomed. Res.* **33**, 186–210.

Clack, J. A., Ahlberg, P. E., Finney, S. M., Dominguez Alonso, P., Robinson, J., and Ketcham, R. A. (2003) A uniquely specialized ear in a very early tetrapod. *Science* **425**, 65–69.

Clarysse, P., Friboulet, D., and Magnin, I. E. (1997) Tracking geometrical descriptors on 3-D deformable surfaces: application to the left-ventricular surface of the heart. *IEEE Trans. Med. Imaging* **16**, 392–404.

Clottes, J. (1995) *Les cavernes de Niaux. Art préhistorique en Ariège.* Paris: Éditions du Seuil.

Cody, D., Moxley, D. M., Davros, W., and Silverman, P. M. (1998) Principles of multislice computed tomography technology. In *Multislice Computed Tomography: A Practical Approach to Clinical Protocols*. Silverman, P. M. (ed.) Philadelphia: Lippincott, Williams & Wilkins, pp. 1–30.

Cole, T. M., and Richtsmeier, J. T. (1998) A simple method for visualization of influential landmarks when using Euclidean distance matrix analysis. *Am. J. Phys. Anthropol.* **107**, 273–283.

Connolly, C. J. (1950) *External Morphology of the Primate Brain*. Thomas Springfield, Ill.

Conroy, G., Weber, G., Seidler, H., Tobias, P., Kane, A., and Brunsden, B. (1998) Endocranial capacity in an early hominid cranium from Sterkfontein, South Africa. *Science* **280**, 1730–1731.

Conroy, G. C., Falk, D., Guyer, J., Weber, G. W., Seidler, H., and Recheis, W. (2002) Endocranial capacity in Sts 71 (*Australopithecus africanus*) by three-dimensional computed tomography. *Anat. Rec.* **258**, 391–396.

Conroy, G. C., and Vannier, M. W. (1984) Noninvasive three-dimensional computer imaging of matrix-filled fossil skulls by high-resolution computed tomography. *Science* **226**, 456–458.

Conroy, G. C., and Vannier, M. W. (1985) Endocranial volume determination of matrix-filled skulls using high-resolution computed tomography. In *Hominid Evolution: Past, Present and Future*. Tobias, P. V. (ed.) New York: Alan Liss, pp. 419–426.

Conroy, G. C., Vannier, M. W., and Tobias, P. V. (1990) Endocranial features of *Australopithecus africanus* revealed by 2D and 3D computed tomography. *Science* **247**, 838–841.

Conroy, G. C., Weber, G. W., Seidler, H., Recheis, W., Zur Nedden, D., and Mariam, J. H. (2000) Endocranial capacity of the Bodo cranium determined from three-dimensional computed tomography. *Am. J. Phys. Anthropol.* **113**, 111–118.

Cooper, D. M., Turinsky, A. L., Sensen, C. W., and Hallgrimsson, B. (2003) Quantitative 3D analysis of the canal network in cortical bone by micro-computed tomography. *Anat. Rec.* **274B**, 169–179.

Cormack, A. M. (1963) Representation of a function by its line integrals with some radiological applications. *J. Appl. Phys.* **34**, 2722–2727.

Cormack, A. M. (1964) Representation of a function by its line integrals with some radiological applications, II. *J. Appl. Phys.* **35**, 2908–2913.

Coward, W. M., and Conathy, D. M. (1996) A Monte Carlo study of the inferential properties of three methods of shape comparison. *Am. J. Phys. Anthropol.* **99**, 369–377.

Cronin, T. W., and King, C. A. (1989) Spectral sensitivity of vision in the mantis shrimp, *Gonodactylus oerstedii*, determined using non-invasive optical techniques. *Biol. Bull.* **176**, 308–316.

Dalton, R. (2000a) Feathers fly over Chinese fossil bird's legality and authenticity. *Nature* **403**, 689–690.

Dalton, R. (2000b) Doubts grow over discovery of fossilized 'dinosaur heart'. *Nature* **407**, 275–276.

Dean, M. C., Stringer, C. B., and Bromage, T. G. (1986) Age at death of the Neanderthal child from Devil's Tower, Gibraltar and the implications for studies of general growth and development in Neanderthals. *Am. J. Phys. Anthropol.* **70**, 301–309.

de Berg, M., van Kreveld, M., Overmars, M., and Schwarzkopf, O. (2002) *Computational Geometry. Algorithms and Applications*. Berlin Heidelberg New York: Springer.

Defleur, A., White, T., Valensi, P., Slimak, L., and Crégut-Bonnoure, E. (1999) Neanderthal cannibalism at Moula-Guercy, Ardèche, France. *Science* **286**, 128–131.

Delattre, A., and Fenart, R. (1960) *L'hominisation du crâne étudiée par la méthode vestibulaire*. Paris: CNRS.

Delibasis, K. S., Matsopoulos, G. K., Mouravliansky, N. A., and Nikita, K. S. (2001) A novel and efficient implementation of the marching cubes algorithm. *Comput. Med. Imaging Graph.* **25**, 343–352.

Dobson, J., and Geelhoed, G. W. (2001) On the Châtelperronian/Aurignacian conundrum: one culture, multiple human morphologies? *Curr. Anthropol.* **42**, 139–140.

Drew, M. S. (2003) *Fundamentals of Multimedia*. Harlow: Prentice Hall.

Dryden, I. L., and Mardia, K. (1998) *Statistical Shape Analysis*. New York: John Wiley & Sons.

Duarte, C., Maurício, J., Pettitt, P. B., Souto, P., Trinkaus, E., van der Plicht, H., and Zilhão, J. (1999) The early upper paleolithic human skeleton from the Abrigo do Lagar Velho (Portugal) and modern human emergence in Iberia. *Proc. Natl. Acad. Sci. U.S.A.*, 7604–7609.

Duchon, J. (1976) Interpolation des fonctions de deux variables suivant le principe de la flexion des plaques minces. *RAIRO Annal. numér.* **10**, 5–12.

Eberly, D. H., and Schneider, P. J. (2002) *Geometric Tools for Computer Graphics*. San Francisco: Morgan Kaufmann.

Ekoule, A. B., Peyrin, F. C., and Odet, C. L. (1991) A triangulation algorithm from arbitrary shaped multiple planar contours. *ACM Trans. Graph.* **10**, 182–199.

Enlow, D. H. (1990) *Facial Growth*. Philadelphia: Saunders.

Fajardo, R. J., and Müller, R. (2001) Three-dimensional analysis of nonhuman primate trabecular architecture using micro-computed tomography. *Am. J. Phys. Anthropol.* **115**, 327–336.

Fajardo, R. J., Ryan, T. M., and Kappelman, J. (2002) Assessing the accuracy of high-resolution X-ray computed tomography of primate trabecular bone by comparisons with histological sections. *Am. J. Phys. Anthropol.* **118**, 1–10.

Falk, D. (1993) Hominid paleoneurology. In *The Human Evolution Source Book*. Ciochon, R. L., and Fleagle, J. G. (eds.) Englewood Cliffs: Prentice Hall, pp. 61–70.

Falk, J., Redstrom, J., and Bjork, S. (1999) Amplifying reality. *Lecture Notes in Computer Science* **1707**, 274–280.

Farin, G. E., and Hansford, D. (1998) *The Geometry Toolbox for Graphics and Modeling*. Wellesley, MA: A.K. Peters.

Feiner, S. K. (2002) Augmented reality: a new way of seeing. *Sci. Am.* **286**, 34–41.

Felsenberg, D., Kalender, W., Sokiranski, R., Ebersberger, J., and Kramer, R. (1988) Reduction of metal artifacts in computed tomography-clinical experience and results. *Electromedica* **56**, 97–104.

Ferson, S. F., Rohlf, F. J., and Koehn, R. K. (1985) Measuring shape variation of two-dimensional outlines. *Syst. Zool.* **34**, 59–68.

Feyerabend, P. (1975) *Against Method: Outline of an Anarchistic Theory of Knowledge*. London: NLB.

Fisher, P. E., Russell, D. A., Stoskopf, M. K., Barrick, R. E., Hammer, M., and Kuzmitz, A. A. (2000) Cardiovascular evidence for an intermediate or higher metabolic rate in an ornithischian dinosaur. *Science* **288**, 503–505.

Flannery, B. P., Deckman, H. W., Roberge, W. G., and D'Amico, K. L. (1987) Three-dimensional microtomography. *Science* **237**, 1439–1444.

Flynn, J. J., Wyss, A. R., Charrier, R., and Swisher, C. C. (1995) An early Miocene anthropoid skull from the Chilean Andes. *Nature* **373**, 603–607.

Foley, J. D., van Dam, A., Feiner, S. K., and Hughes, J. F. (1995) *Computer Graphics. Principles and Practice.* Boston: Addison-Wesley.

Friston, K. (1998) Imaging neuroscience: principles or maps? *Proc. Natl. Acad. Sci. U.S.A.* **95**, 796–802.

Fuchs, H., and Ackerman, J. (1999) Displays for augmented reality: historical remarks and future prospects. In *Mixed Reality: Merging Real and Virtual Worlds. Ohmsha. 1999, Tokyo, Japan.* Ohta, Y., and Tamura, H. (eds.), pp. 31–40.

Galilei, G. (1638) *Discorsi e dimostrazioni matematiche [Mathematical Discourses and Demonstrations].* Leyden: Elsevier.

Gantt, D., Kappelman, J., Ketcham, R., and Colbert, M. (2002) 3d reconstruction of enamel volume in human and gorilla molars. *Am. J. Phys. Anthropol.,* 74–84.

Garrod, D. A. E., Buxton, L. H. D., Smith, G. E., and Bate, D. M. A. (1928) Excavation of a Mousterian rock-shelter at Devil's Tower, Gibraltar. *J. Roy. Anthropol. Inst.* **58**, 33–113.

Gebhardt, A. (2000) *Rapid Prototyping. Werkzeuge für die schnelle Produktentstehung.* München: Hanser.

Gerasimov, M. M. (1945) *Principles of Reconstruction of the Face on the Skull (in Russian).* Moscow: Nauka.

Gerasimov, M. M. (1971) *The Face Finder.* New York: Lippincott.

Gibbons, A. (2002) Glasnost for hominids: seeking access to fossils. *Science* **297**, 1464–1467.

Glassner, A. (1990) *Graphics Gems.* Cambridge, MA: Academic Press.

Golder, W., and Christian, A. (2002) Quantitative CT of dinosaur bones. *J. Comput. Assist. Tomogr.* **26**, 821–824.

Gonzalez, R. C., and Woods, R. E. (2002) *Digital Image Processing.* Upper Saddle River: Prentice Hall.

Goodall, C. R. (1991) Procrustes methods in the statistical analysis of shape. *J. Roy. Stat. Soc. B* **53**, 285–339.

Gould, S. J. (1966) Allometry and size in ontogeny and phylogeny. *Biol. Rev.* **41**, 587–640.

Gray, R. M., and Goodman, J. M. (1995) *Fourier Transforms. An Introduction for Engineers.* Boston, Dordrecht, London: Kluwer Academic Publishers Group.

Griffin, A., McMillin, S., Griffin, C., and Barton, K. (1997) Bioceramic RP materials for medical models. *Proc. 7th Intl. Conf. Rapid Prototyping,* 355–359.

Hankerson, D., Harris, G. A., and Johnson, P. D., Jr. (2003) *Introduction to Information Theory and Data Compression.* Boca Raton: Chapman and Hall/CRC.

Haralick, R. M., and Shapiro, L. G. (1985) Image segmentation techniques. *Comput. Vis. Graph. Image Proc.* **29**, 100–132.

Harders, M., Wildermuth, S., and Szekely, G. (2002) New paradigms for interactive 3D volume segmentation. *J. Vis. Comput. Animat.* **13**, 85–95.

Harville, D. A. (1997) *Matrix Algebra from a Statistician's Perspective.* New York: Springer.

Hauser, O. (1909) *Homo Mousteriensis Hauseri. Schweiz. Vjschr. Zahnheilkunde* **19**, 6–16.

Heberer, G. (1957) Bericht über die Bergung der Skelettreste von Combe Capelle und Le Moustier aus dem Brandschutt des Berliner Museums für Vor- und Frühgeschichte. In *Bericht über die 5. Tagung der Deutschen Gesellschaft für Anthropologie, 5.–7. April 1956 [Homo (suppl.)]* Freiburg i. Br., pp. 67–72.

Hemmy, D. C., Zonneveld, F. W., Lobregt, S., and Fukuta, K. (1994) A decade of clinical three-dimensional imaging: a review. Part I. Historical development. *Invest. Radiol.* **29**, 489–496.

Herman, G. T. (1980) *Image Reconstruction from Projections: the Fundamentals of Computerized Tomography.* New York: Academic Press.

Hjalgrim, H., Lynnerup, N., Liversage, M., and Rosenklint, A. (1995) Stereolithography: potential applications in anthropological studies. *Am. J. Phys. Anthropol.* **97**, 329–333.

Hoffmann, A. (1997) Zur Geschichte des Fundes von Le Moustier. *Acta Praehistor. Archaeol.* **29**, 7–16.

Höhne, K. H., Bomans, M., Riemer, M., Schubert, R., Tiede, U., and Lierse, W. (1992) A volume-based anatomical atlas. *IEEE Comput. Graph. Appl.* **12**, 72–78.

Höhne, K. H., and Hanson, W. A. (1992) Interactive 3D segmentation of MRI and CT volumes using morphological operations. *J. Comput. Assist. Tomogr.* **16**, 285–294.

Holloway, R. L. (1975) Early hominid endocasts: volumes, morphology, and significance for hominid evolution. In *Primate Functional Morphology and Evolution.* Tuttle, R. H. (ed.) The Hague: Mouton, pp. 393–415.

Hounsfield, G. N. (1973) Computerized transverse axial scanning (tomography): Part I. *Br. J. Radiol.* **46**, 1016–1022.

Howells, W. W. (1989) *Skull Shapes and the Map: Craniometric Analyses in the Dispersion of Modern Homo.* Cambridge: Harvard University Press.

Howells, W. W. (1996) Howell's craniometric data on the internet. *Am. J. Phys. Anthropol.* **101**, 441–442.

Huxley, J. S. (1924) Constant differential growth-ratios and their significance. *Nature* **114**, 895–896.

Huxley, J. S. (1932) *Problems of Relative Growth.* London: Methuen.

Ingle, K. A. (2001) *Reverse Engineering.* New York: McGraw-Hill.

Jacko, J. A. (2003) *Human Computer Interaction: Theory and Practice.* Mahwah, NJ: Lawrence Erlbaum Associates.

Jacko, J. A., and Sears, A. (eds.) (2002) *The Human-Computer Interaction Handbook: Fundamentals, Evolving Technologies, and Emerging Applications (Human Factors and Ergonomics).* Mahwah, NJ: Lawrence Erlbaum Associates.

Jackson, J. E. (1991) *A User's Guide to Principal Components.* New York: John Wiley & Sons.

Jacobs, P. F. (1996) *Stereolithography and Other RP&M Techniques.* Dearborn, MI: Society of Manufacturing Engineers.

Jain, A. K., Zhong, Y., and Dubuisson-Jolly, M.-P. (1998) Deformable template models: a review. *Signal Process.* **71**, 109–129.

Janich, K. (1994) *Linear Algebra.* Berlin: Springer.

Jolliffe, I. T. (1986) *Principal Component Analysis.* Berlin, Heidelberg, New York: Springer.

Jones, J. C., Greenberg, W., and Ayers, M. S. (1998) Computed tomographic evaluation of dinosaur egg shell integrity. *Vet. Radiol. Ultrasound* **39**, 133–136.

Jones, K. L. (1988) *Smith's Recognizable Patterns of Human Malformation.* Philadelphia: W.B. Saunders.

Jungers, W. L., and Minns, R. J. (1979) Computed tomography and biomechanical analysis of fossil long bones. *Am. J. Phys. Anthropol.* **50**, 285–290.

Kähler, K., Haber, J., and Seidel, H.-P. (2003) Reanimating the dead: reconstruction of expressive faces from skull data. *Proc. SIGGRAPH* **22**, 555–561.

Kak, A. C., and Slaney, M. (1988) *Principles of Computerized Tomographic Imaging.* New York: IEEE Press.

Kalender, W. A. (1994) Principles and applications of spiral CT. *Nucl. Med. Biol.* **21**, 693–699.

Kalender, W. A. (2000) *Computed Tomography: Fundamentals, System Technology, Image Quality and Applications*. Munich: Publicis MCD.

Kalender, W. A., Hebel, R., and Ebersberger, J. (1987) Reduction of CT artifacts caused by metallic implants. *Radiology* **164**, 576–577.

Kalender, W. A., Polacin, A., and Suss, C. (1994) A comparison of conventional and spiral CT: an experimental study on the detection of spherical lesions. *J. Comput. Assist. Tomogr.* **18**, 167–176.

Kalender, W. A., Vock, P., Polacin, A., and Soucek, M. (1990) Spiral-CT: a new technique for volumetric scans. I. Basic principles and methodology. *Roentgenpraxis* **43**, 323–330.

Kalvin, A. D., Dean, D., and Hublin, J.-J. (1995) Reconstruction of human fossils. *IEEE Comput. Graph. Appl.* **15**, 12–15.

Kass, M., Witkin, A., and Terzopoulos, D. (1988) Snakes: active contour models. *Int. J. Comput. Vis.* **1**, 321–331.

Kendall, D. G. (1981) The statistics of shape. In *Interpreting Multivariate Statistics*. Barnett, V. (ed.) New York: Wiley, pp. 75–80.

Kendall, D. G. (1984) Shape manifolds, Procrustean metrics and complex projective spaces. *Bull. Lond. Math. Soc.* **16**, 81–121.

Kendall, D. G. (1985) Exact distributions for shapes of random triangles in convex sets. *Adv. Appl. Prob.* **17**, 308–329.

Klingenberg, C. P. (1996) Multivariate allometry. In *Advances in Morphometrics – Proceedings of the NATO Advanced Study Institute*. Marcus, L. F., Corti, M., Loy, A., Naylor, G. J. P., and Slice, D. E. (eds.) New York: Plenum, pp. 23–49.

Klingenberg, C. P., Neuenschwander, B. E., and Flury, B. D. (1996) Ontogeny and individual variation: analysis of patterned covariance matrices with common principal components. *Syst. Biol.* **45**, 135–150.

Koch, R. M., Gross, M. H., and Bosshard, A. A. (1998) Emotion edition using finite elements. *Comput. Graph. Forum* **17**, C295–C302.

Kollmann, W. M., and Buchly, W. (1898) Die Persistenz der Rassen und die Reconstruction der Physiognomie praehistorischer Schädel. *Arch. Anthropol.* **25**, 329–359.

Krings, M., Stone, A., Schmitz, R. W., Krainitzki, H., Stoneking, M., and Pääbo, S. (1997) Neandertal DNA sequences and the origin of modern humans. *Cell* **90**, 19–30.

Lambrecht, J. T., Brix, F., and Gremmel, H. (1993) Three-dimensional skull identification via computed tomographic data and video visualization. In *Forensic Analysis of the Skull: Craniofacial Analysis, Reconstruction, and Identification*. Iscan, M. Y., and Helmer, R. P. (eds.) New York: Wiley-Liss, pp. 97–104.

Lang, Z.-P. (2000) *Principles of Magnetic Resonance Imaging: A Signal Processing Perspective*. New York: John Wiley & Sons.

Laughlin, S. B. (1981) Neural principles in the peripheral visual systems of invertebrates. In *Handbook of Sensory Physiology VII/6B*. Land, M. F., Laughlin, S. B., Nässel, D. R., Strausfeld, N. J., and Waterman, T. H. (eds.) Berlin: Springer, pp. 133–280.

Lauterbur, P. C. (1973) Image formation by induced local interactions: examples employing nuclear magnetic resonance. *Nature* **242**, 190–191.

Lavallee, S., Bainville, E., and Bricault, I. (2000) An overview of computer-integrated surgery and therapy. *Crit. Rev. Diagn. Imaging* **41**, 157–236.

Lebedinskaya, G. V., Balueva, T. S., and Veselovskaya, E. V. (1993) Principles of facial reconstruction. In *Forensic Analysis of the Skull*. Iscan, M. Y., and Helmer, R. P. (eds.) New York: Wiley-Liss, pp. 183–198.

Lee, T. Y., and Lin, C. H. (2001) Growing-cube isosurface extraction algorithm for medical volume data. *Comput. Med. Imaging Graph.* **25**, 405–415.

Lele, S. (1993) Euclidean distance matrix analysis (EDMA): estimation of mean form and mean form difference. *Math. Geol.* **25**, 573–602.

Lele, S., and Cole, T. M., III (1996) A new test for shape differences when variance-covariance matrices are unequal. *J. Hum. Evol.* **31**, 193–212.

Lele, S., and Richtsmeier, J. T. (1990) Statistical models in morphometrics: are they realistic? *Syst. Zool.* **39**, 60–69.

Lele, S., and Richtsmeier, J. T. (1991) Euclidean distance matrix analysis: a coordinate-free approach for comparing biological shapes using landmark data. *Am. J. Phys. Anthropol.* **86**, 415–427.

Lele, S., and Richtsmeier, J. T. (1992) On comparing biological shapes: detection of influential landmarks. *Am. J. Phys. Anthropol.* **87**, 49–65.

Lele, S., and Richtsmeier, J. (2001) *An Invariant Approach to the Statistical Analysis of Shapes.* Boca Raton: Chapman and Hall.

Lestrel, P. E. (1989) Some approaches toward the mathematical modeling of the craniofacial complex. *J. Craniofac. Genet. Dev. Biol.* **9**, 77–91.

Lestrel, P. E. (ed.) (1997) *Fourier Descriptors and their Applications in Biology.* Cambridge: Cambridge University Press.

Lieberman, D. E. (1996) How and why humans grow thin skulls: experimental evidence for systemic cortical robusticity. *Am. J. Phys. Anthropol.* **101**, 217–236.

Lieberman, D. E., McBratney, B. M., and Krovitz, G. (2002) The evolution and development of cranial form in *Homo sapiens. Proc. Natl. Acad. Sci. U.S.A.* **99**, 1134–1139.

Lim, J. S. (1991) *Two-Dimensional Signal and Image Processing.* Englewood Cliffs: Prentice Hall.

Lindahl, T. (1997) Facts and artifacts of ancient DNA. *Cell* **90**, 1–3.

Lockwood, C. A., and Kimbel, W. H. (1998) Endocranial capacity of early hominids. *Science* **283**, 9.

Long, L. (1994) *Introduction to Computers and Information Systems.* Englewood Cliffs: Prentice Hall.

Lorensen, W. E., and Cline, H. E. (1987) Marching cubes: a high resolution 3D surface reconstruction algorithm. *Comput. Graph.* **21**, 163–169.

Lounsberg, M., Mann, S., and DeRose, T. (1992) Parametric surface interpolation. *IEEE Comput. Graph. Appl.* **12**, 45–52.

MacLatchy, L., and Muller, R. (2002) A comparison of the femoral head and neck trabecular architecture of Galago and Perodicticus using micro-computed tomography (microCT). *J. Hum. Evol.* **43**, 89–105.

MacLeod, N. (1999) Generalizing and extending the eigenshape method of shape space visualization and analysis. *Paleobiology* **25**, 107–138.

Maisey, J. G. (2001) Remarks on the inner ear of elasmobranchs and its interpretation from skeletal labyrinth morphology. *J. Morphol.* **250**, 236–264.

Malladi, R., and Sethian, J. A. (2002) Fast methods for shape extraction in medical and biomedical imaging. In *Geometric Methods in Bio-Medical Image Processing.* Malladi, R. (ed.) Berlin Heidelberg New York: Springer.

Marcus, L., Bello, E., and García-Valdecasas, A. (eds.) (1993) *Contributions to Morphometrics.* Madrid: Consejo superior de investigaciones científicas.

Marcus, L. F., Corti, M., Loy, A., Naylor, G. J. P., and Slice, D. E. (eds.) (1996) *Advances in Morphometrics.* New York: Plenum Press.

Mardia, K. V., Bookstein, F. L., and Moreton, I. J. (2000) Statistical assessment of bilateral symmetry of shapes. *Biometrika* **87**, 285–300.

Marshall, N. J., Land, M. F., King, C. A., and Cronin, T. W. (1991a) The compound eyes of mantis shrimps (Crustacea, Hoplocarida, Stomatopoda). I. Compound eye structure: the detection of polarized light. *Phil. Trans. Roy. Soc. Lond. B* **334**, 33–56.

Marshall, N. J., Land, M. F., King, C. A., and Cronin, T. W. (1991b) The compound eyes of mantis shrimps (Crustacea, Hoplocarida, Stomatopoda). II. Polychromatic vision by serial and lateral filtering. *Phil. Trans. Roy. Soc. Lond. B* **334**, 57–84.

Masutani, Y., Dohi, T., Yamane, F., Iseki, H., and Takakura, K. (1998) Augmented reality visualization system for intravascular neurosurgery. *Comput. Aided Surg.* **3**, 239–247.

McCarthy, R. C., and Lieberman, D. E. (2001) Posterior maxillary (PM) plane and anterior cranial architecture in primates. *Anat Rec* **264**, 247–260.

McInerney, T., and Terzopoulos, D. (1996) Deformable models in medical image analysis: a survey. *Med. Image Anal.* **1**, 91–108.

McInerney, T., and Terzopoulos, D. (2000) T-snakes: topology adaptive snakes. *Med. Image Anal.* **4**, 73–91.

McMahon, T. A., and Tyler Bonner, J. (1983) *On Size and Life*. New York: Scientific American Books.

Meinguet, J. (1984) Surface spline interpolation: basic theory and computational aspects. In *Approximation Theory and Spline Functions*. Singh, S. P. (ed.) Dordrecht: D. Reidel, pp. 127–142.

Melcher, A. H., Holowka, S., Pharoah, M., and Lewin, P. K. (1997) Non-invasive computed tomography and three-dimensional reconstruction of the dentition of a 2,800-year-old Egyptian mummy exhibiting extensive dental disease. *Am. J. Phys. Anthropol.* **103**, 329–340.

Menzel, R. (1979) Spectral sensitivity and colour vision in invertebrates. In *Handbook of Sensory Physiology, Vol. VII/6A*. Autrum, H. (ed.) Berlin: Springer, pp. 503–580.

Merz, B., Niederer, P., Muller, R., and Ruegsegger, P. (1996) Automated finite element analysis of excised human femora based on precision -QCT. *J. Biomech. Eng.* **118**, 387–390.

Metaxas, D. N. (1997) *Physics-based Deformable Models: Applications to Computer Vision, Graphics and Medical Imaging*. Boston: Kluwer.

Meyers, D., Skinner, S., and Sloan, K. (1992) Surfaces from contours. *ACM Trans. Graph.* **11**, 228–258.

Montani, C., Scateni, R., and Scopigno, R. (1994) A modified look-up table for implicit disambiguation of Marching Cubes. *Visual Comput.* **10**, 353–355.

Morrison, D. F. (1990) *Multivariate Statistical Methods*. New York: McGraw-Hill.

Morrison, N. (1994) *Introduction to Fourier Analysis*. New York: John Wiley & Sons.

Mortenson, M. E. (1999) *Mathematics for Computer Graphics Applications*. New York: Industrial Press.

Mosimann, J. E. (1988) Size and shape analysis. In *Encyclopedia of Statistical Sciences*. Kotz, S., and Johnson, N. L. (eds.) New York: John Wiley & Sons, pp. 219–239.

Moss, M. L., and Young, R. W. (1960) A functional approach to craniology. *Am. J. Phys. Anthropol.* **18**, 281–292.

Mudry, K. M., Plonsey, R. A., and Bronzino, J. D. (eds.) (2003) *Biomedical Imaging*. Boca Raton: CRC Press.

Müller, R., and Rüegsegger, P. (1995) Three-dimensional finite element modelling of non-invasively assessed trabecular bone structures. *Med. Eng. Phys.* **17**, 126–133.

Müller, A., Rüegsegger, P., and Seitz, P. (1985) Optimal CT settings for bone evaluations. *Phys. Med. Biol.* **30**, 401–409.

Murakami, T., Kamimura, A., and Nakajima, N. (2000) Refrigerative stereolithography using sol-gel transformable photopolymer resin and direct masking. In *Solid Freeform and Additive Fabrication – Symposium 2000.* Danforth, S. C., Dimos, D., and Prinz, F. B. (eds.) Warrendale, Pennsylvania, pp. 625–628.

Natterer, F. (1986) *The Mathematics of Computerized Tomography.* New York: John Wiley & Sons.

Nelson, L. A., and Michael, S. D. (1998) The application of volume deformation to three-dimensional facial reconstruction: a comparison with previous techniques. *Forensic Sci. Int.* **94**, 167–181.

Ney, D. R., Fishman, E. K., Magid, D., and Drebin, R. A. (1990) Volumetric rendering of computed tomography data: principles and techniques. *IEEE Comput. Graph. Appl.* **10**, 24–32.

Norton, P. (2003) *Peter Norton's Introduction to Computers.* Boston: McGraw-Hill.

O'Higgins, P., and Williams, N. W. (1987) An investigation into the use of Fourier coefficients in characterizing cranial shape in primates. *J. Zool., Lond.* **211**, 409–430.

Owen, R. (1843) *Lectures on Comparative Anatomy and Physiology of the Invertebrate Animals, Delivered at the Royal College of Surgeons in 1843.* London: Longman, Brown, Green & Longman.

Palmer, A. R. (1994) Fluctuating asymmetry analyses: A primer. In *Developmental Instability: Its Origins and Evolutionary Implications.* Markow, T. A. (ed.) Dordrecht: Kluwer, pp. 335–364.

Path, M., Zollikofer, C. P. E., and Stucki, P. (1998) New approaches in CT artifact suppression – a case study in maxillofacial surgery. In *CAR'98, Computer Assisted Radiology and Surgery (proceedings of the 12th international symposium and exhibition).* Lemke, H. U., Vannier, M. W., Inamura, K., and Farman, A. (eds.) Tokyo, 24–27 June 1998: Elsevier, pp. 830–835.

Patt, Y. N., and Patel, S. J. (2001) *Introduction to Computing Systems: From Bits and Gates to C and Beyond.* Boston: McGraw-Hill.

Perez-Arjona, E., Dujovny, M., Park, H., Kulyanov, D., Galaniuk, A., Agner, C., Michael, D., and Diaz, F. G. (2003) Stereolithography: neurosurgical and medical implications. *Neurol. Res.* **25**, 227–236.

Petzold, R., Zeilhofer, H. F., and Kalender, W. A. (1999) Rapid protyping technology in medicine–basics and applications. *Comput. Med. Imaging Graph.* **23**, 277–284.

Pham, D. L., Xu, C., and Prince, J. L. (2000) Current methods in medical image segmentation. *Annu. Rev. Biomed. Eng.* **2**, 315–337.

Pickering, R. B., Conces, D. J., Braunstein, E. M., and Yurco, F. (1990) Three-dimensional computed tomography of the mummy Wenuhotep. *Am. J. Phys. Anthropol.* **83**, 49–55.

Ponce de León, M. S., and Zollikofer, C. P. E. (1999) New evidence from Le Moustier 1: Computer-assisted reconstruction and morphometry of the skull. *Anat. Rec.* **254**, 474–489.

Ponce de León, M. S., and Zollikofer, C. P. E. (2001) Neanderthal cranial ontogeny and its implications for late hominid diversity. *Nature* **412**, 534–538.

Powell, C. S. (1994) Relinquishing relics. 3-D copies of artifacts could stand in for the real thing. *Sci. Am.* **271**, 46–47.

Pratt, W. K. (2001) *Digital Image Processing.* New York: John Wiley & Sons.

Press, W. H., Teukolsky, S. A., Vetterling, W. T., and Flannery, B. P. (1992) *Numerical Recipes in C.* Cambridge: Cambridge University Press.

Prince, M. R. (2003) *MRI from Picture to Proton.* Cambridge: Cambridge University Press.

Quatrehomme, G., Cotin, S., Subsol, G., Delingette, H., Garidel, Y., Grevin, G., Fidrich, M., Bailet, P., and Ollier, A. (1997) A fully three-dimensional method for facial reconstruction based on deformable models. *J. Forensic Sci.* **42**, 649–652.

Radon, J. (1917) Ueber die Bestimmung von Funktionen durch ihre Integralwerte längs gewisser Mannigfaltigkeiten. *Ber. Saechs. Akad. Wiss.* **29**, 262–277.

Rae, T. C., and Koppe, T. (2000) Isometric scaling of maxillary sinus volume in hominoids. *J. Hum. Evol.* **38**, 411–423.

Rajon, D. A., and Bolch, W. E. (2003) Marching cube algorithm: review and trilinear interpolation adaptation for image-based dosimetric models. *Comput. Med. Imaging Graph.* **27**, 411–435.

Rayfield, E. J., Norman, D. B., Horner, C. C., Horner, J. R., Smith, P. M., Thomason, J. J., and Upchurch, P. (2001) Cranial design and function in a large theropod dinosaur. *Nature* **409**, 1033–1037.

Remmler, D., Olson, L., Duke, D., Ekstrom, R., Matthews, D., and Ullrich, C. G. (1998) Presurgical finite element analysis from routine computed tomography studies for craniofacial distraction: II. An engineering prediction model for gradual correction of asymmetric skull deformities. *Plast. Reconstr. Surg.* **102**, 1395–1404.

Renaud, S., Michaux, J., Jaeger, J. J., and Auffray, J. C. (1996) Fourier analysis applied to *Stephanomys* (Rodentia, Muridae) molars: nonprogressive evolutionary pattern in a gradual lineage. *Paleobiology* **22**, 255–265.

Reyment, R. (1991) *Multidimensional Paleobiology.* Oxford: Pergamon Press.

Reyment, R. A., Blackith, R. E., and Campbell, N. A. (1984) *Multivariate Morphometrics.* New York: Academic Press.

Rhine, J. S., and Campell, H. R. (1980) Thickness of facial tissues in American blacks. *J. Forensic Sci.* **25**, 847–858.

Rhine, J. S., and Moore, C. E. (1984) *Tables of Facial Tissue Thickness of American Caucasoids in Forensic Anthropology.* Albuquerque: UNM.

Richtsmeier, J., Deleon, V. B., and Lele, S. (2002) The promise of geometric morphometrics. *Ybk. Phys. Anthropol.* **45**, 63–91.

Richtsmeier, J. T., and Lele, S. (1993) A coordinate-free approach to the analysis of growth patterns: models and theoretical considerations. *Biol. Rev.* **68**, 381–411.

Rieppel, O. (1989) Homology, topology and typology: the history of modern debates. In *Homology, the Hierarchical Basis of Comparative Biology.* Hall, B. K. (ed.) New York: Academic Press, pp. 63–100.

Robb, R. A. (1998) *Three-Dimensional Biomedical Imaging.* New York: John Wiley & Sons.

Robb, R. A. (1999) 3-D visualization in biomedical applications. *Annu. Rev. Biomed. Eng.* **1**, 377–399.

Rogers, S. W. (1999) Allosaurus, crocodiles, and birds: evolutionary clues from spiral computed tomography of an endocast. *Anat. Rec.* **257**, 162–173.

Rogers Ackermann, R., and Krovitz, G. E. (2002) Common patterns of facial ontogeny in the hominid lineage. *Anat. Rec.* **269**, 142–147.

Rohlf, F. J. (1990) Rotational fit (Procrustes) methods. In *Proceedings of the Michigan Morphometrics Workshop.* Rohlf, F. J., and Bookstein, F. L. (eds.) Ann Arbor: University of Michigan Museum of Zoology, pp. 227–236.

Rohlf, F. J. (1998) On applications of geometric morphometrics to studies of ontogeny and phylogeny. *Syst. Biol.* **47**, 147–158.

Rohlf, F. J. (1999) Shape statistics: Procrustes superimpositions and tangent spaces. *J. Classif.* **16**, 197–225.

Rohlf, F. J. (2000) Statistical power comparisons among alternative morphometric methods. *Am. J. Phys. Anthropol.* **111**, 463–478.

Rohlf, F. J. (2001) Geometric morphometrics in phylogeny. In *Morphology, Shape and Phylogenetics.* Forey, P., and McLeod, N. (eds.) London: Francis & Taylor, pp. 175–193.

Rohlf, F. J., and Archie, J. W. (1984) A comparison of Fourier methods for the description of wing shape in mosquitoes (Diptera: Culicidae). *Syst. Zool.* **33**, 302–317.

Rohlf, F. J., and Bookstein, F. L. (eds.) (1990) *Proceedings of the Michigan Morphometrics Workshop.* Michigan: Michigan Museum of Zoology.

Rohlf, F. J., and Marcus, L. (1993) A revolution in morphometrics. *Trends Ecol. Evol.* **8**, 129–132.

Rohlf, F. J., and Slice, D. (1990a) Extensions of the Procrustes method for the optimal superimposition of landmarks. *Syst. Zool.* **39**, 40–59.

Rowe, T., Ketcham, R. A., Denison, C., Colbert, M., Xu, X., and Currie, P. J. (2001b) Forensic palaeontology: The Archaeoraptor forgery. *Nature* **410**, 539–540.

Rowe, T., McBride, E. F., and Sereno, P. C. (2001a) Dinosaur with a heart of stone. *Science* **291**, 783.

Rüegsegger, P., Koller, B., and Muller, R. (1996) A microtomographic system for the non-destructive evaluation of bone architecture. *Calcif. Tissue Int.* **58**, 24–29.

Ruff, C. B., Trinkaus, E., and Holliday, T. W. (1997) Body mass and encephalization in Pleistocene *Homo. Nature* **387**, 173–176.

Ryan, T. M., and Ketcham, R. A. (2002) The three-dimensional structure of trabecular bone in the femoral head of strepsirrhine primates. *J. Hum. Evol.* **43**, 1–26.

Sailer, H., Haers, P., Zollikofer, C., Warnke, T., Carls, F., and Stucki, P. (1998) The value of stereolithographic models for preoperative diagnosis of craniofacial deformities and planning of surgical corrections. *Int. J. Oral Max. Surg.* **27**, 327–333.

Sarti, A., Mikula, K., Sgallari, F., and Lamberti, C. (2002) Nonlinear multiscale analysis models for filtering of 3D + time biomedical images. In *Geometric Methods in Bio-Medical Image Processing.* Malladi, R. (ed.) Berlin Heidelberg New York: Springer, pp. 702–726.

Schmidt-Nielsen (1984) *Scaling: Why is Animal Size so Important?* Cambridge: Cambridge University Press.

Seeram, E. (2001) *Computed Tomography: Physical Principles, Clinical Applications and Quality Control.* Philadelphia: W.B. Saunders.

Seitz, P., and Rüegsegger, P. (1982) Anchorage of femoral implants visualized by modified computed tomography. *Arch. Orthop. Trauma Surg.* **100**, 261–266.

Sereno, M. I. (1998) Brain mapping in animals and humans. *Curr. Opin. Neurobiol.* **8**, 188–194.

Serre, D. (2002) *Matrices: Theory and Applications.* New York: Springer.

Sherman, W., and Craig, A. (2002) *Understanding Virtual Reality.* Amsterdam: Morgan Kaufmann.

Silverman, P. M., Kalender, W. A., and Hazle, J. D. (2001) Common terminology for single and multislice helical CT. *Am. J. Roentgenol.* **176**, 1135–1136.

Simon, J. C. (2001) *Introduction to Information Systems.* New York: John Wiley & Sons.

Singh, A., Goldgof, D., and Terzopoulos, D. (1988) *Deformable Models in Medical Image Analysis*. Los Alamitos: IEEE Computer Society Press.

Sinha, U., Bui, A., Taira, R., Dionisio, J., Morioka, C., Johnson, D., and Kangarloo, H. (2002) A review of medical imaging informatics. *Ann. N. Y. Acad. Sci.* **980**, 168–197.

Skinner, M. F., and Sperber, G. H. (1982) *Atlas of Radiographs of Early Man*. New York: Alan R. Liss, Inc.

Small, C. G. (1996) *The Statistical Theory of Shape*. New York: Springer.

Soille, P. (2002) *Morphological Image Analysis*. Berlin Heidelberg New York: Springer.

Sokal, R. R., and Rohlf, F. J. (1995) *Biometry*. New York: Freeman.

Soucek, M., Vock, P., Daepp, M., and Kalender, W. A. (1990) Spiral-CT: a new technique for volumetric scans. II. Potential clinical applications. *Roentgenpraxis* **43**, 365–375.

Spitzer, V. M., and Whitlock, D. G. (1998) *Atlas of the Visible Human Male: Reverse Engineering of the Human Body*. Sudbury, MA: Jones and Bartlett Publishers International.

Spoor, C. F., and Zonneveld, F. W. (1995) Morphometry of the primate bony labyrinth: A new method based on high-resolution computed tomography. *J. Anat.* **186**, 271–286.

Spoor, F., Bajpai, S., Hussain, S. T., Kumar, K., and Thewissen, J. G. M. (2002) Vestibular evidence for the evolution of aquatic behaviour in early cetaceans. *Nature* **417**, 133–166.

Spoor, F., Wood, B., and Zonneveld, F. (1994) Implications of early hominid labyrinthine morphology for evolution of human bipedal locomotion. *Nature* **369**, 845–848.

Steidle, C., Klosterman, D., Graves, G., Osborne, N., and Chartoff, R. (1997) Automated fabrication of nonresorbable bone implants using laminated object manufacturing (LOM). *Proc. 42th Int. SAMPE Symposium*, pp. 563–570.

Steidle, C., Klosterman, D., Chartoff, R., Graves, G., and Osborne, N. (1999) Automated fabrication of custom bone implants using rapid prototyping. In *Proc. 44th Int. SAMPE Symposium, Long Beach, CA, May 23–27* (`http://www.udri.udayton.edu/rpdl/papers.htm`).

Stokstad, E. (2000) Paleontology. Learning to dissect dinosaurs–digitally. *Science* **288**, 1728–1732.

Stringer, C. B., Dean, M. C., and Martin, R. D. (1990) A comparative study of cranial and dental development with a recent British sample and among Neandertals. In *Primate Life History and Evolution*. DeRousseau, C. J. (ed.) New York: Wiley-Liss, pp. 115–152.

Stuppy, W. H., Maisano, J. A., Colbert, M. W., Rudall, P. J., and Rowe, T. B. (2003) Three-dimensional analysis of plant structure using high-resolution X-ray computed tomography. *Trends Plant. Sci.* **8**, 2–6.

Sutton, A. (2002) *Textbook of Radiology and Imaging*. Philadelphia: W.B. Saunders.

Sutton, M. D., Briggs, D. E., and Siveter, D. J. (2001) An exceptionally preserved vermiform mollusc from the Silurian of England. *Nature* **410**, 461–463.

Tattersall, I. (1999) The abuse of adaptation. *Evol. Anthropol.* **8**, 115–116.

Taylor, K. T. (2001) *Forensic Art and Illustration*. Boca Raton: CRC Press.

ter Haar Romeny, B. M., Zuiderveld, K. J., Van Waes, P. F., Van Walsum, T., Van Der Weijden, R., Weickert, J., Stokking, R., Wink, O., Kalitzin, S., Maintz, T., Zonneveld, F., and Viergever, M. A. (1998) Advances in three-dimensional diagnostic radiology. *J. Anat.* **193**, 363–371.

Terzopoulos, D., Witkin, A., and Kass, M. (1988) Constraints on deformable models: recovering 3D shape and nonrigid motions. *Artif. Intell.* **36**, 91–123.

Thompson, D. A. W. (1917) *On Growth and Form*. Cambridge: Cambridge University Press.

Thompson, P. M., Woods, R. P., Mega, M. S., and Toga, A. W. (2000) Mathematical/computational challenges in creating deformable and probabilistic atlases of the human brain. *Hum. Brain Mapp.* **9**, 81–92.

Tillier, A.-M. (1982) Les enfants Néanderthaliens de Devil's Tower (Gibraltar). *Zeitschr. Morphol. Anthropol.* **73**, 125–148.

Tyrrell, A. J., Evison, M. P., Chamberlain, A. T., and Green, M. A. (1997) Forensic three-dimensional facial reconstruction: historical review and contemporary developments. *J. Forensic Sci.* **42**, 653–661.

Van Essen, D., Drury, H., Joshi, S., and Miller, M. (1998) Functional and structural mapping of human cerebral cortex: solutions are in the surfaces. *Proc. Natl. Acad. Sci. U.S.A.* **95**, 788–795.

Van Essen, D. C. (2002) Windows on the brain: the emerging role of atlases and databases in neuroscience. *Curr. Opin. Neurobiol.* **12**, 574–579.

Vanezis, M., and Vanezis, P. (2000) Cranio-facial reconstruction in forensic identification–historical development and a review of current practice. *Med. Sci. Law* **40**, 197–205.

Vannier, M. W., Conroy, G. C., Marsh, J. L., Knapp, R. H., and Silcox, M. T. (1985) Three-dimensional cranial surface reconstructions using high-resolution computed tomography. New discoveries on the middle ear anatomy of *Ignacius graybullianus* (Paromomyidae, Primates) from ultra high resolution X-ray computed tomography. *Am. J. Phys. Anthropol.* **67**, 299–311.

van Rietbergen, B. (2001) Micro-FE analyses of bone: state of the art. *Adv. Exp. Med. Biol.* **496**, 21–30.

Wallin, Å. (1991) Constructing isosurfaces from CT data. *IEEE Comput. Graph. Appl.* **11**, 28–33.

Webb, A. G. (2003) *Introduction to Biomedical Imaging*: Hoboken, NJ: Wiley.

Webb, P. A. (2000) A review of rapid prototyping (RP) techniques in the medical and biomedical sector. *J. Med. Eng. Technol.* **24**, 149–153.

Weiss, E. (2003) Understanding muscle markers: aggregation and construct validity. *Am. J. Phys. Anthropol.* **121**, 230–240.

Welch, T. A. (1984) A technique for high-performance data compression. *IEEE Computer* **17**, 8–19.

White, T. (2003) Early hominids - diversity or distortion? *Science* **299**, 1994–1997.

Williams, F. l. E. (2000) Heterochrony and the human fossil record: comparing Neandertal and modern human craniofacial ontogeny. In *Neanderthals on the Edge*. Stringer, C. B., Barton, R. N. E., and Finlayson, J. C. (eds.) Oxford: Oxbow Books, pp. 257–267.

Wilting, J. E., and Timmer, J. (1999) Artefacts in spiral CT images and their relation to pitch and subject morphology. *Eur. J. Radiol.* **9**, 316–322.

Wind, J. (1984) Computerized X-ray tomography of fossil hominid skulls. *Am. J. Phys. Anthropol.* **63**, 265–228.

Wind, J., and Zonneveld, F. W. (1985) Radiology of fossil hominid skulls. In *Homind Evolution: Past, Present and Future*. Tobias, P. V. (ed.) New York: Alan R. Liss, pp. 437–442.

Yasuda, T., Yokoi, S., Ohshita, H., and Toriwaki, J.-I. (1992) 3D visualization of an ancient Egyptian mummy. *IEEE Comput. Graph. Appl.* **12**, 13–17.

Zhao, S., Robertson, D. D., Wang, G., Whiting, B., and Bae, K. T. (2000) X-ray CT metal artifact reduction using wavelets: an application for imaging total hip prostheses. *IEEE Trans. Med. Imaging* **19**, 1238–1247.

Zhou, Z., Clarke, J. A., and Zhang, F. (2002) *Archaeoraptor*'s better half. *Nature* **420**, 285.

Zollikofer, C. P. E. (2002) A computational approach to paleoanthropology. *Evol. Anthropol. Suppl.* **1**, 64–67.

Zollikofer, C. P. E., and Ponce de León, M. S. (1995) Tools for rapid prototyping in the biosciences. *IEEE Comp. Graph. Appl.* **15**, 48–55.

Zollikofer, C. P. E., and Ponce de León, M. S. (2001a) The brain and its case: computer-based case studies on the relation between software and hardware in living and fossil hominids. In *Humanity from African Naissance to Coming Millennia*. Tobias, P. V., Raath, M. A., Moggi-Cecchi, J., and Doyle, G. A. (eds.) Florence: Florence University Press, pp. 379–384.

Zollikofer, C. P. E., and Ponce de León, M. S. (2001b) Computer-assisted morphometry of hominoid fossils: The role of morphometric maps. In *Phylogeny of the Neogene Hominoid Primates of Eurasia*. De Bonis, L., Koufos, G., and Andrews, P. (eds.) Cambridge: Cambridge University Press, pp. 50–59.

Zollikofer, C. P. E., and Ponce de León, M. S. (2002) Visualizing patterns of craniofacial shape variation in *Homo sapiens*. *Proc. Roy. Soc. B* **269**, 801–807.

Zollikofer, C. P. E., Ponce de León, M. S., and Martin, R. D. (1998) Computer-assisted paleoanthropology. *Evol. Anthropol.* **6**, 41–54.

Zollikofer, C. P. E., Ponce de León, M. S., Martin, R. D., and Stucki, P. (1995) Neanderthal computer skulls. *Nature* **375**, 283–285.

Zollikofer, C. P. E., Ponce de León, M. S., Vandermeersch, B., and Lévêque, F. (2002a) Evidence for interpersonal violence in the St. Césaire Neanderthal. *Proc. Natl. Acad. Sci. U.S.A.* **99**, 6444–6448.

Zollikofer, C. P. E., Ponce de León, M. S., Esteves, F., Tecelao Silva, F., and Pacheco Dias, R. (2002b) The computer-assisted reconstruction of the skull. In *Portrait of the Artist as a Child. The Gravettian Human Skeleton from the Abrigo do Lagar Velho and its Archeological Context*. Zilhão, J., and Trinkaus, E. (eds.) Lisbon: Instituto Português de Arqueologia, pp. 326–341.

Zonneveld, F. W. (1994) A decade of clinical three-dimensional imaging: a review. Part III. Image analysis and interaction, display options, and physical models. *Invest. Radiol.* **29**, 716–725.

Zonneveld, F. W., and Noorman van der Dussen, M. F. (1992) Three-dimensional imaging and model fabrication in oral and maxillofacial surgery. *Oral Maxillofac. Surg. Clin. N. Am.* **4**, 19–33.

Zonneveld, F. W., and Vijerberg, G. P. (1984) The relationship between slice thickness and image quality in CT. *Medicamundi* **29**, 104–117.

Zonneveld, F. W., and Wind, J. (1985) High-resolution computed tomography of fossil hominid skulls: a new method and some results. In *Hominid Evolution: Past, Present and Future*. Tobias, P. V. (ed.) New York: Alan R. Liss, pp. 427–436.

Zonneveld, F. W., and Fukuta, K. (1994) A decade of clinical three-dimensional imaging: a review. Part II. Clinical applications. *Invest. Radiol.* **29**, 574–589.

# INDEX

Accuracy, 59, 216, 277–279, 299. *See also* Resolution; Precision
Active contours, *see* Deformable models
Address, *see* Computer(s)
Alligator, 78, 165–166
Allometry
  definition, 232
  equation, 299–300
  visualization, 254, 257
Analysis
  Euclidean distance matrix (EDMA), *see* Euclidean distance matrix analysis
  multivariate, 235, 237
  of outlines, *see* Outline analysis
  of shape, *see* Shape space
Anatomy
  space, 166–169
  system of coordinates, 167–168, 300
Animation, 43, 148, 209
Applicaton program, 100
Archaeoraptor, 193
Archimedes, 225
ASCII (American Standard Code for Information Interchange), *see* Data format(s)
Asymmetry, *see* Fossil reconstruction
Augmented reality, *see* Enhanced reality
Autapomorphic, *see* Morphology, characters
Axes, anatomic, 299–300. *See also* Principal components

Backprojection, 73–74, 87, 105–107. *See also* Computed tomography
Baudrillard, J., 155
Bauplan, 189
Big-endian, *see* Data encoding, endian-ness
Binary system, *see* Number system(s)

Biomedical
  data, 50–56
  imaging, *see* Computed tomography; Magnetic resonance imaging
Bit, 23. *See also* Pixel(s), depth
Bitmap, *see* Data format(s)
Bone
  thickness, 274
  trabecular structure, 83
  X-ray density, 160
Boundary detection, *see* Edge detection
Brain mapping, 56, 273
Brightness, 35
Byte, *see* Data encoding

CAD, *see* Computer-aided design
Camera
  photographic, 59, 97, 132, 140, 277
  virtual, 139–143
Centroid size
  definition, 231
  formula, 303
Chamisso, A. von, 129
Channel(s), *see* Data format(s),
Character(s), *see* Data format(s); Morphology
Charge-coupled device (CCD), 60, 277
Chip, *see* Computer(s)
CIE (Commision Internationale de l'Éclairage), 37
CMYK (cyan-magenta-yellow-black), *see* Data format(s)
Color(s)
  additive and subtractive, 38–39
  cube, 36
  data, *see* Data format(s)
  look-up table (LUT), 32, 41-42
  receptor(s), *see* Color vision
  triangle, 35
  vision, 33–39. *See also* Visual system